AF556984

STUDIES IN COMPUTATIONAL MATHEMATICS 10

Editors:

C.K. CHUI
Stanford University
Stanford, CA, USA

P. MONK
University of Delaware
Newark, DE, USA

L. WUYTACK
University of Antwerp
Antwerp, Belgium

An imprint of Elsevier Science

2003

Amsterdam – Boston – Heidelberg – London – New York – Oxford – Paris
San Diego – San Francisco – Singapore – Sydney – Tokyo

BEYOND WAVELETS

Grant V. WELLAND
University of Missouri – St. Louis
Department of Mathematics and Computer Science
St. Louis, USA

An imprint of Elsevier Science
2003

Amsterdam – Boston – Heidelberg – London – New York – Oxford – Paris
San Diego – San Francisco – Singapore – Sydney – Tokyo

ELSEVIER SCIENCE Inc.
360 Park Avenue South
New York, NY 10010-1710, USA

© 2003 Elsevier Science Inc. All rights reserved.

This work is protected under copyright by Elsevier Science, and the following terms and conditions apply to its use:

Photocopying
Single photocopies of single chapters may be made for personal use as allowed by national copyright laws. Permission of the Publisher and payment of a fee is required for all other photocopying, including multiple or systematic copying, copying for advertising or promotional purposes, resale, and all forms of document delivery. Special rates are available for educational institutions that wish to make photocopies for non-profit educational classroom use.

Permissions may be sought directly from Elsevier's Science & Technology Rights Department in Oxford, UK: phone: (+44) 1865 843830, fax: (+44) 1865 853333, e-mail: permissions@elsevier.com. You may also complete your request on-line via the Elsevier Science homepage (http://www.elsevier.com), by selecting 'Customer Support' and then 'Obtaining Permissions'.

In the USA, users may clear permissions and make payments through the Copyright Clearance Center, Inc., 222 Rosewood Drive, Danvers, MA 01923, USA; phone: (978) 7508400, fax: (978) 7504744, and in the UK through the Copyright Licensing Agency Rapid Clearance Service (CLARCS), 90 Tottenham Court Road, London W1P 0LP, UK; phone: (+44) 207 631 5555; fax: (+44) 207 631 5500. Other countries may have a local reprographic rights agency for payments.

Derivative Works
Tables of contents may be reproduced for internal circulation, but permission of Elsevier Science is required for external resale or distribution of such material.
Permission of the Publisher is required for all other derivative works, including compilations and translations.

Electronic Storage or Usage
Permission of the Publisher is required to store or use electronically any material contained in this work, including any chapter or part of a chapter.

Except as outlined above, no part of this work may be reproduced, stored in a retrieval system or transmitted in any form or by any means, electronic, mechanical, photocopying, recording or otherwise, without prior written permission of the Publisher.
Address permissions requests to: Elsevier's Science & Technology Rights Department, at the phone, fax and e-mail addresses noted above.

Notice
No responsibility is assumed by the Publisher for any injury and/or damage to persons or property as a matter of products liability, negligence or otherwise, or from any use or operation of any methods, products, instructions or ideas contained in the material herein. Because of rapid advances in the medical sciences, in particular, independent verification of diagnoses and drug dosages should be made.

First edition 2003

Library of Congress Cataloging in Publication Data
A catalog record from the Library of Congress has been applied for.

British Library Cataloguing in Publication Data
A catalogue record from the British Library has been applied for.

Academic Press
An Elsevier Science Imprint
525 B Street, Suite 1900, San Diego, California 92101-4495, USA
http://www.academicpress.com

ISBN: 0 12 743273 6
ISSN: 1570 579 X (Series)

Typeset by Alden Press, Oxford
Printed in Great Britain by MPG Books Ltd, Bodmin, Cornwall

♾ The paper used in this publication meets the requirements of ANSI/NISO Z39.48-1992 (Permanence of Paper).
Printed in The Netherlands.

PREFACE

The themes of classical wavelets include terms such as compression and efficient representation. Important features which play a role in analysis of functions in two variables are dilation, translation, spatial and frequency localization and singularity orientation. Singularities of functions in more than one variable vary in dimensionality. Important singularities in one dimension are simply points. In two dimensions zero and one dimensional singularities are important. A smooth singularity in two dimensions may be a one dimensional smooth manifold. Smooth singularities in two dimensional images often occur as boundaries of physical objects. Efficient representation in two dimensions is a hard problem and is addressed in the first six chapters. The next two chapters return to problems of one dimension where new important results are given. The final two chapters represent a transition from harmonic analysis to statistical methods and filtering theory but the goals remain consistent with those of earlier chapters. We have chosen to title "Beyond Wavelets". We could have used the title, "Pursuing the Promise of Wavelets". We briefly describe each chapter.

The lead chapter, "Digital Ridgelet Transform based on True Ridge Functions" by David Donoho and Georgina Flesia addresses the problem of analyzing the structure of a function of two real variables. It extends work of Donoho and an associated group of co-workers. Special credit is due to Emmanuel Candès. Donoho and Candès have constructed a system called curvelets which gives high-quality asymptotic approximation of singularities. Passage from their continuum study to one appropriate for applications requires development of digital algorithms to implement concepts of the continuum study faithfully. A less obvious proposal than a standard tensor product basis was made earlier by Donoho emphasizing "wide-sense" ridgelets with localization properties in radial and angular frequency domains. Wide-sense ridgelets are no longer of strict ridge form but allow the possibility of an orthonormal set of elements. The theory is related to that of the Radon transform and to rotation and scaling of images. At the continuum level these are natural but for digital data issues are problematic. In this chapter a definiton of digital ridgelet transform is given. The digital transform has structural relationships strongly analogous to those of the continuum case. The transform takes a n-by-n array of data in Cartesian coordinates and expands it by a factor of 4 in creating a coefficient array. This leaves room for further improvements.

Chapter 2 is a companion chapter to Chapter 1 and continues the study of digital implementation of ridgelets with ridgelet packets. The two principal approaches given are the frequency-domain approach and the Radon approach. In the first approach a recursive dyadic partition of the polar Fourier domain produces a collection of rectangular tiles followed by a tensor basis of windowed sinusoids in the angular and radial variables for each tile. In the Radon approach transformation to the Radon domain is followed by using wavelets in the angular variable and wavelet packets in the second Radon variable. The Radon isometry is important in this case. The notion of pseudopolar Fast Fourier Transform and a pseudo Radon isometry called the normalized Slant Stack are discussed and used. In both cases analysis of image data relies on directionally oriented waveforms. The wavelet packet and the local sinusoidal packet bases are generalizations of the original wavelet systems of elements. Ridgelet packets which follow in the spirit of these systems are highly orientation selective and bear much the same relationship to ridgelets as do wavelet packets to wavelets.

In Chapter 3, François Meyer and Raphy Coifman create brushlets to address the problem of describing an image with a library of steerable wavelet packets. By careful design of the window of the local Fourier basis, brushlets with very fast decay are obtained. They note that other directionally oriented filter banks have been constructed which a redundancy factor of 2 or 4. This presents a major hurdle to computing a sparse image representation. By use of a construction in the Fourier domain they create wavelet packets which are complex valued functions with a phase. A key ingredient of the construction is a window used for local Fourier analysis. The window is required to have very fast decay.

Do and Vetterli study image representation in Chapter 4. An observation that the curvelet transform is defined in the frequency domain leads to the question: "Is there as spatial domain scheme for refinement which at each generation, doubles the spatial resolution as well as the angular resolution?" They propose a filter bank construction that effectively deals with piecewise smooth images with smooth contours. The resulting image expansion is a frame composed of contour segments, which are named contourlets. Their work leads to an effective method to implement the discrete curvelet transform.

Chan and Zhou open discussion of the ENO-wavelet construction in Chapter 5, by discussing oscillations which emulate the classical Gibbs' phenomenon. It has be discovered that the wavelet Gibbs' phenomenon is generated by using difference filters across boundaries of discontinuity. ENO is the acronym for the phrase essential non-oscillatory which represents an approach for suppression of unwanted oscillations encountered at discontinuities. Rigorous approximation error bounds are found to depend on the smoothness of function away from discontinuities when the ENO approach is used. Several applications of the ENO method are given which include function approximation, image compression and signal denoising.

An explicit model for Bayesian reconstruction of tomographic data is given by S. Zhao and H. Cai in Chapter 6. Their approach to image analysis is based on an interesting analogy to classical mechanics. The intensity of each pixel of an image is modelled by a transverse motion of a "pixtron". The energy for Bayesian tomographic reconstruction is interpreted as the total kinetic energy of the collection of pixtrons and log-likelihood is interpreted as potential energy restricting motion of pixtrons. Finally, the use of the minimization of a log-posterior is analogous to the principle of least action of classical mechanics. The analogy allows them to show that a Gaussian Markov random field prior can viewed as the kinetic energy of free motion of pixtrons. The analogy leads to a novel image prior for Bayesian tomographic reconstruction based on level-set evolution of an image driven by the mean curvature motion. Their methods are accompanied by applications to brain slice images which demonstrate algorithms produced by the model.

Chui and Stöckler give extensive description of recent developments of spline wavelets and frames in Chapter 7. Splines have many of the natural features required in the original design of I. Daubechies for wavelets which result in beautiful formulas. Vanishing moments reflect smoothness. Design of wavelet frames with vanishing moments requires a series of new ideas. The authors explain why early design approaches fail to create wavelets with higher orders of vanishing moments and then provide steps to recover vanishing moments. The method involves the notion of vanishing moment recover functions. The theory is extended in the direction of tight spline-wavelet frames with arbitrary knot sequences that allow stacked knots. Knot Stacking provides local increase in smoothness and can be applied at the boundaries of bounded intervals and half line segments. This gives greater flexibility overcoming standard rigid design features of classical wavelets in which supports are closely tied to the dilation factor of wavelet families. Multi-wavelets represent a special case of this more general construction.

Chapter 8, "Affine, Quasi-affine and Co-affine Wavelets", by Washington University the group of researchers, is devoted to fully understanding results of Ron and Shen. Dilations and translation are two characteristic operators used to define the wavelet pyramid. The question studied asks whether the order in which dilation and translation are applied is important. A subset of the affine group, used in the wavelet definition, is the set translations followed by dilation. A second subset of the affine group is the set for which dilation is applied first which is followed by translation. The effects are dramatically different. Ron and Shen found that by reversing the order of these operators at a 'half-way' point in the wavelet pyramid results in a different set of functions and yet they are sufficient to solve the representation problem. This chapter is devoted to understanding this phenomenon and it is discovered that the choice of Ron and Shen is essentially optimal.

Bènichou and Saito search for relations between the related criteria in Chapter 9. Two studies motivate them. Olshausen and Field pioneered an approach to imaging which investigates representation of natural images emphasizing sparsity of representation using a large library of photographs of natural images and computer experiments to derive a set of basis elements for efficient representation. Bell and Sejnowski conducted similar studies in which statistical independence was the major criterion. The pair of studies suggests both the basis derived for sparse representation and the basis derived under the independence criterion produce elements efficient for capture of edges, orientation and location; all features prominently studied by image researchers. Their study is based on a modest goal that begins with an artificial stochastic process, the spike process, from which they obtain theorems which give precise conditions on the sparsity and statistical independence criteria to select the same basis for the spike process.

S. Akkarakaran and P.P. Vaidyanathan provide a new direction from previous work in Chapter 10. Standard filter banks fall under the theory of design and uniform filter banks. A nonuniform filter bank is one whose channel decimation rates need not all be equal. Most nonuniform filter bank designs result in approximation or near-perfect reconstruction which leaves open theoretical issues for nonuniform filter banks. Their study is restricted to filter banks with integer decimation rates. A set, S, of integers satisfies maximal decimation if the reciprocals of the integers sum to unity. They only study filter banks with integer decimation rates. Their study searches for necessary and sufficient conditions on S for existence of a perfect-reconstruction filter bank belonging to some class which uses S as its set of decimators. They present examples with conditions which are either sufficient or necessary but unfortunately different. They focus on rational filter banks and strengthen known necessary conditions providing an important step to solving the problem. However, the basic problem remains unresolved. Necessary and sufficient conditions remain unknown. Thus they open an important problem and provide insight toward solving it.

This volume is a product which was conceived during a conference funded by the National Science Foundation and the Conference Board of Mathematical Sciences at which David Donoho was the principal speaker in May of 2000 at the University of Missouri - St. Louis. The title "Beyond Wavelets" is due to David Donoho. I thank the NSF and the University of Missouri - St. Louis and the support staff of the Mathematics Department there. A very special thanks is extended to David Donoho for his continued support and understanding.

Many contributed to the success of that conference and to the original idea to develop "Beyond Wavelets". I give thanks to Charles Chui, Raphy Coifman, Ingrid Daubechies, and Joachim Stöckler and Shiying Zhao. I thank the contributors to the volume both for their efforts and understanding. I take responsibility for the delays encountered and beg your forgiveness. Many more deserve to be mentioned to whom I extend my thanks anonymously.

Grant Welland
St. Louis, MO
February, 2003.

CONTENTS

Beyond Wavelets
G. V. Welland (Editor)
© 2003 Elsevier Science (USA) All rights reserved

1

DIGITAL RIDGELET TRANSFORM BASED ON TRUE RIDGE FUNCTIONS

D.L. DONOHO AND A.G. FLESIA

Department of Statistics, Stanford University
Sequoia Hall, 390 Serra Mall, Stanford, CA 94305-4065
donoho@stat.stanford.edu flesia@stat.stanford.edu

Abstract

We study a notion of ridgelet transform for arrays of digital data in which the analysis operator uses true ridge functions, as does the synthesis operator. There are fast algorithms for analysis, for synthesis, and for partial reconstruction. Associated with this is a transform which is a digital analog of the orthonormal ridgelet transform (but not orthonormal for finite n). In either approach, we get an overcomplete frame; the result of ridgelet transforming an $n \times n$ array is a $2n \times 2n$ array. The analysis operator is invertible on its range; the appropriately preconditioned operator has a tightly controlled spread of singular values. There is a near-parseval relationship.

Our construction exploits the recent development by Averbuch et al. (2001) of the Fast Slant Stack, a Radon transform for digital image data; it may be viewed as following a Fast Slant Stack with fast 2-d wavelet transform. A consequence of this construction is that it offers discrete objects (discrete ridgelets, discrete Radon transform, discrete Pseudopolar Fourier domain) which obey inter-relationships paralleling those in the continuum ridgelet theory (between ridgelets, Radon transform, and polar Fourier domain).

We make comparisons with other notions of ridgelet transform, and we investigate what we view as the key issue: the summability of the kernel underlying the constructed frame. The sparsity observed in our current implementation is not nearly as good as the sparsity of the underlying continuum theory, so there is room for substantial progress in future implementations.

1.1 INTRODUCTION

1.1.1 Ridgelets on the Continuum

Recently, several theoretical papers have called attention to the potential benefits of analyzing continuum objects $f(x, y)$ with $(x, y) \in \mathbf{R}^2$ using new bases/frames called *ridgelets* [3], [4] and [12]

A *ridge function* $\rho(x, y) = r(ax + by)$, that is to say, it is a function of two variables which is obtained as a scalar function $r(t)$ of a synthetic scalar variable $t = ax+by$ [20]. Geometrically, the level sets of such a function are lines $ax+by = t$ and so the graph of such a function, viewed as a topographic surface, exhibits ridges. The function $r(t)$ is the profile of the ridge function as one traverses the ridge orthogonally to its level sets.

In Candès' thesis [3], a ridgelet is a function $\rho_{a,b,\theta}(x, y) = \psi((\cos(\theta)x + \sin(\theta)y - b)/a)/a^{3/2}$ where $\psi(t)$ is a wavelet – an oscillatory function obeying certain moment conditions and smoothness conditions. The continuous Ridgelet transform $R_f(a, b, \theta) = \langle f, \rho_{a,b,\theta}\rangle$ is defined on functions f in L^1 and extends by density to L^2. This transform obeys a parseval relation and an exact reconstruction formula. Candès also showed that discrete decompositions were possible, so that for L^2 spaces of compactly supported functions one could develop a *frame of ridgelets* – a discrete family $(\psi_{a_n,b_n,\theta_n}(x))$ serving the role of an approximating system.

The "classic ridgelets" of Candès are not in $L^2(\mathbb{R}^2)$, being constant on lines $t = x_1 \cos\theta + x_2 \sin\theta$ in the plane. This fact seems responsible for certain technical difficulties in the deployment and interpretation of discrete systems based on Candès' notion of ridgelet. In [12] Donoho proposed to broaden the concept of ridgelet somewhat, allowing "wide-sense" ridgelets to be functions obeying certain localization properties in a radial frequency $\times$ angular frequency domain. Under this broader conception, ridgelets no longer are of strict ridge the form $\rho_{a,b,u}(x)$, so the elegant simplicity of formulation is lost. However, in exchange, it becomes possible to have an orthonormal set of "wide-sense" ridgelets. These orthonormal ridgelets are believed to be appropriate L^2-substitutes for ridge functions, and to fulfill the goal of a constructive and stable system which *although not based on true ridge functions* are believed to play operationally the same role as ridge functions, compare [12, 13].

For either classic ridgelets or orthonormal ridgelets, the central issue is that such systems should behave very well at representing functions with *linear singularities.* As a prototype, consider the mutilated Gaussian :

$$g(x_1, x_2) = 1_{\{x_2>0\}} e^{-x_1^2-x_2^2}, \qquad x \in \mathbf{R}^2 \,. \tag{1.1.1}$$

See Figure 1.1. This is discontinuous along the line $x_2 = 0$ and smooth away from that line. Due to the singularity along the line, this function has coefficients of relatively slow decay in both wavelet and Fourier domains, so it requires large numbers of wavelets or sinusoids to represent accurately. The rate of convergence of best N-term superpositions of wavelets or sinusoids cannot be faster than $O(N^{-1})$. On the other hand, g can be represented by relatively few ridgelets: the rate of

Figure 1.1. 'Half Dome'– a Mutilated Gaussian

convergence of appropriate N-term superpositions of ridgelets or ortho-ridgelets can be faster than $O(N^{-m})$ for any $m > 0$. And the situation is the same for any rotation or translation of g so that the line $x_2 = 0$ becomes a line $\cos(\theta)x + \sin(\theta)y = t$.

While perfectly straight singularities are rare, many two-dimensional objects concern imagery with edges, which may be regarded as curved singularities. While ridgelets *per se* do not provide the right tool for such curved singularities, Candès and Donoho have used ridgelets to construct a system called *curvelets* which gives high-quality asymptotic approximations to such singularities. Curvelets are ridgelets that have been dilated and translated and subjected to a special space/frequency localization explained in [6]. The rate of convergence of an appropriate N-term superpositions of curvelets is nearly $O(N^{-2})$ in squared error, whereas the comparable behavior for classical systems would by $O(N^{-1})$ or worse.

1.1.2 Discretization of Ridgelets

The conceptual attractiveness of this theoretical work drives us to consider the problem of translating it (if possible) from continuum concepts, useful in theoretical discussion, to algorithmic concepts capable of widespread application. It is *initially* by no means obvious how to do this or whether it can really be done. The theory of ridgelets is closely related to the theories of Radon transformation, and of rotation and scaling of images, all of which seem natural and simple on the continuum, and for which it is widely believed that there is no simple, inevitable definition for digital data.

A number of prior attempts at defining a digital ridgelet transform have been made; these will be discussed in detail further below.

In this paper, we propose a definition of digital ridgelet transform with several desirable properties. We believe that this definition is based on a clear understanding of the fundamental opportunities and limitations posed by data on a Cartesian grid, and has clear superiority over some other notions of discrete ridgelet transform which are, in our view, false starts.

Our definition offers:

- *Analysis and synthesis by true ridge functions.* The underlying analysis and synthesis functions depend on (u, v) as $\rho(u + bv)$ or $\rho(v + bu)$. This means that the transform is geometrically faithful, and avoids wrap-around artifacts.
- *Exact reconstruction formula.* There is an iterative algorithm which in the limit gives exact reconstruction from the ridgelet transform.
- *Near-Parseval Relationship.* There is a variant of the DRT, which we call the (pseudo-) Ortho-Ridgelet Transform, in which the energy in coefficient space is equal to the energy in original space, to within a few percent.
- *Fast algorithm.* There is a fast algorithm requiring only $O(N \log(N))$ flops for data sampled in an n by n grid, where $N = n^2$ is the total number of data.
- *Continuum analogies.* The transform and related objects have structural relationships bearing a strong analogy with all the principal relationships that exist in the continuum case, between ridgelet transform, Radon transform, and Polar Fourier transform.
- *Cartesian data structures.* The transform takes data on a Cartesian grid and creates a rectangular coefficient array indexed according to a semi-direct product of simple integer indices measuring scale, location, and orientation.
- *Overcompleteness.* The transform takes an n-by-n array and expands it by a factor of 4 in creating the coefficient array.

We also compare properties of this DRT with its continuum counterpart, and with other discrete counterparts, particularly as regards sparse representation of objects with discontinuities along lines. We point out certain conceptual and practical advantages of the new transform, over, for example, the Z_p^2 transform proposed by Do and Vetterli [8], and certain advantages over straightforward discretizations of the Fourier plane proposed by Donoho [9] and Starck et al. [22].

Our current implementation provides a frame whose kernel does not have, in our view, sufficient sparsity to provide in the digital setting all the quantitative advantages offered by the continuum theory, leaving ample room for further improvements.

1.2 DIGITAL RIDGELETS

Let $\psi_{j,k}(t) \equiv \psi_{j,k}(t; m)$ be the periodic discrete Meyer wavelet for the m point discrete circle $-m/2 \leq t < m/2$ with indices $J_0 \leq j < \log_2(m)$, and $0 \leq k < 2^j$; this is studied in, for example, Kolaczyk's thesis [18]. This is actually defined as the discrete inverse Fourier transform

$$\psi_{j,k}(t) = \sum_{w=-m/2}^{m/2-1} c_w^{j,k} \exp((i2\pi/m)wt)$$

of a certain complex sequence $(c_w^{j,k})$ which can be derived, e.g. using arguments in [1]. Since the formula makes sense for all t and not only for integers in the range

$-m/2 \leq t < m/2$, *the periodic discrete Meyer wavelet is unambiguously defined not just at integral* t, *but in fact for all real* t. Figure 1.2 displays a Meyer Wavelet of degree 2.

We will also have use for *fractionally-differentiated Meyer wavelets*, defined as follows. For a certain sequence (δ_w)

$$\delta_w = \begin{cases} \sqrt{2w/m} & w \neq 0 \\ \sqrt{1/4m} & w = 0 \end{cases}$$

we apply this as a multiplier to the Fourier coefficients of $\psi_{j,k}$, getting

$$\tilde{\psi}_{j,k}(t) = \sum_{w=-m/2}^{m/2-1} \delta_w \cdot c_w^{j,k} \cdot \exp((i2\pi/m)wt).$$

(Equivalently, we could define $\tilde{\psi}_{j,k} = \tilde{\Delta} \star \psi_{j,k}$, where $\star$ denotes m-point circular convolution and $\tilde{\Delta}$ is the inverse discrete Fourier transform of (δ)). This is equally well viewed as a trigonometric polynomial defined at all t. Figure 1.2 displays a fractionally-differentiated Meyer wavelet. For reasons that will be clear later, we also call the $\tilde{\psi}_{j,k}$ *normalized wavelets.* In this paper we consider images as n by

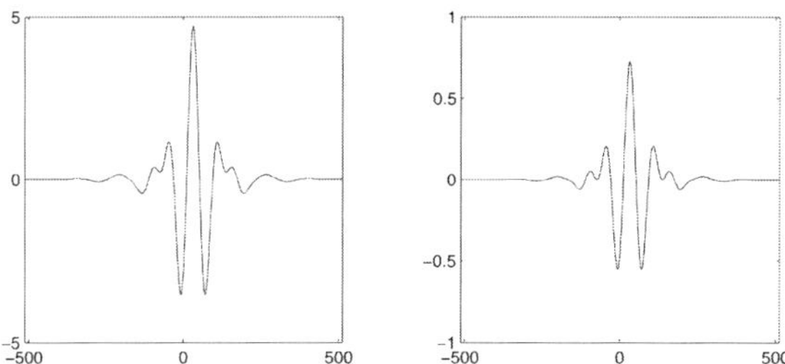

Figure 1.2. Left side: Meyer Wavelet of degree 2. Right side: Fractionally differenciated Meyer wavelet of degree 2

n arrays indexed by coordinates (u, v) ranging in the square $-n/2 \leq u, v < n/2$ centered at $(0, 0)$. Let $\theta^s_{\ell;n}$ be defined so that

$$\tan(\theta^1_{\ell;n}) = 2\ell/n, \quad -n/2 \leq \ell < n/2; \qquad \text{cotan}(\theta^2_{\ell;n}) = 2\ell/n, \quad n/2 \leq \ell < n/2.$$

The lines $v = \tan(\theta^1_{\ell;n})u + t$ we speak of as 'basically horizontal lines' and the lines $u = \text{cotan}(\theta^2_{\ell;n})v + t$ we speak of as 'basically vertical lines'. Each family of lines is equispaced in slope, rather than angle. Figure 1.3 illustrates this family of angles.

Definition 1.2.1 Let n be given. A digital ridgelet $\rho_{j,k,s,\ell}$ is an n by n array built as ridge functions from Meyer wavelets by the formula

$$\rho_{j,k,s,\ell}(u, v) = \psi_{j,k}(u + \tan(\theta^s_\ell)v), \qquad s = 1,$$

and

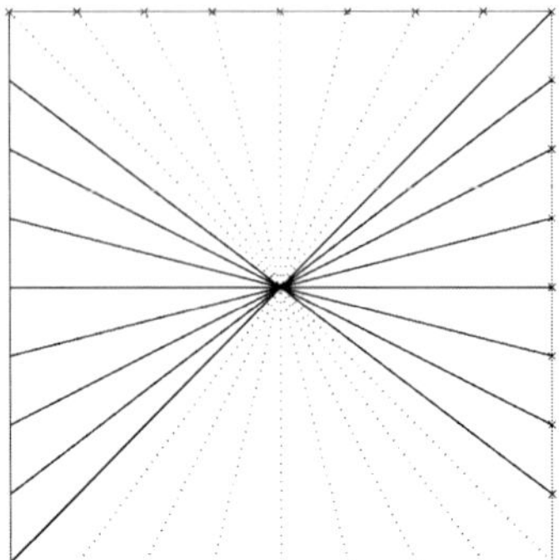

Figure 1.3. Lines in frequency space corresponding to pseudopolar angles

$$\rho_{j,k,s,\ell}(u,v) = \psi_{j,k}(v + \mathrm{cotan}(\theta_\ell^s)u), \qquad s = 2,$$

where the parameter m underlying the definition obeys $m = 2n$. We also call digital ridgelet any function built as ridge functions from fractionally-differentiated Meyer wavelets by the formula

$$\tilde{\rho}_{j,k,s,\ell}(u,v) = \tilde{\psi}_{j,k}(u + \tan(\theta_\ell^s)v), \qquad s = 1,$$

and

$$\tilde{\rho}_{j,k,s,\ell}(u,v) = \tilde{\psi}_{j,k}(v + \mathrm{cotan}(\theta_\ell^s)u), \qquad s = 2.$$

These definitions in fact guarantee that the resulting objects $\rho_{j,k,s,\ell}(u,v)$ and $\tilde{\rho}_{j,k,s,\ell}(u,v)$ are digital samplings of true continuum ridge functions. We note that there are $m = 2n$ wavelets and $2 \times n$ angles $\theta^s_{\ell,n}$. For future use, we write $\lambda = (j,k,s,\ell)$ for quads occurring in this definition, and Λ for the set of all $4n^2$ quads.

Figure 1.4 gives a few examples of such ridgelets.

Figure 1.4. Digital Ridgelets

Definition 1.2.2 The digital ridgelet analysis operator applied to an $n \times n$ image $(I(u,v) : -n/2 \leq u, v < n/2)$ is the array with $4n^2$ entries

$$RI = (\langle I, \rho_\lambda \rangle : \lambda \in \Lambda)$$

We also call digital ridgelet analysis the corresponding normalized operator

$$\tilde{R}I = (\langle I, \tilde{\rho}_\lambda \rangle : \lambda \in \Lambda)$$

In either case, we conventionally think of the DRT as a $2n$ by $2n$ array, as in Figure 1.5, which shows the analysis of an object with linear singularity. Figure 1.6(a) gives a map of the coefficient space.

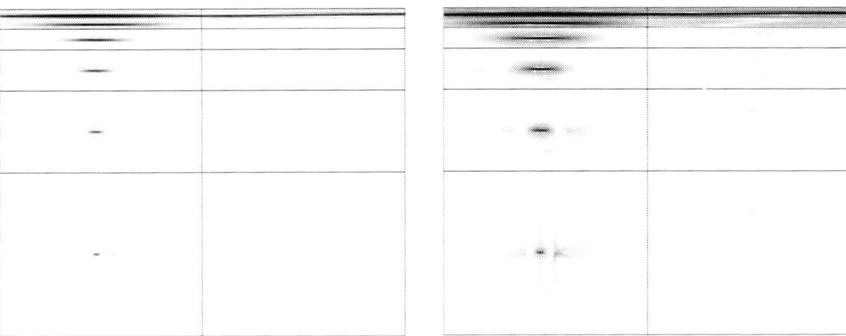

Figure 1.5. Ridgelet analysis of an object with a linear singularity: Left side:Amplitude Map of Ridgelet coefficients. Right side: Amplitude Map of Ridgelet coefficients on a square root scale

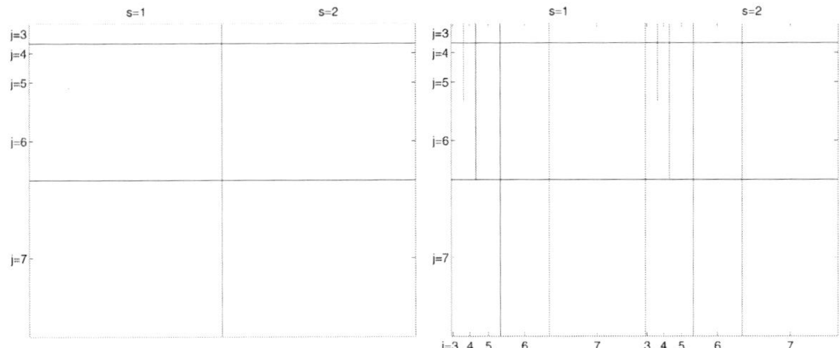

Figure 1.6. Map of coefficient space: Left side: Ridgelets, Right side: Ortho-Ridgelets

Definition 1.2.3 The digital ridgelet synthesis operator takes a $2n$ by $2n$ coefficient array $(\alpha_\lambda : \lambda \in \Lambda)$ into an $n \times n$ array

$$R^\star \alpha = \sum_\lambda \alpha_\lambda \rho_\lambda.$$

We also call digital ridgelet synthesis the corresponding normalized operator

$$\tilde{R}^\star \alpha = \sum_\lambda \alpha_\lambda \tilde{\rho}_\lambda.$$

The notation $R^\star$ is meant to suggest the adjoint operation, and in fact $R^\star$ is precisely the formal adjoint of R.

The first main result is that these transforms are in a sense invertible, and so exact reconstruction is possible in principle.

Theorem 1.2.4 *The operators R and $\tilde{R}$ are one-one and so invertible on their range.*

The second main result is that these transforms are rapidly computable.

Theorem 1.2.5 *The operators R, $\tilde{R}$, $R^\star$ and $\tilde{R}^\star$ can all be computed exactly in order $O(N \log(N))$ exact arithmetic operations, where $N = n^2$ is the number of entries in the n by n image.*

The next 'result' is really a distillation of computational experience.

Empirical Fact. *The normalized transforms $\tilde{R}$, and $\tilde{R}^\star$ have their nonzero singular values within about 10% of each other. The generalized inverse $\tilde{R}^\dagger$ can be computed to seven digits accuracy in 4 iterations of a conjugate gradient solver.*

As a corollary of this empirical result, we have that *the system $\tilde{\rho}_\lambda$ makes a frame, with the ratio of frame bounds empirically smaller than 1.10* . Thus the system $(\tilde{\rho}_\lambda)$ behaves nearly as well as would a tight frame or ortho basis.

It will also be important to consider a discrete analog of orthonormal ridgelets. Let $W_{i,\ell}(u)$ denote a discrete orthonormal Cohen-Daubechies-Feauveau-Jawerth boundary adjusted wavelet for the discrete interval $-n/2 \le u < n/2$. There are n of these wavelets, with indices $0 \le i < \log_2(n)$ and $0 \le \ell < 2^i$. For real sequences (W_u) and (V_u) indexed by $-n/2 \le u < n/2$, let $\{W, V\}$ denote the inner product $\sum_u W_u V_u$.

Definition 1.2.6 Given the normalized discrete Ridgelet transform array RI, we define the (pseudo-) Ortho Ridgelet transform array UI by taking the wavelet transform along the angular variable of $\tilde{R}I$:

$$(UI)_{j,k;i,\ell} = \{RI_{j,k;s,\cdot}, W_{i,\ell}(\cdot)\}.$$

Figure 1.7 shows the ortho-ridgelet transform of the same Halfdome object as in Figure 1.5. It will be evident that the display is more sparse; the transform in θ has compressed the laterally elongated features in Figure 1.5 into more point-like features. Figure 1.6(b) gives a map of the coefficient space M for this transform.

There is, of course, a Riesz representer for each coefficient of UI. For later use, let $\mu = (j, k; i, \ell)$ denote the tuple indexing UI, and let ν_μ denote the Riesz representer of $(UI)_\mu$, i.e. the vector obeying

$$(UI)_\mu = \langle I, \nu_\mu \rangle.$$

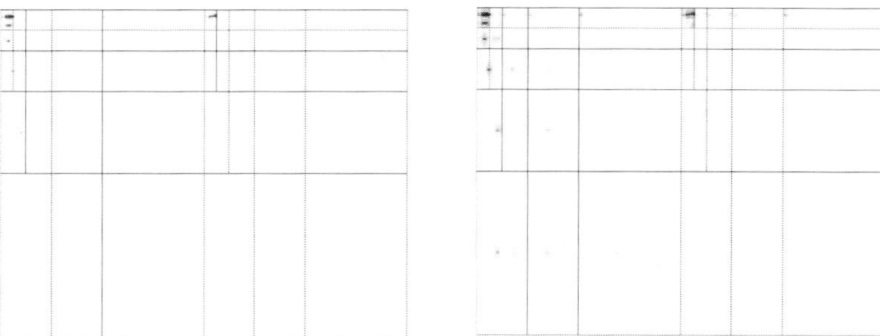

Figure 1.7. Ortho-Ridgelet analysis of an object with a linear singularity: Left side:Amplitude map of Ortho-Ridgelet Coefficients. Right side:Ortho-Ridgelet Coefficients on a square root scale

Figure 1.8 gives an example of such a representer, which we will call a (pseudo-) ortho ridgelet. Just like the ortho ridgelets for the continuum, these are no longer true ridge functions; they crudely behave like fragments of ridgelets windowed by circular windows depending on i. Now U and $\tilde{R}$ are related by an orthonormal

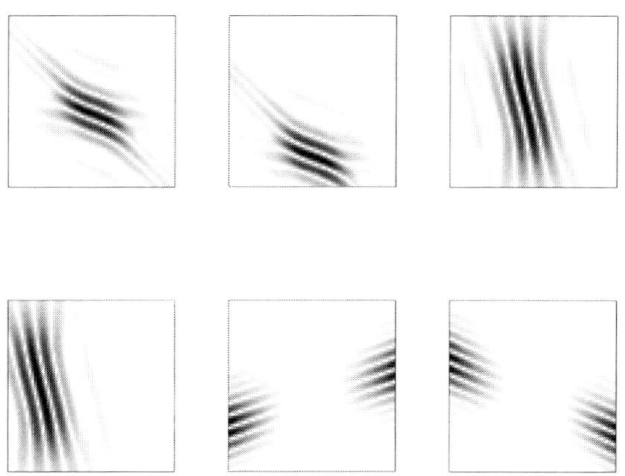

Figure 1.8. Some (pseudo-) Ortho-Ridgelets

transformation of the space of $2n$ by $2n$ arrays, so it follows that all the injectivity and frame bounds properties obeyed by $\tilde{R}$ follow for U as well.

Corollary 1.2.7 *The $4n^2$ elements $(\nu_\mu : \mu \in M)$ make a frame with frame bounds empirically within about 10% of each other.*

1.3 RELATION TO FAST SLANT STACK

The article [2] defined the notion of *Fast Slant Stack* a certain kind of discrete Radon transform which is intimately connected to our notion of Ridgelet transform.

Given an array $I(u,v)$, a slope a with $|a| \leq 1$, and an offset z, we initially define the Radon transform associated with the basically horizontal line $y = ax + b$ via

$$\mathrm{Radon}(\{y = ax + b\}, I) = \sum_u \tilde{I}^1(u, au + b).$$

Thus, we are summing at n values $(u, au + b)$ along the line $y = ax + b$. Since a is not integral, the ordinates at which we are summing do not in general lie in the original pixel grid on integer pairs. Therefore, the values we are summing come not from the original image I, but instead an interpolant $\tilde{I}^1(u, y)$, which takes integer values in the first argument, and real values in the second argument.

The interpolation "in y only", is performed as follows. Letting $m = 2n$, we define the Dirichlet kernel of order m by

$$D_m(t) = \mathrm{cotan}(\pi t/m)\sin(\pi t)/m + i\sin(\pi t)/m.$$

We then set

$$\tilde{I}^1(u, y) = \sum_{v=-n/2}^{n/2-1} I(u, v) D_m(y - v).$$

We note that D_m is an interpolating kernel, so that

$$I(u, v) = \tilde{I}^1(u, v), \qquad -n/2 \leq u, v < n/2.$$

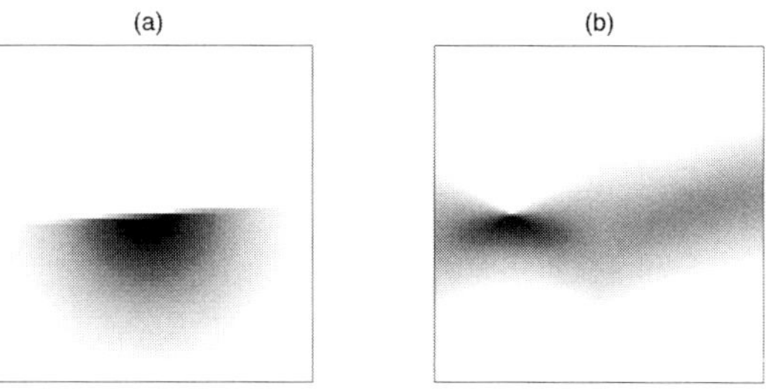

Figure 1.9. Left: Half Dome. Right: Slant Stack of Half Dome

In the case of basically vertical lines, we define the Radon transform similarly, interchanging roles of x and y:

$$\mathrm{Radon}(\{x = ay + b\}, I) = \sum_u \tilde{I}^2(av + b, v),$$

with the interpolant defined analogously:

$$\tilde{I}^2(x, v) = \sum_{u=-n/2}^{n/2-1} I(u, v) D_m(x - u).$$

It is convenient to also have θ to represent the angle associated to the slope s.

Definition 1.3.1 The Slant Stack operator S is defined by

$$(SI)(t, \theta) = \text{Radon}(\{y = \tan(\theta)x + t\}, I),$$

and, for $\theta \in [\pi/4, 3\pi, 4)$,

$$(SI)(t, \theta) = \text{Radon}(\{x = \text{cotan}(\theta)y + t\}, I).$$

where the intercept ranges through $-n \leq t < n$, and the angles vary through $\theta^1_{\ell,n} = \arctan(2\ell/n)$, $-n/2 \leq \ell < n/2$ and $\theta^2_{\ell,n} = \pi/4 + \arctan(2\ell/n)$.

We note that while I is n by n, SI may be regarded as a $2n$ by $2n$ array. Figure 1.9 depicts the slant stack of the Half Dome image. Recall the fractional

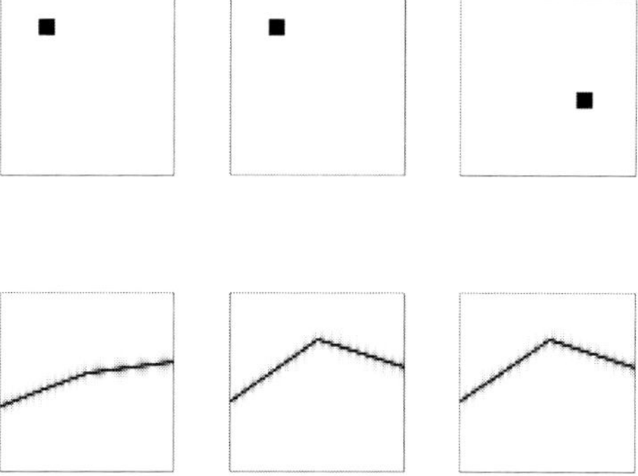

Figure 1.10. The Slant Stack of a point is a broken line. First row: Images with nonzero entry at a single point. Second row corresponding Radon Transform, with break in slope at n corresponding to $\theta = \pi/4$.

differentiation operator introduced in Section 2. We can apply this to t-slices of S to produce a normalized Slant Stack operator:

$$\tilde{S}I(\cdot; s, l) = \tilde{\Delta} \star \tilde{S}I(\cdot; s, l)$$

The following properties of the Slant Stack operator have been established in [2].

Theorem 1.3.2 *The operator S is one-one and hence invertible on its range.*

Theorem 1.3.3 *The operators S and $S^\star$ can each be computed in order $O(N \log(N))$ flops.*

Theorem 1.3.4 *The operator $\tilde{S}$ has n^2 nonzero singular values; the ratio of the largest to smallest is bounded independently of n.*

The next 'result' summarizes the computational experience in [2].

Empirical Fact. *The normalized transform $\tilde{S}$, and adjoint $\tilde{S}^\star$ have all their nonzero singular values within about 10% of each other. The generalized inverse $\tilde{S}^\dagger$ can be computed to seven digits accuracy in 4 iterations of a conjugate gradient solver.*

We can now exhibit the relevance of the Slant Stack to our notion of Ridgelet transform. To do so, we let $\langle\langle , \rangle\rangle$ denote the inner product purely in the t-variable.

Theorem 1.3.5 *The Digital Ridgelet Transform is the 1-dimensional Meyer-wavelet transform, in t, of the Slant Stack Radon Transform:*

$$(RI)(j,k,s,l) = \langle\langle SI(\cdot;s,l), \psi_{j,k}(\cdot)\rangle\rangle.$$

The normalized Digital Ridgelet Transform is the 1-dimensional Meyer-wavelet transform, in t, of the normalized Slant Stack Radon Transform:

$$(\tilde{R}I)(j,k,s,l) = \langle\langle \tilde{S}I(\cdot;s,l), \psi_{j,k}(\cdot)\rangle\rangle.$$

With this equivalence established, it is clear that all the results of the last section follow immediately from the results quoted here. Since the Meyer wavelet transform is an isometry, we have

$$\|SI\|_2 = \|RI\|_2, \qquad \|\tilde{S}I\|_2 = \|\tilde{R}I\|_2.$$

All the norm bounds and norm ratios for the ridgelet transform and for the normalized Slant Stack transform are identical. Since the 1-dimensional Meyer transform costs $O(n\log(n))$ flops, and we perform this once for each column of S the conversion from the $2n \times 2n$ slant stack domain to ridgelet domain costs a total of $O(n^2\log(n))$ or $O(N\log(N))$ flops.

It remains to prove Theorem 1.3.5. Let $[,]$ denote the inner product in the $2n \times 2n$ slant domain, and let $1_{s_0,l_0}$ denote the Kronecker sequence indexed by (s,ℓ) which has a 1 in position $s = s_0$, $l = l_0$ and is zero elsewhere. Then

$$\langle\langle SI(\cdot;s,l), \psi_{j,k}(\cdot)\rangle\rangle = [SI, \psi_{j,k} \otimes 1_{s,l}].$$

Now, by the definition of adjoint,

$$[SI, \psi_{j,k} \otimes 1_{s,l}] = \langle I, S^*(\psi_{j,k} \otimes 1_{s,l})\rangle.$$

Finally, we arrive at the key observation:

$$S^*(\psi_{j,k} \otimes 1_{s,l}) = \rho_{j,k;s,l},$$

a *digital ridgelet is the (slant-) Radon backprojection of a wavelet living in a single θ-slice.* In turn, this follows because *trigonometric interpolation is exact on Meyer wavelets,*

$$\psi_{j,k}(x) = \sum_{t=-n}^{n-1} D_m(x-t)\psi_{j,k}(t).$$

Indeed, S^* involves application of the trigonometric interpolation operator to each column of $\psi_{j,k} \otimes 1_{s,l}$ and this gives exactly the same result at a given $(v = \tan(\theta)u + z)$ as applying the formula for $\psi_{j,k}$ directly at that point. Theorem 3.5 is established.

1.4 STRUCTURAL ANALOGIES

There are several analogies between the discrete ridgelet analysis proposed here and continuum ridgelet analysis. We believe that these analogies further support the correctness of our notion of digital Ridgelet analysis.

1.4.1 Two Continuum Radon Transforms

To understand our analogies it is important to introduce an important variant of the traditional continuum Radon transform – the continuum slant stack.

It is convenient in this paper to denote the Radon transform operator using the letter "X", as R has already been taken by ridgelet transform, and X suggests X-ray. Set

$$(Xf)(t,\theta) = \int f(x)\delta(x_1\cos\theta + x_2\sin\theta - t)\;dx\;, \tag{1.4.1}$$

where $\theta \in [0, 2\pi)$ and $t \in \mathbb{R}$.

Suppose we define, for $\theta \in [-\pi/4, \pi/4)$

$$Y^1 f(t,\theta) = \int f(x,y)\delta(t - x - y\tan(\theta))dxdy, \tag{1.4.2}$$

and for $\theta \in [\pi/4, 3\pi/4)$

$$Y^2 f(t,\theta) = \int f(x,y)\delta(t - \operatorname{cotan}(\theta)x - y)dxdy \tag{1.4.3}$$

and encapsulate these in a single object Yf defined by

$$Yf(t,\theta) = \begin{cases} Y^1 f(t,\theta) & \theta \in [-\pi/4, \pi/4) \\ Y^2 f(t,\theta) & \theta \in [\pi/4, 3\pi/4) \end{cases}.$$

Then, if f is a small pointlike 'bump', a display of Yf will look like a broken line, with a break at the transition angle $\theta = \pi/4$.

The continuum Slant stack transform originated in seismics [7, 23]. Another field in which this continuum transform has been (independently) developed is medical tomography, where Yf is called the Linogram [14, 15], in reference to the fact that points map under Y into broken lines, whereas in the usual Radon transform, points map into sinusoids; because of this in medical tomography, the usual Radon transform is sometimes called the sinogram.

We remark that the continuum Slant stack and the continuum Radon transform contain the same information: for $\theta \in [-\pi/4, \pi/4)$,

$$(Xf)(t \cdot \cos(\theta), \theta) = (Yf)(t, \theta);$$

a similar relationship holds for $\theta \in [\pi/4, 3\pi/4)$.

It should be evident that the continuum slant stack has a close relationship to the discrete slant stack; hence, the above relationship between the continuum slant stack and the continuum Radon transform, provides a connection between the discrete slant stack and the continuum Radon, albeit with a certain amount of relabelling.

This observation is responsible for the several analogies described in this section.

1.4.2 Analogies between Polar and Pseudopolar Fourier Domains

To understand still better the underlying relationships, note that there are actually *three* important domains associated with traditional ridgelet analysis in the continuum case: the ridgelet domain, the Radon domain, and the so-called *polar Fourier* domain. To complete our understanding of the situation, we need to know about all three.

The polar Fourier transform is defined in terms of the usual Fourier transform by simple cartesian-to-polar transformation:

$$\tilde{F}(\omega, \theta) = \hat{f}(\omega \cdot \cos(\theta), \omega \cdot \sin(\theta))$$

Through the well-known projection-slice theorem [7], many facts about the Radon domain can be translated isometrically into facts about the polar Fourier domain, and vice versa. The projection-slice theorem says that the one-dimensional Fourier transform in t of a fixed θ-slice of the Radon transform yields precisely a slice of the Polar Fourier transform at the same θ. So, if $f(x, y)$ is a function in $L^2(\mathbf{R}^2)$ with Radon transform $Xf(t, \theta)$ and polar Fourier transform $\tilde{F}(\omega, \theta)$ and if F_1 denotes 1-dimensional Fourier transform in the first variable

$$(F_1 Xf)(\omega, \theta) = \tilde{F}(\omega, \theta). \tag{1.4.4}$$

The digital ridgelet transform obeys comparable relationships, between ridgelet domain, (slant-) Radon domain, and *pseudopolar Fourier domain.* The pseudopolar domain is discussed in detail in [2]. It offers a notion of polar Fourier domain better adapted to digital data. The digital Fourier domain is viewed as a sequence of squares, not circles, and the "radial shells" picked out by the (pseudo-) "radial" variable are squares.

Define the ordinary 2-d Fourier transform of I by

$$\hat{I}(\xi_1, \xi_2) = \sum_{u,v} I(u, v) \exp\{-i(u\xi_1 + v\xi_2)\}.$$

The discrete pseudopolar Fourier transform P of the digital image I is defined for $-n \leq k < n$ and $-n/2 \leq \ell < n/2$ by sampling the $2-D$ Fourier transform at the collection of pseudopolar grid points illustrated Figure 1.11. Formally,

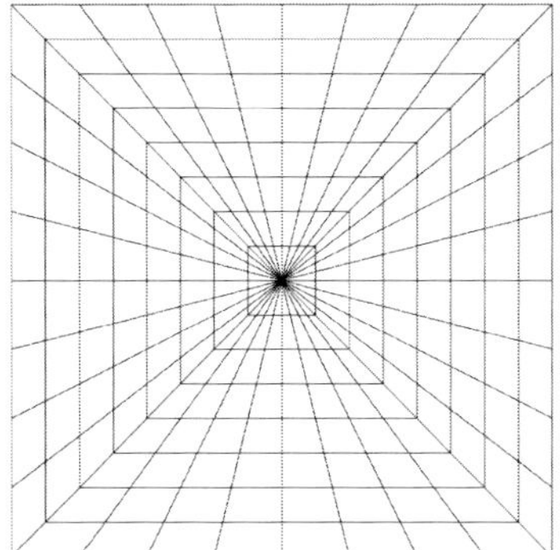

Figure 1.11. Pseudopolar grid points

$$(PI)(k,s,\ell) = \begin{cases} \hat{I}(\pi k/n \tan(\theta_\ell^s), \pi k/n) & s = 1 \\ \hat{I}(\pi k/n, \operatorname{cotan}(\theta_\ell^s)\pi k/n) & s = 2 \end{cases}$$

The frequencies of evaluation in this relation do not lie in a Cartesian grid.

A discrete projection-slice theorem exists relating the (slant-) Radon transform and the pseudopolar Fourier transform. The discrete projection-slice theorem [2] says that if we take the 1-dimensional Fourier transform in t of the slant-Radon data $SI(t, \theta)$, we get samples of the two-dimensional Fourier transform of I.

$$(F_1 Sf)(\omega, \theta) = P(\omega, \theta), \tag{1.4.5}$$

where now F_1 denotes discrete Fourier transform of length $2n$ in the first variable and $\omega = \pi k/n$, $-n \leq k < n$ and $\theta = \theta_{\ell;n}^s$. It follows that the (slant-) transform of digital data is isometric to the pseudopolar Fourier Transform. We remark that this situation parallels a similar relationship for the continuum-domain Slant stack transform Yf introduced above.

1.4.3 Analogies between Radon Isometries

An important consequence of (1.4.4) and (1.4.5): a simple postprocessing of the Radon transform creates an isometry. Indeed, as the Fourier transform is an isometry, we have that the norm of f is conveniently measured in the polar Fourier domain by

$$\|f\|_2^2 = \frac{1}{2} \cdot \int_{-\infty}^{\infty} \int_0^{2\pi} |\tilde{F}(\omega, \theta)|^2 |\omega| d\omega d\theta.$$

Indeed,

$$\begin{aligned} \int_{-\infty}^{\infty} \int_0^{2\pi} |\tilde{F}(\omega, \theta)|^2 |\omega| d\omega d\theta &= 2 \int_0^{\infty} \int_0^{2\pi} |\hat{f}(r \cdot \cos(\theta), r \cdot \sin(\theta))|^2 r dr d\theta \\ &= 2\|\hat{f}\|_2^2 \\ &= 2\|f\|_2^2. \end{aligned}$$

It follows that the object

$$\bar{F}(\omega,\theta) = |\omega|\tilde{F}(\omega,\theta)$$

is isometric with f, and equally that the new object

$$\bar{X}(\cdot,\theta) = F_1^{-1}\bar{F}(\cdot,\theta)$$

is a Radon-like object isometric with f, which we call the *Radon isometry*. Now this object has an alternate definition. Let $\overline{\Delta}$ denote the convolution operator on smooth $L^2(\mathbf{R})$ functions g defined on the Fourier side by the multiplier relation

$$(\overline{\Delta} \star g)(\omega) = |\omega|\hat{g}(\omega), \qquad \omega \in \mathbf{R}.$$

It follows that

$$\bar{X}f(\cdot,\theta) = \overline{\Delta} \star Xf(\cdot,\theta)$$

so that simple postprocessing of X by the Fourier multiplier $|\omega|$ creates an isometry. Since $\overline{\Delta}\star\overline{\Delta} = -(\frac{d}{dt})^2$, $\overline{\Delta}$ is in an obvious sense a fractional differentiation operator.

The Radon isometry $\bar{X}$ bears a strong analogy with the normalized Slant Stack $\tilde{S}$. Indeed, the fact that $\bar{X}$ is an isometry of L^2, makes it comparable to a matrix operator having all its singular values equal to one, while $\tilde{S}$ has all its singular values within a reasonable percentage of each other. Moreover, $\bar{X} = (\overline{\Delta} \otimes I)X$ – i.e. a fractional differential operator is applied to the Radon transform; while $\tilde{S} = (\tilde{\Delta} \otimes I)S$ – a discrete analog of fractional operator is applied to the Slant stack.

1.4.4 Analogies between Ortho-Ridgelet Analyses

In the continuum case, the ridgelet orthobasis is built essentially as follows: we create an orthobasis of wavelets $(W_\lambda(t,\theta))$ living in the Radon domain $(t,\theta) \in \mathbf{R} \times [0, 2\pi)$. We then apply the inverse Radon Isometry $\bar{X}^{-1}$ to the wavelet basis, getting $\rho_\lambda = \bar{X}^{-1}W_\lambda$. In words, (ortho-) ridgelet analysis is wavelet analysis composed with radon isometry.

In the discrete case, we do something entirely similar. We create an orthobasis of wavelets living in the (slant-) Radon domain $-n \le t < n$, $\theta^1_{\ell;n} = \tan(2\ell/n)$, $\theta^2_{\ell;n} = \cotan(2\ell/n)$, by taking the direct product $\psi_{j,k} \otimes w^\varepsilon_{i,l}$. We then apply the normalized (slant-) operator $S^\dagger$ to this, getting (pseudo-) ortho-ridgelets $\tau_{j,k;i,\ell}$.

In the digital ridgelet case, we have something similar, but based on concentric squares rather than concentric circles, see Figure 1.12(b).

1.4.5 Analogies Between Frequency-Domain Tilings

For additional insights, we can consider the ortho-ridgelets in the frequency domain rather than the Radon domain. The typical member $\rho_\lambda(x_1,x_2)$ of the ortho-ridgelets basis for $L^2(\mathbf{R}^2)$ can be defined in the frequency domain by

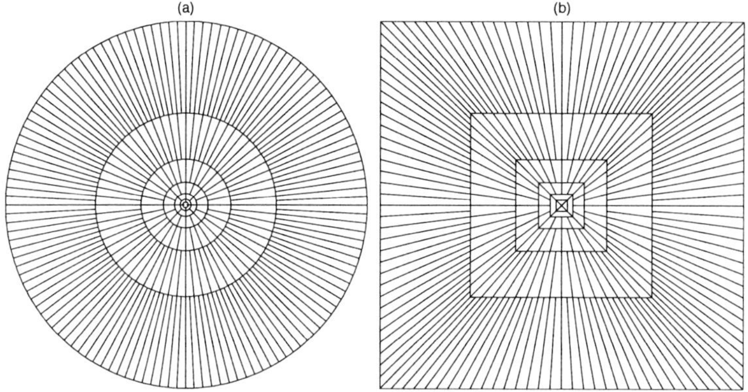

Figure 1.12. Ridgelet tiling and Digital Ridgelet tiling

$$\hat{\rho_\ell}(\xi) = |\xi|^{-\frac{1}{2}}(\hat{\psi_{j,k}}(|\xi|)w^{\varepsilon}_{i,\ell}(\theta) + \hat{\psi_{j,k}}(-|\xi|)w^{\varepsilon}_{i,\ell}(\theta+\pi))/2\ . \tag{1.4.6}$$

Here $(\psi_{j,k}(t) : j \in \mathbb{Z}, \mathrm{k} \in \mathbb{Z})$ denotes an orthonormal basis of Meyer wavelets for $L^2(\mathbb{R})$, and $(w^0_{i_0\ell}(\theta),\ \ell = 0,\dots,2^{i_0}-1;\ w^1_{i,\ell}(\theta),\ i \geq i_0,\ \ell = 0,\dots,2^i-1)$ an orthonormal basis for $L^2[0,2\pi)$ made of periodized Lemarié scaling functions $w^0_{i_0\ell}$ at level i_0 and periodized Meyer wavelets $w^1_{i\ell}$ at levels $i \geq i_0$.

Now the 1-dimensional Fourier transform of a wavelet $\hat{\psi_{j,k}}(\lambda)$ is a windowed sinusoid, with window supported in $\lambda \in [\frac{2}{3}\pi 2^j, \frac{8}{3}\pi 2^j]$. Hence, the Fourier transform of this ridgelet lives in a dyadic annulus of radius $\approx 2^j$. The wavelet $w^{\varepsilon}_{i,\ell}(\theta)$ is mostly concentrated to a window near $\theta_{i,\ell} = 2\pi\ell/2^i$. Hence, the Fourier transform of the ortho ridgelet lives in an angular wedge. Combining these remarks, we have the pattern illustrated in Figure 1.12(a), which we call the ridgelet tiling of the frequency plane.

1.5 EXAMPLE: HALFDOME

We now use the HalfDome object to illustrate the relationships we have just discussed, and to underline the plausibility of ridgelet analysis. Figure 1.13 shows the HalfDome object in Fourier space both with and without the Ridgelet tiling superposed. It is evident that the energy in the object is concentrated near a highly elongated sausage-shaped feature oriented at 90° to the orientation of the discontinuity in the HalfDome object in original space. We of course hope that the ridgelet tiling is well-adapted to the underlying distribution of energy in the object under study – i.e. that only a few tiles are required to cover the bulk of the energy in the object and that only a few coefficients per tile are needed to represent the part of the object overlapping that tile. It is evident from the distribution of the energy in frequency space that there is a strong possibility that things will work out as hoped.

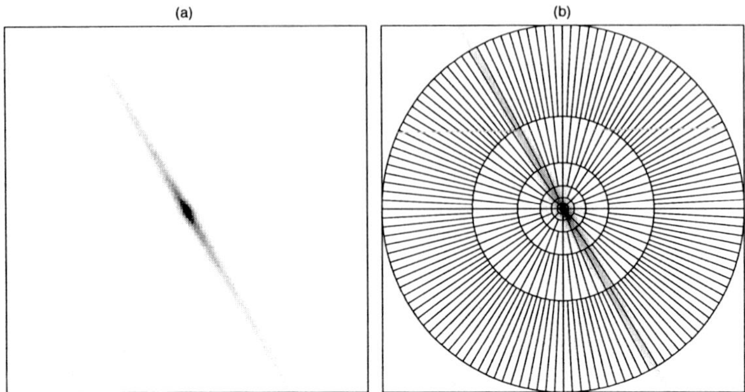

Figure 1.13. Fourier Transform of Halfdome on log scale. (a) without (b) with ridgelet tiling

We now consider the situation in the Radon domain. Figure 1.14 reveals that the Slant Stack of the HalfDome object is smooth away from a singularity at a single (s, l, t) value. Now, the ridgelet coefficients of HalfDome are roughly speaking such wavelet coefficients of the Slant Stack. It makes sense that a $2-d$ wavelet transform of this object will have sparse coefficients, with the large coefficients concentrated at indices associated with spatial positions near the singularity. Hence, we expect the ridgelet transform to be sparse. However, to be rigorously correct, we must remember that the ridgelet coefficients arise from the wavelet transform of the normalized Slant Stack, which is portrayed in Figure 1.14.

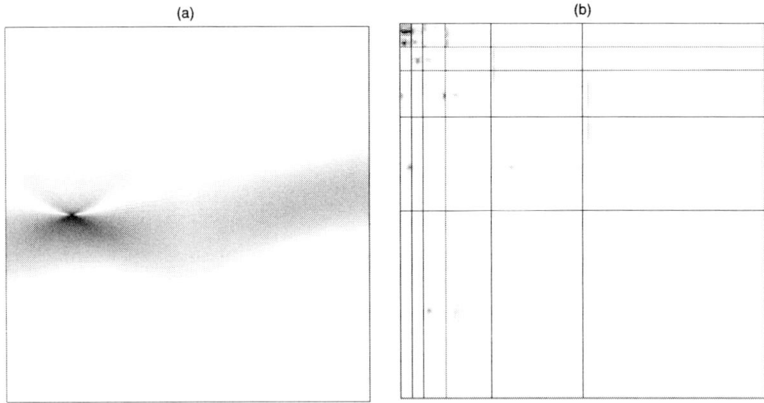

Figure 1.14. a) Normalized Slant Stack of HalfDome and b) Wavelet Transform of it

The normalized Slant Stack exhibits - like the unnormalized Slant Stack - smoothness away from a singularity at a single point. Consequently, it is clear that we should indeed have sparsity of the 2-D wavelet transform and hence sparsity in the ridgelet domain, as indeed we observe in Figure 1.5. For extra clarity, we show the wavelet transform with nonlinear transformation.

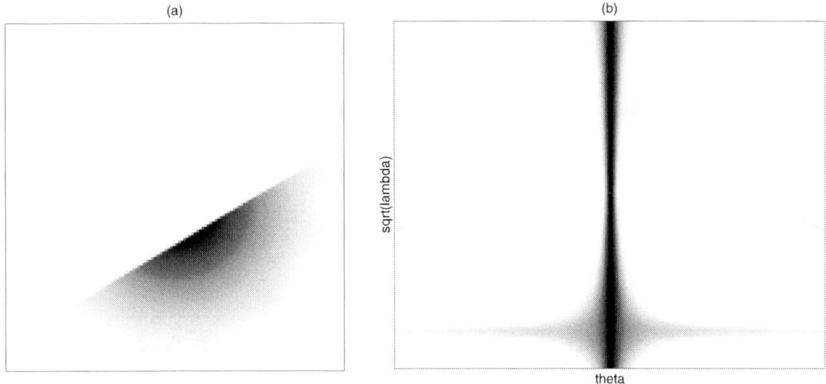

Figure 1.15. Left: Half Dome, Right:Pseudopolar FT of HalfDome

Continuing in our chain of equivalences, suppose we now take the discrete Fourier transform in t of each t-slice of the normalized Slant Stack. Then we are viewing the (normalized) pseudopolar Fourier transform of HalfDome; this is displayed below. In this pseudo-polar coordinate system, the object of interest is well-concentrated in a few tiles. Again, we see reason to expect a sparse representation.

1.6 SPARSITY OF THE FRAME KERNEL

Both ridgelet transforms defined here are expansive: they transform an n by n array into a $2n$ by $2n$ coefficient array. Equivalently, each set of analysis and synthesis functions, having $4n^2$ elements for representing objects with n^2, is overcomplete.

As a result, it becomes important to study the frame kernel $G(\mu, \nu) = \langle \rho_\mu, \rho_\nu \rangle$, particularly as regards the sparsity of its rows kernel. If the ridgelet system were orthonormal, each row would have a single nonzero element on the diagonal, and hence be extremely sparse. In the overcomplete case, what we can hope for is that the nonzero elements in each row, once rearranged in order of decreasing amplitude, decay rapidly.

Owing to the obviously high inner products of true ridge functions oriented in neighboring directions, we should not expect rapid decay of the coefficients if we work in the ridgelet system, but should look instead to the ortho-ridgelet system.

To check the sparsity condition, we consider the following computational experiment: we synthesize a single ortho-ridgelet and then analyze that ridgelet, inspecting the resulting transform coefficients for rapid decay.

We give examples here in two cases.

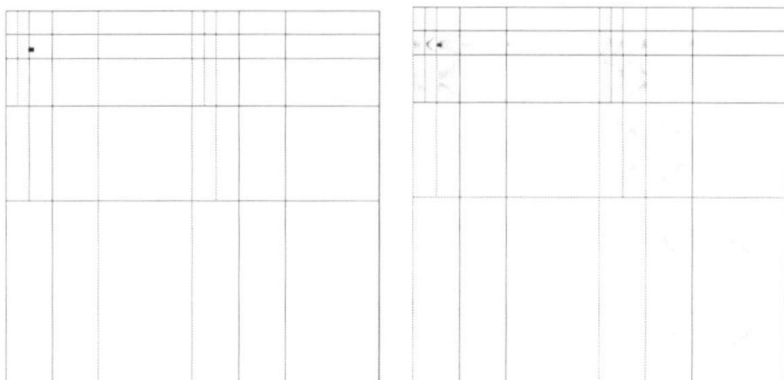

Figure 1.16. Ortho-Ridgelet Synthesis and Analysis Planes: Left side: OrthoRidgelet Synthesis. Right side:OrthoRidgelet Analysis on a square root scale

1.6.1 Analysis of a Coarse-scale ridgelet

Figure 1.16 shows the ortho-ridgelet synthesis plane – evidently a delta sequence peaking at the appropriate place in coefficient space –, and the ortho-ridgelet analysis plane. The analysis plane is considerably more spread out than the synthesis plane. Figure 1.17 presents a close-up of the principal subband.The delta in the synthesis plane is replaced by a blob in the analysis plane. Another display of

Figure 1.17. Close-up of principal subband of Ridgelet Analysis Plane

behavior in the analysis plane simply plots the sizes of coefficients in decreasing order, as shown in Figure 1.18. Evidently, there are few "big" coefficients, followed by a decaying tail.

1.6.2 Remarks on Decay

While the analysis plane is visually quite sparse, in fact the degree of sparsity exhibited by the transform is disappointing. This is caused by the fact – seen in Figure 1.18– that, after the initial quite rapid drop-off of coefficient amplitudes, there is a rather slow decay in the tail region. This sort of phenomenon is well known in the harmonic analysis of nonperiodic signals, and it is an interesting open question whether some variant on the procedure could repair this slow decay.

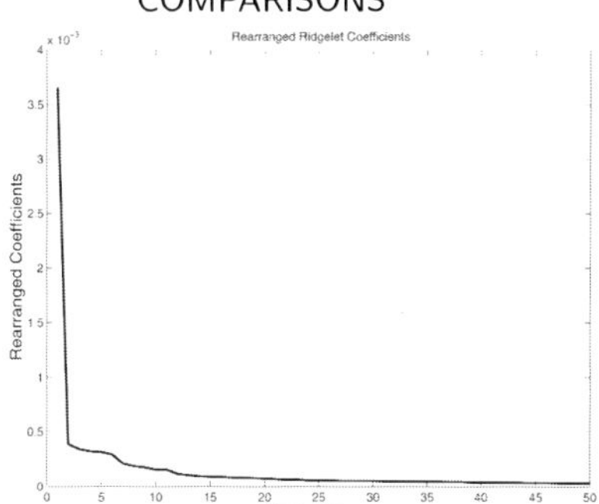

Figure 1.18. Decreasing Rearrangement of Ridgelet Analysis Plane

1.6.3 Edge Effects

It seems important to note that the kernel decay becomes worse the more the ridgelet in question is concentrated near the edges of the image. We consider a ridgelet at a fine scale that happens to live near the corner of the image domain, and again show its ten-term reconstruction, see Figure 1.19. We again show the analysis and synthesis planes. The analysis plane is dramatically more spread out than before, exhibiting long stripes, see Figure 1.20. Figure 1.21 presents a close-up of the principal subband.

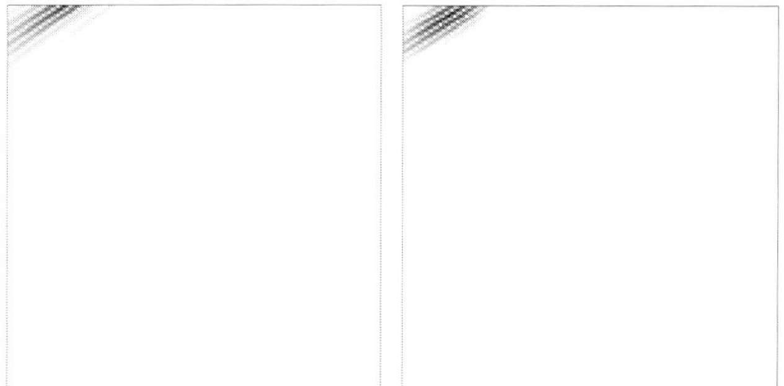

Figure 1.19. Left, Corner Ridgelet, and right, 10-term Reconstruction

1.7 COMPARISONS

1.7.1 Comparison with Z_p^2-ridgelets

Do and Vetterli [8] have proposed a method of ridgelet analysis based on the use of the Radon transform for Z_p^2, the cartesian product of two copies of the integers mod p, where p is a prime. At a formal level, this has much to recommend it, including orthogonality. Essentially, one applies the Z_p^2 Radon transform, and then take the wavelet transform in the 't' direction.

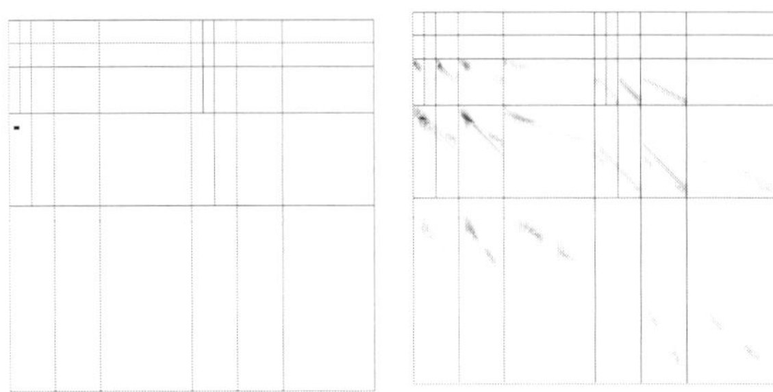

Figure 1.20. Corner Ridgelet: Synthesis and Analysis Planes

Figure 1.21. Corner Ridgelet: Principal subband of Ridgelet Analysis Plane

Unfortunately, the Z_p^2 Radon transform integrates over 'lines' which are defined algebraically rather than geometrically, and so the points in a 'line' can be rather arbitrarily and randomly spread out over the spatial domain. In consequence, the Z_p^2 ridgelets that are defined in this way have a rather strange apppearance. Typical examples are shown on Figure 1.22. Such 'ridge functions' have their support

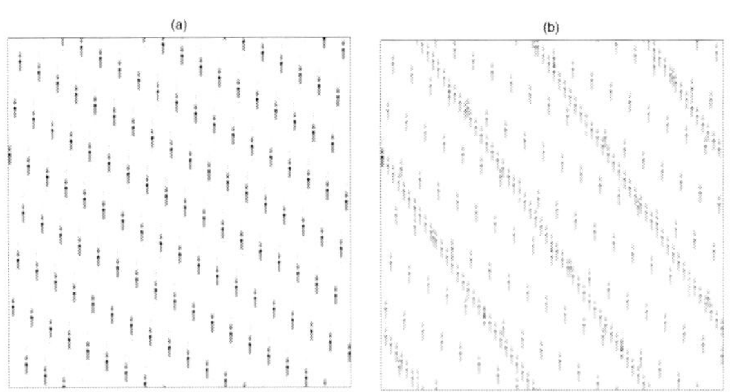

Figure 1.22. Examples of Z_p^2 Ridgelets

scattered haphazardly throughout the image plane, and resemble neither a a traditional ridge function, nor any spatially coherent object. As a consequence of this behavior, partial reconstructions in the Z_p^2 system have errors which look very much like textured random noise. Figure 1.23 compares reconstruction of Half-Dome by 50 Z_p^2 ridgelet coefficients with reconstruction by 50 ridgelets based on the transform developed here. The additional noisiness is clear. Figure 1.24 con-

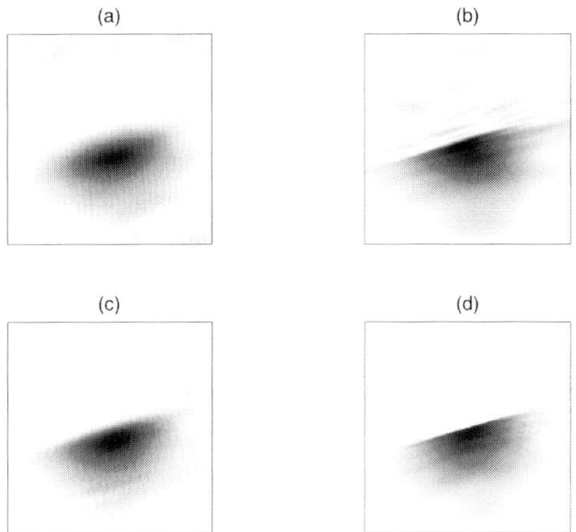

Figure 1.23. Reconstruction of HalfDome by Z_p^2 ridgelets, and digital ridgelets. (a) Reconstruction by 50 Z_p^2 coefficients, (b) Reconstruction by 50 DR coefficients, (c) Reconstruction by 100 Z_p^2 coefficients, (d) Reconstruction by 100 DR coefficients

siders an object used by Do and Vetterli [8] in their paper on Z_p^2 ridgelets, and compares reconstruction by 50 Z_p^2 ridgelet coefficients with reconstruction by 50 ridgelets based on the transform developed here. Again, the additional noisiness in Z_p^2 reconstruction is clear.

In summary, the Z_p^2 approach simply is not based on a geometrically faithful notion of ridgelet, and suffers from textural artifacts.

1.7.2 Comparison with earlier ridgelets

We briefly mention three other notions of ridgelet transform we know about.

In [9], an initial attempt was made to construct a discrete ridgelet transform operating in order $N\log(N)$ flops, where $N = n^2$ is the image size. The idea was based on approximate cartesian-to-polar resampling. One would overlay a true polar grid on the discrete Fourier transform, and approximately evaluate the discrete Fourier transform at polar grid points by interpolation from nearby cartesian grid points. It was shown that by this device one could obtain a Frame, provided the polar grid were sufficiently dense. However, this approach had three seeming drawbacks. First, it might be that it would require a very high degree of oversampling to obtain the Frame property. Second, the kind of interpolation required would involve considerable arithmetic for each desired polar grid point, with frequent accesses to data in a near a corresponding pixel location, which, on

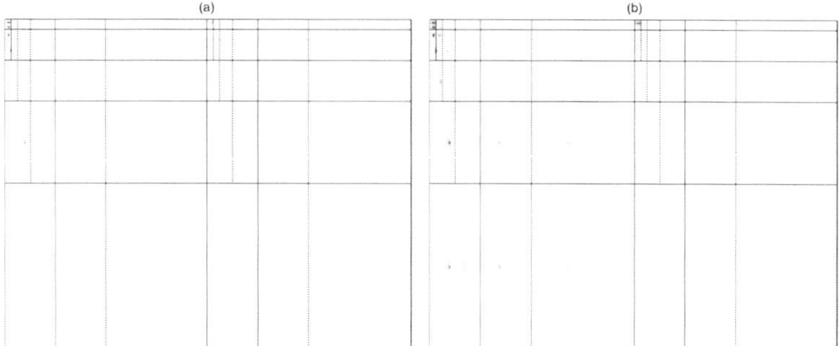

Figure 1.28. Ridgelet Analysis of BandPassed HalfDome. Left side:Amplitude Map of Ridgelet coefficients. Right side:Amplitude Map of Ridgelet coefficients on a square root scale

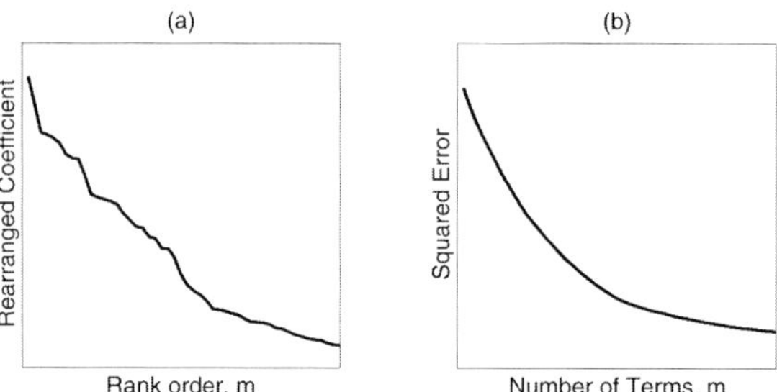

Figure 1.29. Left side: Decreasing Rearrangement of Ridgelet Analysis Plane, Right side: m-term aproximation errors

the image de-noising application described there, this factor does not seem to be very important.

Yoel Shkolnisky has pointed out another possible variation on our approach, discussed in [2] and forthcoming work. In extending the image we zero pad out to length $m = 2n + 1$ rather than $2n$, obtaining a corresponding expression for D_m which is purely real, namely

$$D_m(t) = \frac{\sin(\pi t)}{m \cdot \sin(\pi t/m)}.$$

This fixes a slightly inelegant property of the choice $m = 2n$, namely that the ridgelets with the highest radial index j are not purely real – they have a small imaginary component, owing to the small imaginary component of the kernel D_{2n}. With $m = 2n + 1$ this component goes away.

1.8 DISCUSSION

We have described a notion of ridgelet transform which is able to synthesize or analyze using true ridge functions and which has various exact reconstruction

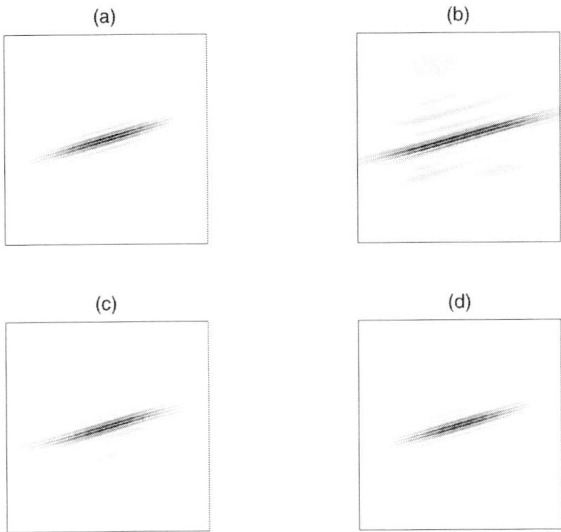

Figure 1.30. Reconstructions from 16, 50 and 100 OR coefficients: (a) Band Passed Half Dome, (b) Reconstructed by 16 coefficients, (c) reconstructed by 50 coefficients, (d) reconstructed by 100 coefficients.

and frame properties. It also obeys a series of relationships with a notion of digital Radon transform and a notion of digital polar Fourier transform which are precisely analogous to corresponding relationships that exist in the frequency domain. At its heart, the method is based on the use of pseudopolar FFT and Fast Slant Stack described in [2].

The principal disappointment of the existing implementation is the relatively slow decay of the ortho-ridgelet coefficients of a ortho-ridgelet. Figure 19 shows that, after an initially steep decline in coefficients, a kind of flat 'background' sets in. The slow decay is reminiscent of the behavior one would see from Gibbs phenomena in fourier analysis of discontinuities, or from critical sampling in Gabor analysis. It is not hard to see why this behavior obtains, and to see that it is *intrinsically* tied to our central assumption – the use of ridge functions in a digital setting. Figure 1.31 illustrates the ridgelet analysis process; Figure 1.32 illustrates its adjoint, the ridgelet synthesis process. The ridgelet analysis process works as follows: an image is extended to twice its length, then sheared, then projected, then wavelet analyzed. The ridgelet synthesis process works 'in reverse': a delta sequence in the wavelet coefficient domain is inverted into a wavelet, which is then backprojected into a ridge function, which is then sheared into a tilted ridge function, *which is then mutilated.*

This last step – mutilation – is the adjoint of extension by zero-padding, meaning that digital ridgelets, when viewed as an array, amount to a series of columns containing 1-d wavelets which have been brutally truncated from an $n \times 2n$ array to fit in an $n \times n$ array. If the wavelets were not mutilated, their inner products would decay rapidly with separation in index space; but the mutilation spoils the

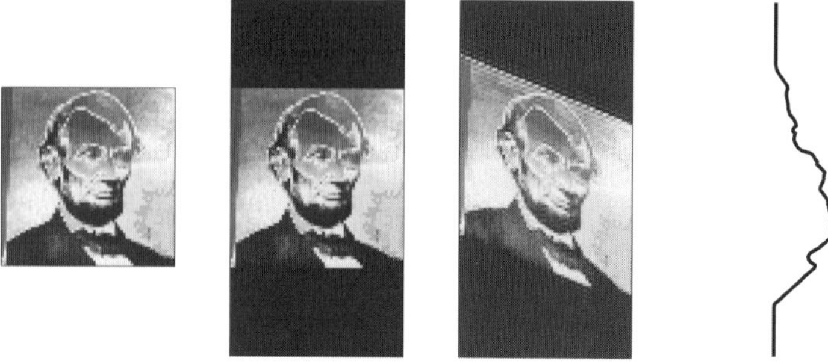

Figure 1.31. Stages of Radon analysis: padding, shearing, projection

Figure 1.32. Stages of Radon synthesis (right to left): backprojection, shearing, mutilation

decay property. From this point of view, it is rather obvious what to try next. One should develop analysis and synthesis transforms not based on ridge functions, but instead based on windowed ridge functions. We have not explored this proposal in detail here simply because although a straightforward and obvious extension, it violates the initial assumption marking the origin of this project: the use of true ridge functions.

ACKNOWLEDGEMENT

This work has been partially supported by AFOSR MURI 95–P49620–96–1–0028, by National Science Foundation grant DMS 98–72890 (KDI), and by DARPA ACMP BAA 98-04. The authors would like to thank Amir Averbuch, Emmanuel Candès, Raphy Coifman, Jean-Luc Starck, Yoel Shkolnisky, and Martin Vetterli for helpful comments, preprints, and references. AGF would like to thank the Statistics Department at UC Berkeley for its hospitality.

REFERENCES

[1] P. Auscher, G. Weiss and M.V. Wickerhauser, Local sine and cosine bases of Coifman and Meyer and the construction of smooth wavelets. *Wavelets: A tutorial in Theory and Applications* Academic Press: Boston, 237–256.

[2] A. Averbuch, R.R. Coifman, D.L. Donoho, M. Israeli and J. Waldén *Fast Slant Stack: A notion of Radon Transform for Data in a Cartesian Grid which is Rapidly Computible, Algebraically Exact, Geometrically Faithful and Invertible*, to Appear: SIAM J.Sci. Comp.

[3] E. Candès *Ridgelets: Theory and Applications.* Ph.D. Thesis, Department of Statistics, Stanford University, 1998.

[4] E.J. Candès, and D. L. Donoho *Ridgelets: The key to High-Dimensional Intermittency?* in Phil. Trans. R. Soc. Lond. A. **357** 1999, 2495–2509.

[5] E. Candès and D. Donoho *Curvelets and Curvilinear integrals* (1999) Technical Report, Department of Statistics, Stanford University. *http://www-stat.stanford.edu/~donoho/Reports/2000/Curves-Curvelets.ps.*

[6] E. Candès and D.L. Donoho. Curvelets: a surprisingly effective nonadaptive representation of objects with edges. *Curve and Surface Fitting: Saint-Malo 1999* Albert Cohen, Christophe Rabut, and Larry L. Schumaker (eds.) Vanderbilt University Press, Nashville, TN. ISBN 0-8265-1357-3.

[7] S. R. Deans *The Radon transform and some of its applications.* New York: Wiley & Sons.

[8] M. Do and M. Vetterli. *Orthonormal finite Ridgelet transform for image compression.* To appear Proc. of IEEE International Conference on Image Processing, ICIP-2000.

[9] D. L. Donoho. Fast Ridgelet Transform in Dimension 2. Available: *http://www-stat.stanford.edu/~donoho/Reports/1997/FRT.pdf*,1997.

[10] D.L.Donoho, Fast Ridgelet Transform via Digital Polar Coordinate Transform. Available: *http://www-stat.stanford.edu/~donoho/Reports/ 1998/FRTvD-PCT.pdf*, 1998.

[11] D.L. Donoho. Tight Frames of k-Plane Ridgelets and the Problem of Representing d-dimensional singularities in $\mathbf{R}^n$. *Proc. Nat. Acad. Sci. USA*, **96** 1998, 1828–1833.

[12] D.L. Donoho. *Orthonormal Ridgelets and Linear Singularities.* SIAM J. Math Anal. **31** no 5 2000, 1062-1099.

[13] D.L. Donoho *Ridge Functions and Orthonormal Ridgelets.* Journ. Approx. Thry. **111** 2001, 143–179.

[14] P. Edholm and G.T. Herman.*Linograms in image reconstruction from projections.* IEEE Trans. Medical Imaging, **MI-6(4)** 1987, 301–307.

[15] P. Edholm, G.T. Herman, and D.A. Roberts. *Image reconstruction from linograms: Implementation and evaluation.* IEEE Trans. Medical Imaging, **MI-7(3)**1988, 239–246.

[16] I.M. Gel'fand and G.E. Shilov *Generalized Functions: Properties and Operations.* Academic Press, 1964.

[17] S. Helgason, *Groups and Geometric analysis: integral geometry, invariant differential operators, and spherical functions.* Series title: Pure and applied

mathematics. Orlando: Academic Press,1984.

[18] E. Kolaczyk *Wavelet-Vaguelette Decomposition of certain Homogeneous Linear Inverse problems.* Ph.D. Thesis, Department of Statistics, Stanford University, 1994.

[19] P. G. Lemarié Y. and Meyer. *Ondelettes et bases Hilbertiennes*, Revista Matematica Iberoamericana **2** 1986, 1-18.

[20] B.F. Logan L.A. and Shepp *Optimal reconstruction of a function from its projections.* Duke Math. J. **42** 1975, no. 4,645–659 .

[21] F. Matus and J. Flusser.*Image representations via a Finite Radon Transform.* IEEE Trans. Pattern Ana. Machine Intell. **15** 1993, 996-1006.

[22] J.L. Starck, E. Candes and D.L. Donoho *The Curvelet Transform for Image Denoising*, to appear IEEE Transactions on Image Processing.

[23] Oz Yilmaz, *Seismic Data Processing* (SEG Investigations in Geophysics N 2.) 2000.

Beyond Wavelets
G. V. Welland (Editor)
© 2003 Elsevier Science (USA) All rights reserved

2

DIGITAL IMPLEMENTATION OF RIDGELET PACKETS

A.G. FLESIA, H. HEL-OR,
A. AVERBUCH, E.J. CANDÈS,
R.R. COIFMAN AND D.L. DONOHO

Department of Statistics, Stanford University
Sequoia Hall, 390 Serra Mall, Stanford, CA 94305-4065
donoho@stat.stanford.edu flesia@stat.stanford.edu

Department of Computer Science, Haifa University
Haifa 31905, Israel
H. Hel-Or, *hagit@cs.haifa.ac.il*

Computer Science Department, Tel Aviv University
Tel Aviv 69978, Israel
A. Averbuch, *amir@math.tau.ac.il*

Applied and Computational Mathematics, California Institute of Technology
1200 E. California Boulevard, MC 217-50 Pasadena, CA 91125, USA
E.J. Candès, *emmanuel@acm.caltech.edu*

Department of Mathematics, Yale University
PO Box 208283 New Haven, CT 06520-8283, USA
R.R. Coifman, *coifman@math.yale.edu*

Abstract

The Ridgelet Packets library provides a large family of orthonormal bases for functions $f(x, y)$ in $L^2(dxdy)$ which includes orthonormal ridgelets as well as bases deriving from tilings reminiscent from the theory of wavelets and the study of oscillatory Fourier integrals. An intuitively appealing feature: many of these bases have elements whose envelope is strongly aligned along specified 'ridges' while displaying oscillatory components across the main 'ridge'.

There are two approaches to constructing ridgelet packets; the most direct is a frequency-domain viewpoint. We take a recursive dyadic partition of the polar Fourier domain into a collection of rectangular tiles of various widths and lengths. Focusing attention on each tile in turn, we take a tensor basis, using windowed sinusoids in θ times windowed sinusoids in r. There is also a Radon-domain approach to constructing ridgelet packets, which involves applying the Radon isometry and then, in the Radon plane, using wavelets in θ times wavelet packets in t, with the scales of the wavelets in the two directions carefully related.

We discuss digital implementations of the two continuum approaches, yielding many new frames for representation of digital images $I(i, j)$. These rely on two tools: the pseudopolar Fast Fourier Transform, and a pseudo Radon isometry called the normalized Slant Stack; these are described in Averbuch et al. (2001). In the Fourier approach, we mimic the continuum Fourier approach by partitioning the pseudopolar Fourier domain, building an orthonormal basis in the image space subordinate to each tile of the partition. On each rectangle of the partition, we use windowed sinusoids in θ times windowed sinusoids in r. In the Radon approach, we operate on the pseudo-Radon plane, and mimic the construction of orthonormal ridgelets, but with different scaling relationships between angular wavelets and ridge wavelets. Using wavelet packets in the ridge direction would also be possible.

Because of the wide range of possible ridgelet packet frames, the question arises: what is the best frame for a given dataset? Because of the Cartesian format of our 2-D pseudopolar domain, it is possible to apply best-basis algorithms for best anisotropic cosine packets bases; this will rapidly search among all such frames for the best possible frame according to a sparsity criterion – compare N. Bennett's 1997 Yale Thesis. This automatically finds the best ridgelet packet frame for a given dataset.

2.1 INTRODUCTION

There is considerable scientific and technological interest in representing 2-dimensional objects $f(x, y)$ – images – in terms of oscillatory $2 - d$ waveforms with anisotropic envelopes.

For example, it is widely believed in the study of computational and biological vision that early vision depends on the analysis of image data by directionally oriented waveforms [19, 35, 20, 21]. It is also believed that representations by oriented oscillatory waveforms are important for efficient representation of real image data, [2, 30, 29], and to texture simulation [22].

These scientific and engineering interests interact with the agenda of computational harmonic analysis, where an important goal in dealing with 2-D and higher-dimensional objects is to construct representations which have a wide range of orientations, aspect ratios, locations, scales, and oscillation numbers. For example, one would like the ability to decompose an object $f(x)$, $x \in R^d$, stably and compactly, into a sum of oscillatory waveforms with anisotropic envelopes:

$$\psi_{A,b,\xi}(x) = e^{i\xi'(x-b)} \exp\{-(x-b)'A(x-b)\}.$$

If such an ability were available, it would encompass in one framework most of the benefits of wavelets, Gabor transforms, ridgelets, and many other specific decompositions.

Unfortunately, no effective, stable decomposition into general collections of such atoms is available today. The present paper does make some progress in this direction in the two-dimensional case $d = 2$.

To put our approach in context, recall that an important stage in the development of computational harmonic analysis was the generalization from wavelet bases to wavelet packet and cosine packet bases [11]. Coifman and Meyer recognized that one could develop a very intuitive picture of basis construction around the notion of tilings of the time-frequency domain, and this picture would explain how to go beyond wavelets, creating a very wide variety of orthonormal bases with various interesting properties. In their approach, one had available a general family of tilings of the time-frequency plane based on recursive dyadic partitioning, and one could build a basis corresponding to each such tiling by using some simple tools deployed within an appropriately general framework. Reducing this insight to very intuitive terms, these constructions allow bases made of waveforms which, like wavelets, were localized in time and scale, but which instead of the fixed, small number of oscillations offered by wavelets, had a variable number of oscillations, controllable by the user.

In this paper, we describe a system of orthonormal bases and frames for $L^2(dxdy)$ as well as frames for two-dimensional digital data $I(i, j)$. The system we describe, *Ridgelet Packets*, has many points in common with ridgelet analysis, while allowing for a more complex and oscillatory structure. Thus, the system contains bases where the basis elements are, like ridgelets, highly orientation-selective, but which have greater degrees of oscillation either along or across the direction of primary orientation. We also describe algorithms for two-dimensional digital data which can adaptively construct frames which are well-adapted to sparse representation of given digital data. The tools can be viewed as a contribution towards solving the problem proposed in paragraph one above: automatically decomposing objects into compact collections of anisotropic, oscillating waveforms. They bear much the same relationship to ridgelets as do wavelet packets to wavelets.

One idea driving our construction is indicated in Figure 2.1 below. We illustrate the tiling underlying the orthonormal ridgelet basis, and two other tilings which we have labeled 'wavelet-like' and 'FIO'. These tilings of Fourier space are all similar in that they all involve the subdivision of the radial axis dyadically. They differ in the degree of angular subdivision that they apply. For ridgelets, the frequency band $2^j < r < 2^{j+1}$ is subdivided into 2^j tiles angularly, while for the FIO tiling, the same frequency band is divided into $2^{j/2}$ tiles, and for the wavelet-like tiling, the frequency band is divided in a constant number of tiles, irrespective of the frequency range. (Inspirations for studying tilings of these forms include: for the ridgelet tiling – Candès' thesis [8, 7]; for wavelet-like tiling – work in representation of natural images, such as Watson's article on the Cortex transform [35], and Field's articles on statistics of natural images [19, 20]; for the FIO-tiling – work applying 'Second Dyadic Decomposition' in harmonic analysis going back to Fefferman's analysis of Bochner-Riesz summability; see [32, 31].)

After tiling the frequency domain according to one of these schemes, we can create a basis where each basis element is associated with a specific frequency

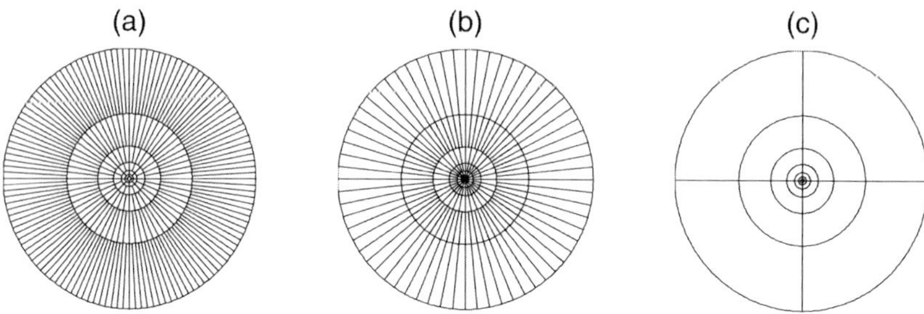

Figure 2.1. Three Tilings of the Frequency Domain: (a) Ridgelet tiling; (b) FIO tiling; (c) Wavelet tiling

domain tile. If the tile is very narrow in the angular sense, then the basis element, viewed in the original spatial (x, y) domain, will exhibit a very high degree of orientational preference. If the tile is very anisotropic, the spatial envelope of the basis element, viewed in the spatial domain, will have a very anisotropic envelope, and so a long and thin tile (long in r and thin in θ) will give rise to basis elements which are long and thin as well.

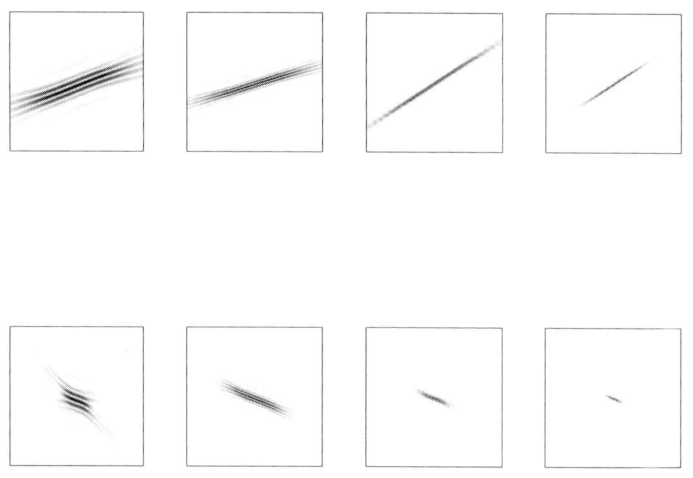

Figure 2.2. First Row, Left to Right: Basis Elements associated with Ridgelet tiling, from coarse to fine ridge scale, with same angular position. Second Row, Left to Right: Basis Elements associated with FIO tiling, from coarse to fine ridge scale, with same angular position.

We have illustrated this concept in Figure 2.2, the first row depicts basis elements associated with the Ridgelet tiling (a) from Figure 2.1, that belongs to four consecutive scales in r, plotted from left to right as r increases, and same angular direction θ. The second row depicts basis elements associated with the FIO tiling (b) from Figure 2.1, chosen from the same scales in r as the Ridgelet basis elements. We can see that the ridgelet basis elements are generally thinner and longer than corresponding FIO-derived elements.

In discussing such tilings, we have already claimed that we can construct a basis adapted to each such tiling. To see how this can work, recall that [15] introduced orthonormal ridgelets, defined as follows. Let $(\psi_{j,k}(t) : j \in \mathbf{Z}, k \in \mathbf{Z})$ be an orthonormal basis of Meyer wavelets for $L^2(\mathbf{R})$, and let $(w^0_{i_0\ell}(\theta),\ \ell=0,\ldots,2^{i_0}-1;\ w^1_{i,\ell}(\theta),\ i \geq i_0,\ \ell=0,\ldots,2^i-1)$ be an orthonormal basis for $L^2[0,2\pi)$ made of periodized Lemarié scaling functions $w^0_{i_0,\ell}$ at level i_0 and periodized Meyer wavelets $w^1_{i\ell}$ at levels $i \geq i_0$. Let $\hat{\psi_{j,k}}(\omega)$ denote the Fourier transform of $\psi_{j,k}(t)$, and define ridgelets $\rho_\lambda(x)$, $\lambda = (j,k;i,\ell,\varepsilon)$ as functions of $x \in \mathbf{R}^2$ using the frequency-domain definition

$$\hat{\rho}_\lambda(\xi) = |\xi|^{-\frac{1}{2}}(\hat{\psi_{j,k}}(|\xi|)w^\varepsilon_{i,\ell}(\theta) + \hat{\psi_{j,k}}(-|\xi|)w^\varepsilon_{i,\ell}(\theta+\pi))/2\ . \tag{1.1}$$

Here the indices run as follows: $j,k \in \mathbf{Z}$, $\ell = 0,\ldots,2^{i-1}-1$, $i \geq i_0$; and, if $i > i_0$, $i \geq j$. Also, if $i > i_0$ and $i > j$, then necessarily $\varepsilon = 1$. Let Λ denote the set of all such indices λ. In that article, it was shown that this collection of functions makes an orthonormal set for $L^2(\mathbf{R}^2)$.

From (1.1), and the definition of 1-D Meyer wavelets in the frequency domain as having the form $e^{it_{j,k}\lambda}w(\lambda/2^j)/2^{j/2}$ of a windowed sinusoid, we can see that the construction is essentially based on selecting a tile, using a windowed sinusoid in $r = |\xi|$ and using a wavelet in θ which peaks near that tile in θ.

In the end, the construction is based on the use of a tensor basis in (r,θ) for the space of functions localized (near) the tile.

In this paper we will carefully describe a general construction that contains the idea behind the ridgelets construction as a special case. The result will be a family of bases with a variety of interesting space-frequency localization properties. For example, by modifying the ridgelet construction using wavelet packets in the ridge direction and wavelets in the angular direction, one induces on real space elements which are oscillatory ridgelets in the sense that they have angular localization features similar to the orthonormal ridgelet family – being nearly ridge functions – while being oscillatory in the ridge direction, see Figure 2.3. On the other hand, by using wavelets in the ridge direction together with cosine packets in the angular direction one can produce effects which are more like brush strokes – bundles of line elements with given orientation, position, and textural cross-section.

We will also develop a digital realization of these ideas, based on the so-called pseudopolar FFT. In our digital realization, we replace the polar Fourier plane (which we can view as based on concentric circles of different radii) by a pseudopolar system, based on concentric squares of different pseudo-radii. The result is that we have tilings as illustrated in Figure 2.4.

We will also describe an adaptive algorithm, based on tree pruning, that can rapidly search through a wide variety of such bases, and select one that can provide the sparsest representation of the digital data. An example of the output of this algorithm is provided in Figure 2.5 below.

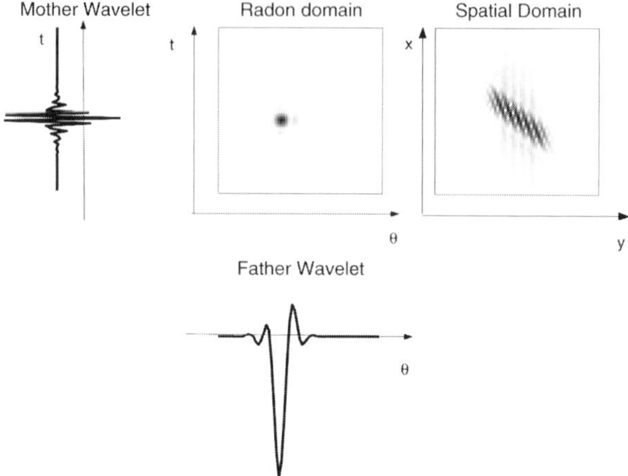

Figure 2.3. Construction of a Ridgelet Packet. A wavelet packet in the ridge direction and father wavelet Symmlet 8 in the angular direction are combined by tensor product, creating a waveform in the normalized Radon domain; this is backprojected into the spatial domain, creating an oriented waveform there

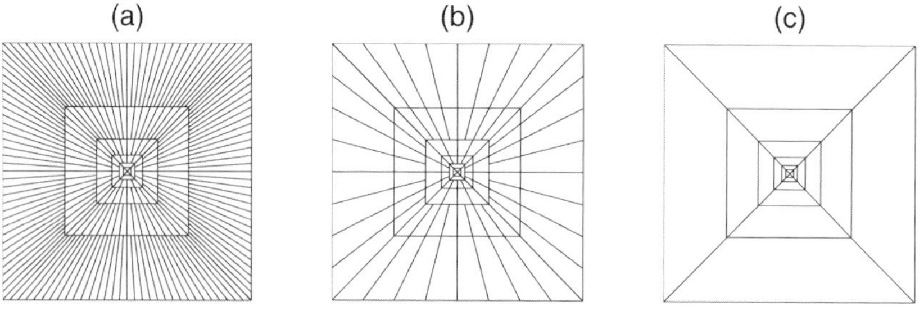

Figure 2.4. Three Tilings of the Digital Frequency Domain: (a) Digital Ridgelet; (b) Digital FIO; (c) Wavelet-like

Finally, we will describe a software package for calculating with ridgelet packets which is available as part of the Beamlab software distribution at *www.beamlab.net.*

2.2 FOURIER PRELIMINARIES

A function $f(x, y)$ in $L^2(dxdy)$, has a Fourier transform $\hat{f}(\xi)$ which, by Parseval/Plancherel, is isometric to it: $\|f\|_2^2 = (\frac{1}{2\pi})^2\|\hat{f}\|_2^2$ (we will typically use the term isometric in the broad sense, meaning preservation of norm up to multiplication by a fixed scalar multiple). Another isometry is given by the Cartesian-to-polar transformation $\hat{f} \mapsto \overline{F}$ defined by

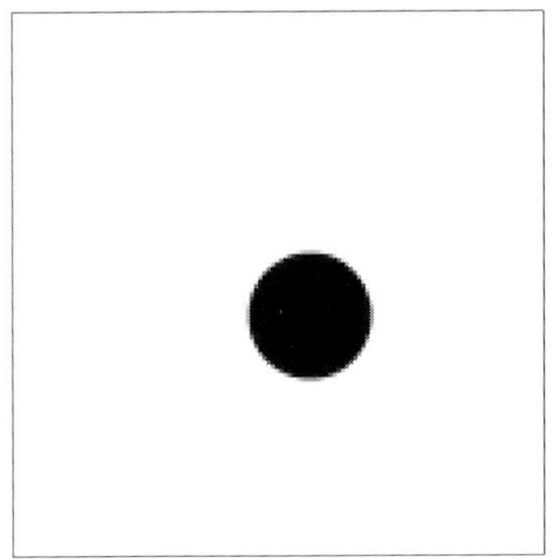

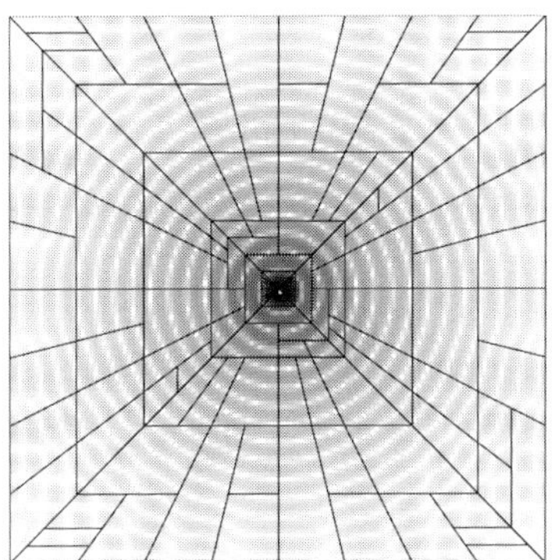

Figure 2.5. Disc-shaped object and Adapted Tiling. (a) Disc-shaped object. (b) Adaptive Tiling of the Fourier Plane, overlaid on the Fourier Transform of the disc. It is interesting to note that owing to the Cartesian structure of the data, the Fourier Transform of the disc does not have radial symmetry near the corners. This may explain the 'unexpected' high-frequency splits.

$$\overline{F}(\lambda, \theta) = |\lambda|^{1/2} \hat{f}(\lambda \cdot \cos(\theta), \lambda \cdot \sin(\theta)), \tag{2.1}$$

is an isometry from $L^2(d\xi_1 d\xi_2)$ to $L^2(\mathbf{R} \times [0, 2\pi))$. The composition of these two isometries, $f \mapsto \overline{F}$ maps the spatial domain into the polar Fourier domain, is again an isometry, and motivates the following

Definition 2.2.1 By **Polar Fourier Transform** we mean the mapping $\tilde{F} = \Pi(f)$ defined by

$$\tilde{F}(\lambda, \theta) = \hat{f}(\lambda \cdot \cos(\theta), \lambda \cdot \sin(\theta)),$$

whereas the **Polar Fourier Isometry** $\overline{F} = \overline{\Pi}(f)$ is defined by (2.1).

Suppose now that we take a tiling of the Fourier domain, such as one of the tilings of the introduction, and consider its equivalent in polar coordinates, as illustrated in Figure 2.6.

We then construct an orthonormal basis for functions on the polar strip which is the 'gluing together' of individual orthonormal bases for functions (nearly) localized to the individual tiles. Finally, we invert the Polar Fourier Isometry, producing a basis for $L^2(dx_1 dx_2)$.

It is well understood how to construct interesting bases based on tilings in dimension 1 [11, 36]. If we want to divide the frequency axis into intervals $\{I\}$ and then construct bases for the space of functions localized near those intervals, we construct smooth windows w_I localized near those intervals and obeying special properties and multiply those by sinusoids $\phi_{k,I}$ producing windowed sinusoids $w_I(t)\phi_{k,I}(t)$.

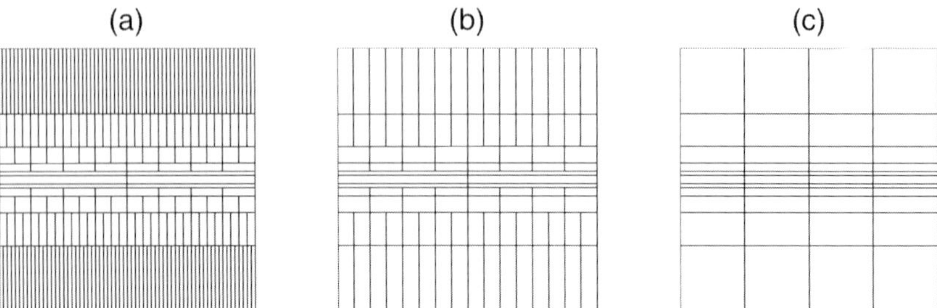

Figure 2.6. Tilings of Figure 1 Translated into Polar Form

One can adapt this idea to the two-dimensional case, produce bases associated with tilings of the polar strip by a tensor product construction. We use a similar cosine packets idea in θ, partitioning the interval $[0, 2\pi)$ into a sequence of intervals $\{J\}$ and consider the tensor cosine packet as the candidate basis element

$$C_{J,\ell,I,k}(\theta,\lambda) = w_J(\theta)\phi_{\ell,J}(\theta) \cdot w_I(\lambda)\phi_{k,I}(\lambda).$$

The collection of all such elements gives (if the windows and sinusoids are set up appropriately) an orthobasis for the polar frequency domain $L^2(\mathbf{R} \times [0, 2\pi))$.

We would next attempt to use this basis to induce a frame in the spatial domain. We might be tempted to identify each $C_{J,m,I,k}$ as the image, under polar Fourier isometry, of an object $\gamma_{J,m,I,k}$ in $L^2(dx_1 dx_2)$; and then to invert the isometry, getting $\gamma_{J,m,I,k} = \overline{\Pi}^*(C_{J,m,I,k})$. The inspiration is clear. Several details are less clear:

- Can this procedure really be carried out: is such a tensor product really in the range of the polar Fourier isometry? Strictly speaking, the answer is no; some modification is required. The information in the polar Fourier strip is redundant:

$$\overline{F}(\lambda,\theta) = \overline{F}(-\lambda,\theta+\pi);$$

  and if, in addition f is real-valued,

$$\overline{F}(\lambda,\theta) = \overline{F}(\lambda,\theta+\pi)^*.$$

  The basis elements $C_{J,m,I,k}(\theta,\lambda)$ will have to be modified to obey these constraints. A suitable modification is:

$$\tilde{C}_{J,m,I,k}(\theta,\lambda) = (C_{J,m,I,k}(\theta,\lambda) + C_{J,m,I,k}(\theta+\pi,-\lambda))/2$$

- Even after suitable modifications, does this lead to an interpretable basis and interpretable basis elements? Candès and Donoho, in preliminary calculations, have some theoretical insights on this matter. The digital implementations described later below provide an empirical tool for studying the resulting bases.

- The above construction does not correspond to the ortho-ridgelet definition (1.1); so apparently there is more than one way to operationalize a given set of tilings.
- There is no well-defined polar Fourier transform for digital data; it is unclear whether the ideas underling this construction have serious implications for digital data.

Because of these observations, it will turn out that the polar Fourier viewpoint tells only part of the story. In broad terms, there are two viewpoints, one in the polar Frequency domain and one in the (closely related) Radon domain. It will be very instructive in the coming few sections to develop a Radon viewpoint, which provides full or partial answers to all the open issues above.

2.3 RADON PRELIMINARIES

For a smooth function $f(x) = f(x_1, x_2)$ of rapid decay, let Rf denote the Radon transform of f, the integral along a line $L_{(\theta,t)}$, expressed using the Dirac mass δ as

$$(Rf)(t,\theta) = \int f(x)\delta(x_1 \cos\theta + x_2 \sin\theta - t)\ dx\ , \tag{3.1}$$

where we permit $\theta \in [0, 2\pi)$ and $t \in \mathbf{R}$. To create a space of such objects, we let $[\ ,\ ]$ denote the pairing

$$[F,G] = \frac{1}{4\pi}\int_0^{2\pi}\int_{-\infty}^{\infty} F(t,\theta)G(t,\theta)^* dt\ d\theta\ , \tag{3.2}$$

and by $L^2(dt\ d\theta)$ norm we mean $\|F\|^2 = [F,F]$. We let R^+ denote the adjoint of the Radon transform so that for all sufficiently nice $G \in A$ and all sufficiently nice $f \in L^2(dx)$,

$$[Rf, G] = \langle f, R^+G\rangle, \tag{3.3}$$

which leads to

$$(R^+G)(x) = \frac{1}{4\pi}\int_0^{2\pi} G(x_1\cos\theta + x_2\sin\theta, \theta)\ d\theta\ . \tag{3.4}$$

Define the Riesz order-1/2 fractional differentiation operator Δ^+ by

$$(\Delta^+ f)(t) = \frac{1}{2\pi}\int_{-\infty}^{\infty} e^{it\omega}\hat{f}(\omega)|\omega|^{\frac{1}{2}}\ d\omega\ . \tag{3.5}$$

This unbounded operator is well-defined on functions which are sufficiently smooth [formally, the domain $D(\Delta^+) = \{f : \int_{-\infty}^{\infty} |\hat{f}(\omega)|^2\ |\omega| d\omega\}$]. In particular, it is well-defined on every 1-D Meyer wavelet $\psi_{j,k}$, owing to $\text{supp}(\hat{\psi_{j,k}}) \subset \{\omega : |\omega| \in [\frac{2}{3}\pi 2^j,\ \frac{8}{3}\pi 2^j]\}$. Moreover, on the appropriate domain, it is self-adjoint.

Definition 2.3.1 The **Radon Isometry** $\overline{R}$ is the operator densely defined by the formal expression

$$\overline{R}(f) = (\Delta^{+} \otimes I)R[f].$$

The **adjoint Radon Isometry** $\overline{R}^{*}$ is the operator densely defined by the formal expression

$$\overline{R}^{*}(G) = R^{+}[(\Delta^{+} \otimes I)G].$$

Now note that Rf has the antipodal symmetry

$$(Rf)(-t, \theta + \pi) = (Rf)(t, \theta) \ . \tag{3.6}$$

We adopt the convention that F (and G and variants) typically will denote a function on $\mathbf{R} \times [0, 2\pi)$ obeying the same antipodal symmetry:

$$F(-t, \theta + \pi) = F(t, \theta) \ . \tag{3.7}$$

Let A be the closed subspace of $L^2(dt\ d\theta)$ of functions F obeying (3.7). Let $P_A F$ be the orthoprojector from $L^2(dt\ d\theta)$ onto A, defined by

$$(P_A F)(t, \theta) = (F(t, \theta) + F(-t, \theta + \pi))/2 \ . \tag{3.8}$$

2.4 THE RIDGELET CONSTRUCTION, AND ITS PROPERTIES

The Radon machinery can be used to construct orthonormal ridgelets very efficiently. Define the operator of reflection of functions of one variable $(Tf)(t) = f(-t)$ and the operator of translation by half a period by $(Sg)(\theta) = g(\theta + \pi)$. Note that the space A consists of objects invariant under $T \otimes S$; (3.7) can be rewritten $(T \otimes S)F = F$. In fact, $P_A = (I + T \otimes S)/2$. Set now, for $j, k \in \mathbf{Z}$, and $i \geq i_0$, $\ell = 0, \ldots, 2^{i-1} - 1$, $\varepsilon \in \{0, 1\}$

$$W_\lambda(t, \theta) = P_A(\psi_{j,k} \otimes w_{i,\ell}^{\varepsilon}) \ , \tag{4.1}$$

where $\lambda = (j, k; i, \ell, \varepsilon)$. For later reference, we spell this out:

$$W_\lambda(t, \theta) = (\psi_{j,k}(t) w_{i,\ell}^{\varepsilon}(\theta) + \psi_{j,k}(-t) w_{i,\ell}^{\varepsilon}(\theta + \pi))/2 \ . \tag{4.2}$$

It was shown in [15] that the W_λ provide an orthobasis for A. In order to obtain orthonormality with respect to the scalar product $[,]$, a particular normalization was imposed. In that normalization, $\|\psi_{j,k}\|_{L^2(\mathbf{R})} = \sqrt{2}$, and $\|w_{i,\ell}^{\varepsilon}\|_{L^2[0,2\pi]} = 2\sqrt{\pi}$. In a sense, the $(W_\lambda : \lambda \in \Lambda)$ constitute a "tensor wavelet basis which has been antipodally symmetrized".

If we now let ρ_λ denote the ridgelet elements defined in the Introduction, we note that

$$\rho_\lambda = \overline{R}^{*}[W_\lambda], \tag{4.3}$$

this says that an orthonormal ridgelet is *isometric to a wavelet in Radon-space which has been antipodally-symmetrized.*

The ortho-ridgelet construction may be viewed as transferring a basis from Radon space to real space via an isometry $\overline{R}^*$. If we reflect on the details of the above construction, we notice that the basis we used was not completely arbitrary: it had to consist of elements both in the domain of $\overline{R}$ and the range of $\overline{R}^*$. Now the range and domain both consist of functions in A, and the easiest elements in both range and domain to describe are functions which are bandpass in t, i.e. functions with support in the frequency domain contained in a compact set separated from the origin.

These ideas imposed the following restrictions on the construction.

- The basis on Radon space was a basis for A rather than $L^2(dtd\theta)$. This meant that its elements had to obey an antipodal symmetry requirement, or equivalently that an element W of the basis had to obey the invariance $P_A W = W$.
- In order to construct such a basis, we started with an orthonormal basis for $L^2(dtd\theta)$ and operated on it by P_A, creating a tight frame with antipodal symmetry. But as turned out, the tight frame was actually an orthobasis, owing to two special *closure properties* of these families we used; closure under reflection about the origin in the ridge direction:

$$\psi_{j,k}(-t) = \psi_{j,1-k}(t) \ , \tag{4.4}$$

  and closure under translation by half a cycle in the angular direction:

$$w_{i,\ell}^{\varepsilon}(\theta + \pi) = w_{i,\ell+2^{i-1}}^{\varepsilon}(\theta) \ . \tag{4.5}$$

  The closure property (4.4) would ***not*** hold for certain other prominent wavelet families, such as Daubechies' compactly supported wavelets. The significance of the closure properties was that for certain pairs $(W_\lambda, W_{\lambda'})$, $P_A W_\lambda = P_A W_{\lambda'}$, so that the induced frame consisted of many identical pairs. Systematically removing one element from each such pair, and rescaling the other element, we obtained an orthonormal basis.
- An 'absence of low-frequencies' restriction was imposed: the basis in the ridge-direction consisted entirely of bandpass elements, i.e. elements with frequency-domain support in an octave band disjoint from the origin.

If we weakened these conditions, the following would still be true.

- We can always start from an orthobasis for $L^2(\mathbf{R})$ and apply the projector P_A, getting a tight frame for A. We can then apply the isometry to this, getting a tight frame for real space.
- If, in addition, the original basis obeyed appropriate closure under ridge-reflection and angular translation, the tight frame in real space can be decimated by a factor of two to form an ortho-basis.
- The condition that the low-frequency terms be absent from all basis elements is simply a regularity condition on the outcome of the procedure. If certain elements in the basis have support near the origin in frequency space, then the construction can still take place; however, some of the corresponding frame elements will have poor decay.

In other words, the construction is quite general, but it might lead to a redundant set with redundancy two and it might lead to a basis where certain elements do not exhibit good spatial decay.

2.5 RIDGELET PACKET CONSTRUCTION

We now propose a class of tight frames based on the remarks just given. In certain cases, these can be subsampled to form orthobases.

2.5.1 General Procedure

We begin with a general set of ingredients:

- An orthonormal basis $(U_\mu(t))$ for $L^2(\mathbf{R})$ for the ridge direction. If the elements are bandlimited, we call this a bandlimited basis. If the elements obey the closure condition
$$U_\mu(-t) = U_{\mu'}(t)$$
the basis will be called a basis *closed under reflection.*
- A basis $(V_\nu(\theta))_\nu$ for $L^2[0, 2\pi)$ in the angular direction. If the elements obey the closure condition
$$V_\nu(\theta + \pi) = V_{\nu'}(\theta)$$
the basis will be called a basis *closed under translation.*
- A collection of antipodally-symmetric functions A will be constructed from the two families of bases. Letting $\lambda = (\mu, \nu)$ group the indices in each of the variables,
$$W_\lambda(t, \theta) = P_A[U_\mu \otimes V_\nu]; \tag{5.1}$$
as the result of applying an orthonormal projector to the orthonormal basis, the W_λ make a tight frame for A.
- A collection of functions ρ_λ will be induced by the isometry $\overline{R}^*$:
$$\rho_\lambda = \overline{R}^*(W_\lambda) \qquad \forall \lambda.$$
As an isometry of a tight frame, the ρ_λ make a tight frame for their span. In fact their span is all of $L^2(\mathbf{R}^2)$.

We note the following.

First, there is a simple expression for the element ρ_λ
$$\hat{\rho}_\lambda(\xi) = |\xi|^{-1/2}(\hat{U}_\mu(|\xi|) \cdot V_\nu(\theta) + \hat{U}_\mu(-|\xi|) \cdot V_\nu(\theta + \pi))/2$$
valid for $\xi = (|\xi| \cos(\theta), |\xi| \sin(\theta))$.

Second, if the elements U_μ are bandpass, with C^∞ Fourier transforms, and if the elements V_ν are C^∞ then the elements ρ_λ are likewise bandpass with smooth Fourier transforms; it follows that they are C^∞ with spatial rapid decay.

Third, the general procedure described above has been stated for tensor product bases $U_\mu \otimes V_\nu$. In general, there is no reason to restrict ourselves in this way. More generally, we may allow a semi-direct product

$$W_\lambda = P_A[U_\mu \otimes V_{\nu|\mu}], \qquad \forall\lambda = (\mu, \nu) \tag{5.2}$$

where the basis $(V_{\nu|\mu} : \nu)$ depends on μ. The orthoridgelet basis defined in the Introduction in fact has this form, as can be seen from the constraint $i \geq j$. By and large, the freedom enabled by the rule (5.2) will only be exercised in a limited way, as exemplified by the way it is exercised in the ridgelet orthobasis; the coarsest-scale of resolution may be adjusted to the properties of the corresponding ridge element.

2.5.2 Bases of Ridgelet Packets

While in principle, *any* pair of bases may be used for the above construction, we are interested here in those bases deriving from applying certain principles of time-frequency localization [13, 33].

Definition 2.5.1 We call **ridgelet packet basis** a basis constructed by the above procedure, where the basis U_μ is chosen from a **wavelet packets dictionary** and the basis V_ν (or $V_{\nu|\mu}$ if rule (5.2) is used) is chosen from a **wavelet packets dictionary** or a **cosine packets dictionary**.

When the basis in the angular direction is chosen from the wavelet packets dictionary, we will sometimes speak of the *Radon-domain approach* to defining ridgelet packets, whereas when the basis in the angular direction is chosen from the cosine packets dictionary, we will speak of the *polar Fourier-domain approach.* This distinction reflects the structure of the underlying algorithms in the two situations, as we will discuss later. Admittedly, this is artificial to some extent, since the Radon and Fourier domains are related in 1-1 fashion, but we find the distinction helpful.

2.5.3 Radon Approach: Wavelets in both Ridge and Angular Directions

The orthonormal ridgelet basis is built using wavelets in both the ridge and angular directions. Other bases can be built within this framework, by simply varying the base resolution level of the angular wavelets as a function of the resolution of the ridge wavelets.

In the ortho-ridgelet case, we start with ridge wavelets $\psi_{j,k}(t)$ for $j, k \in \mathbf{Z}$ and with angular wavelets $w^{\varepsilon}_{i,\ell}(\theta)$. The key decision is that we limit $i \geq j$, and we have $\varepsilon = 1$ for $i > j$, while $\varepsilon \in \{0, 1\}$ for $i = j$.

To interpret these choices, focus on the situation where $i = j$ and $\varepsilon = 0$. Hence we are looking at a tensor product based on the male-gendered wavelet at scale j, $w^0_{j,\ell}(\theta) \cdot \psi_{j,k}(t)$ Note that for $\varepsilon = 0$, $w^{\varepsilon}_{j,\ell}(\theta)$ is a "bump", integrating to $2^{j/2}$. The tensor product is thus localized near $\theta = \ell/2^j$, and has each constant-θ, varying-t profile proportional to the wavelet $\psi_{j,k}$.

- *Wilson-like Basis.* If we use breakpoints $\{1, 2, 3, \dots\}$, we get a partition into intervals

$$I_j = (-j, -(j-1)] \cup [j-1, j), \tag{5.3}$$

and we obtain in this way elements familiar to those who understand [1] and who have studied the construction of the Wilson basis [14]. In effect, the basis elements are windowed sinusoids of frequency roughly j, exponentially localized near a position proportional to k in the time domain.

Other examples of the construction may seem more exotic:

- *Intermediate Coherence Length.* Suppose we use breakpoints

$$\{1, 2, 4, 6, 8, 12, 16, 20, 24, 28, 32, ...\},$$

where in general the $2j$-th and $2(j+1)$-th initial intervals $[2^j, 2^{j+1})$ and $[2^{j+1}, 2^{j+2})$ are recursively subdivided $j/2$ times, yielding a family of $2^j/2$ subintervals. Then we obtain a basis where the typical elements supported near high frequency ω have a frequency localized in a band of width about $\sqrt{\omega}$ and a time localization, according to the Heisenberg principle, to a correspondingly short interval of length about $1/\sqrt{\omega}$. This says that the time coherence of effects at frequency ω is not as short as in the wavelet system, where it is proportional to $1/\omega$, nor as long as in the Gabor system, where coherent effects last for about one unit of time.
- *Increasing Coherence Length.* If we use breakpoints

$$\{1, 2, 3, 4, 4\frac{1}{2}, 5, 5\frac{1}{2}, 6, 6\frac{1}{2}, 7, 7\frac{1}{2}, 8, 8\frac{1}{4}, 8\frac{1}{2}, 8\frac{3}{4}, 9, ...\},$$

where in general the j-th initial dyadic interval $[2^j, 2^{j+1})$ is subdivided dyadically through $2j-2$ complete generations, then we obtain a basis where the typical elements supported near high-frequency ω have a frequency localized in a band of width about $1/\sqrt{\omega}$ and a time localization, according to the Heisenberg principle, to a correspondingly short interval of length about $\sqrt{\omega}$. This says that the time coherence of effects at frequency ω is not as short as in the Gabor system, where coherent effects last for about one unit of time, nor as long as in the Fourier system, where coherent effects last for infinite time.

With any of these choices, we can then subdivide the angular variable in a fashion subordinate to the ridge frequency variable, according to the same principle as in the ortho ridgelet basis. Let $V_{\nu|\mu}$ be simply the periodized Meyer wavelet as in the ortho ridgelet basis – under a low frequency constraint to be determined below – and let W_μ be a wavelet packet basis based on a different partition than the dyadic wavelet partition. Consider for example the Wilson-like basis partition (5.3) based on integer breakpoints. Choose the low-frequency constraint on $w_{i,\ell}^{\varepsilon}$ so that $i \geq j$ i.e. so that the angular scale is finer than the ridge frequency. It results that for $j > 0$, the ρ_λ are bandlimited and of rapid decay.

For each ρ_λ we have from (4.3) the formula

$$\rho_\lambda = \overline{R}^*[W_\lambda]$$

which gives the explicit formula

$$\rho_\lambda(x) = \frac{1}{4\pi}\int (U_\mu^+(x_1\cos(\theta) + x_2\sin(\theta))w_{i,\ell}^\varepsilon(\theta))/2d\theta$$
$$+\frac{1}{4\pi}\int (U_\mu^+(x_1\cos(\theta+\pi) + x_2\sin(\theta+\pi))w_{i,\ell}^\varepsilon(\theta+\pi))/2d\theta.$$

Now roughly speaking, U_μ^+, with $\mu = (j,k)$ is a sinusoid of frequency j, say localized to an interval of length ≈ 1 situated near $t \approx k$. Hence, the ridge function $U_\mu^+(x_1\cos(\theta)+x_2\sin(\theta))$ is localized near $x_1\cos(\theta)+x_2\sin(\theta) = k$. Similarly $w_{i,\ell}^\varepsilon(\theta)$ is localized near $\theta = \theta_{i,\ell} = 2\pi\ell/2^i$. It follows that the integrand is large for x in a range where $x \approx (k\cos(\theta_{i,\ell}), k\sin(\theta_{i,\ell}))$, so we may expect that for $\varepsilon = 0$ and $i = i_0(\mu)$, the function ρ_λ concentrates near $x \approx (k\cos(\theta_{i,\ell}), k\sin(\theta_{i,\ell}))$. For $\varepsilon = 1$ and $i > i_0$, one must argue by cancelation, which is more subtle.

2.5.5 Polar Fourier Approach: Wavelet $\otimes$ Cosine Packet

Let now (U_μ) be simply the standard Meyer wavelet basis for $\mathbf{R}$, just as in the ortho-ridgelet basis (1.1). Let $V_{\nu|\mu}$ however, be a cosine packet basis based on a recursive dyadic partition of the angle domain. Consider for example, a partition based on dividing the angular domain into 2^j equal sectors. Use the cosine packets subordinate to this partitioning. In the polar Fourier domain, things are very simple, because Meyer wavelets are the Fourier transforms of cosine packets in the frequency domain. Hence we have cosine packets in λ times cosine packets in θ. Hence, bivariate cosine packets are being used, subordinate to a recursive dyadic partition.

For each ρ_λ we have the explicit formula

$$\rho_\lambda(x) = \frac{1}{4\pi}\int (\psi_{j,k}^+(x_1\cos(\theta) + x_2\sin(\theta))V_\nu(\theta))/2d\theta$$
$$+\frac{1}{4\pi}\int (\psi_{j,k}^+(x_1\cos(\theta+\pi) + x_2\sin(\theta+\pi))V_\nu(\theta+\pi))/2d\theta.$$

Now, roughly speaking, $\psi_{j,k}^+$ is a wavelet of scale 2^{-j}, localized near $t \approx t_{j,k} = k/2^j$. Hence, the ridge function $\psi_{j,k}^+(x_1\cos(\theta) + x_2\sin(\theta))$ is localized near $x_1\cos(\theta) + x_2\sin(\theta) = t_{j,k}$. Similarly V_ν is localized to an interval $J_{m,\ell}$. It follows that the integrand is large for x in a range where $|x| \approx t_{j,k}$, and $\theta \in J_{m,l}$. We may expect that the function ρ_λ is large in the neighborhood where $x \approx \pm(|t_{j,k}|\cos(\theta_{i,\ell}), |t_{j,k}|\sin(\theta_{i,\ell}))$. Knowing the exact shape of the support requires additional insight.

Now we make the more detailed assumption that V_ν is a sinusoid in θ of frequency $2\pi\cdot k_1$ localized near the interval $J_{m,\ell}$. This allows us to study the details of ρ_λ on its support. For large $|t_{j,k}|$ and large m, the integrand is approximately of the form

$$\psi_{j,k}^+(x_1\cos(\theta_{m,l}) + x_2\sin(\theta_{m,l}))w_{J_{m,l}}(\theta)\phi_{k_1}(\theta).$$

Hence it has approximately the form of a wavelet function in the ridge direction and the form of a localized sinusoid in the transverse direction.

2.6 IMPLEMENTATION ON DIGITAL DATA

Ridgelet Packets bases for digital data can be constructed based on an adaptation of a circle of ideas associated to digital implementation of the Radon transform, polar Fourier transform, and ridgelet transform [3, 17].

2.6.1 Fast Slant Stack

Averbuch et al. (2001) [3] describe a realization of the Radon transform suited for n-by-n image data, called Fast Slant Stack, claiming that the transform is geometrically accurate and can be implemented by a fast algorithm. The geometric accuracy, for example, implies that the backprojection of a point in Radon space is a true ridge function, i.e. a true object of the form $\psi(x + sy)$, where $\psi(\cdot)$ is delta-like.

This scheme has been deployed by Donoho and Flesia [17] to produce a discrete ridgelet transform based on true ridge functions. In our work for this paper, we have used the same scheme to provide a digital implementation of ridgelet packets.

2.6.2 Pseudopolar FFT

Underlying the Fast Slant Stack is a notion of digital polar transform Fourier called pseudopolar FFT in [3].

The key point is to view the digital Fourier domain not as a cartesian grid, but instead as a special pointset as shown in Figure (2.7). Then define the *pseudopolar Fourier transform* as the evaluation of the Fourier transform

$$\hat{I}(\xi) = \sum_{x_1,x_2=0}^{n-1} I(x_1, x_2) \exp\{-(x_1\xi_1 + x_2\xi_2)\}$$

at the $4n^2$ points of this pointset. The pointset can be viewed as a set of "concentric squares" stacked inside each other (like Chinese boxes), with equispaced points along the boundary of the box. The half-width of a side functions as a pseudo radius, and the arclength along the perimeter of the box functions as a pseudo angular variable.

As shown in [3], the evaluation of the Fourier sum on this set of gridpoints can be performed in order $N \log(N)$ flops, where $N = n^2$ is the total number of pixels. The underlying ideas that allow rapid evaluation of these specific gridpoints date back to work of Pasciak [27], Edholm and Herman [18], and Lawton [24], working variously in Medical Imaging and in Synthetic Aperture Radar.

The resulting set of pseudopolar values may be viewed as a $2n$ by $2n$ array: $2n$ points on each line through the origin, and $2n$ lines through the origin, grouped in columns as different lines through the origin, in rows as different 'radii'. We define the pseudopolar FFT $P(I)$ to be the transform from n by n arrays to $2n$ by $2n$ arrays produced in this way.

Note that the pseudopolar grid samples the region near the origin more finely than the region near the boundary. In fact the spacing between samples on line segments varies inversely with distance of the segment from the origin. Define the

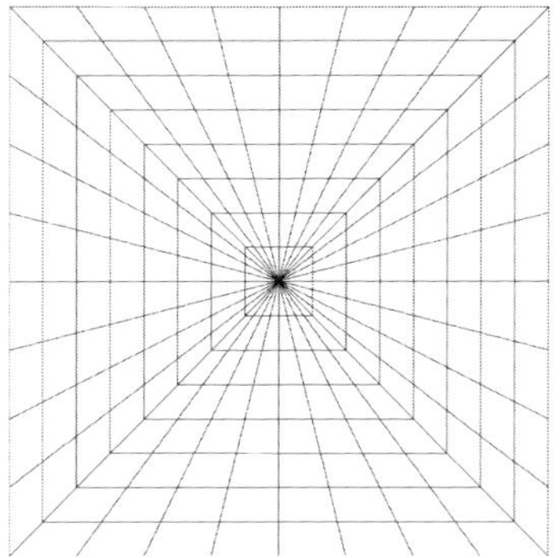

Figure 2.7. Pseudopolar Fourier Grid

normalized pseudopolar FFT $\overline{P}(I)$ to be the result of applying a simple rescaling of entries in $P(I)$ according to the square root of the local sample spacing in the pseudopolar grid at the corresponding grid point. Since $P(I)$ is a discrete analog of $F(r,\theta) = \hat{f}(r\cos(\theta), r\sin(\theta))$, sampled at specific points in (r,θ), the definition of $\overline{P}(I)$ is very analogous to defining in the continuum case $\overline{F}(r,\theta) = r^{1/2}\hat{f}(r\cos(\theta), r\sin(\theta))$. Recall that $f \mapsto \overline{F}$ is an isometry from $L^2(dxdy)$ to $L^2(drd\theta)$; we can't get quite so much in the digital case: Instead we have

$$C_1\|I\|_2 \leq \|\overline{P}(I)\|_2 \leq C_2\|I\|_2, \tag{6.1}$$

where empirically, $C_2/C_1 < 1.1$.

Note that if we had $C_1 = C_2$ then, up to normalization, $\overline{P}$ would be an ℓ^2 isometry. In that sense, the mapping $I \mapsto \overline{P}(I)$ is *a digital analog of the polar Fourier Isometry.*

2.6.3 Digital Radon Domain

If we apply a 1-dimensional inverse FFT to each column of the 2-D pseudopolar FFT array, we create a new $2n$-by-$2n$ matrix. This matrix is a digital Radon transform of I; each column gives the sums of (an interpolant of) I along a family of equispaced parallel lines, where the slope of the lines in that family is indexed by the column index (which provide a pseudo-angular variable) [3]. Call the overall mapping $S(I)$ the slant stack.

If we apply instead a 1-dimensional inverse FFT to each column of the 2-D *normalized* pseudopolar FFT array, we create another new $2n$-by-$2n$ matrix. This matrix is a *preconditioned* digital Radon transform of I. Call the overall transform mapping $\overline{S}(I)$ the normalized slant stack. Because of the near-isometry property of $\overline{P}(I)$, we have

$$C_1\|I\|_2 \leq \|\overline{S}(I)\|_2 \leq C_2\|I\|_2.$$

Here C_1 and C_2 are the same as in (6.1). Again, if $C_1 = C_2$ then, up to normalization, then $\overline{S}(I)$ would be an ℓ^2 isometry. In that sense, the mapping $I \mapsto \overline{S}(I)$ is *a digital analog of the Radon Isometry.*

2.6.4 Strategy for Digital Implementation

We have now introduced a set of digital-data friendly tools which are analogous to the continuum-domain tools discussed in earlier sections. In a sense, we have built up a 'dictionary' to translate between the continuum domain and the digital domain. The dictionary is summarized in the table below.

Continuum Concept	Symbol	Digital Concept	Symbol
Polar Fourier Transform	$F(\lambda, \theta)$	Pseudopolar FFT	$P(I)$
Polar Fourier Isometry	$\overline{F}(\lambda, \theta)$	Normalized Pseudopolar FFT	$\overline{P}(I)$
Radon Transform	$R(t, \theta)$	Slant Stack	$S(I)$
Radon Isometry	$\overline{R}(t, \theta)$	Normalized Slant Stack	$\overline{S}(I)$

Our strategy for digital implementation of ridgelet packets is to use this dictionary to substitute digital concepts for continuum concepts in the original definitions. Thus, if a certain orthobasis for the continuum case used wavelets 'in each direction' followed by dual Radon isometry, then we propose to work with the normalized slant stack, and use a discrete wavelet basis 'in each direction', followed by the adjoint normalized slant stack 'to return to the spatial domain'.

A remark about this strategy: before heading a great distance down this path, it is important to know that there is a 'proof of concept' which shows that at least in a special case, the strategy provides decent results. In this case, the proof of concept has been provided by the algorithm for the digital ridgelet transform in [17]. In constructing that transform, the authors have followed the strategy suggested above and carefully documented the properties of the digital domain transforms that result.

A second remark: while the continuum approach leads to the definition of various bases, in the discrete case we will only get frames, for a simple reason. The normalized pseudopolar FFT and the normalized slant stack are both transforms from $n \times n$ arrays to $2n \times 2n$ arrays. Hence the strategy must in general lead to overcomplete systems (frames) rather than orthobases. However, the frames generated in this manner can be expected to have good frame bounds, owing to the closeness of C_1 and C_2 in (6.1).

2.6.5 Digital Ridgelet Packets

Our implementation strategy leads to the following general schema:

Definition 2.6.1 A **Digital Ridgelet Packet transform** in dimension two is a transform of n-by-n data defined as follows.

[DRPT1]

The n-by-n digital array is transformed into a digital Radon domain via the fast normalized slant stack algorithm in [3], which gives a $2n$-by-$2n$ array.

[DRPT2]These arrays are then transformed according to some specific combination of wavelet packets and cosine packets in each of the two directions (angular vs. ridge), where the combination of bases in the two directions is made according to a direct product or a semidirect product.

The **Inverse Digital Ridgelet Packet transform** in dimension two is a transform returning from the Ridgelet Packet domain as follows.

[IDRPT1] The Ridgelet packet coefficients are transformed back into the Radon domain by inverting the transform in step [DRPT2] above.

[IDRPT2]The Radon domain data are transformed back into the original digital spatial domain by inverting the normalized slant stack transform in step [DRPT1] above, using the algorithm in [3].

This algorithm has the following general characteristics for an image of size n by n with $N = n^2$ pixels.

- Storage Space: The algorithm requires permanent storage of order $O(N)$, and temporary storage of comparable size.
- Complexity: The forward transform algorithm requires $O(N \cdot \log N)$ flops, and the inverse transform algorithm requires $O(C(\varepsilon) \cdot N \cdot \log N)$ flops, where $C(\varepsilon)$ depends on the relative accuracy required. $C(10^{-6}) \approx 7$.

For certain purposes it may be useful to apply the adjoint of the forward digital ridgelet packet transform. This is obtained by a two-step procedure

[ADRPT1]The Ridgelet packet coefficients are transformed back into the digital Radon domain by applying the adjoint of the transform in step [DRPT2] above. If the transform is orthogonal (as would be typical) this is the same as inverting the transform.

[ADRPT2] The Radon domain data are transformed back into the original digital domain by applying the adjoint of the transform in step [DRPT1] above, using the adjoint algorithm in [3].

The adjoint uses $O(N \log(N))$ flops and $O(N)$ space.

2.6.6 Digital Implementation

We have developed a digital implementation of these ideas as part of a MATLAB toolbox called *BeamLab*. This toolbox contains tools for Ridgelet, beamlet, and curvelet analysis, and is available for download from
`http://www.beamlab.net`
where further information is available.

BeamLab contains scripts which can reproduce all the figures in the article you are reading. It also contains a directory named *RP_FCP* which implements ridgelet packets based on anisotropic cosine packets in the pseudopolar Fourier plane, and a directory named *RP_RWW* which implements ridgelet packets based on applying wavelets and wavelet packets in the Radon plane.

2.6.7 Examples of Digital Implementation

We now demonstrate the implementation indicated above by displaying some basis functions and the associated coefficient functionals.

2.6.8 Synthesis from Tiles

What do ridgelet packets look like? Given the emphasis of this paper on tilings, it might seem most natural to try many different tilings, and some specific basis functions associated with each. We suggest that another approach would be more directly informative: to study individual tiles.

Indeed, the system of tilings underlying our constructions has a relatively few different shapes of tiles: they are all oriented parallel to the axes in (r, θ) space, and their widths and heights vary through a dyadic set $(2^i, 2^j)$. They are located at various places in the (r, θ) plane, but most of this variation is easily visualized, as either a rescaling or a rotation.

In short, we propose that to get a good understanding of the system, we should fix a certain (r, θ) and consider objects associated to tiles at a range of different aspect ratios: $(2^i, 2^{i+h})$, for varying h, or else $(2^{j+h}, 2^j)$ for varying h. In this way we see the effect of varying the tile shape on the basis elements.

Carrying out this proposal, we begin by working in the Fourier plane with Cosine packet bases. Figure 2.8 gives an example of several basis functions obtained from picking a tile extending from 1/4 to 1/2 of Nyquist in the radial frequency variable, and considering various widths ranging from very narrow to very broad. In these figures, we tried to keep the base frequency as low as possible, so that we explore the shape of the envelope rather than the oscillations within the envelope.

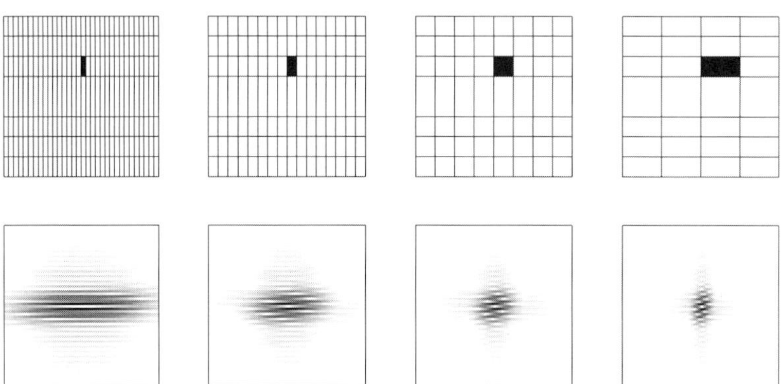

Figure 2.8. Basis Functions Derived from Tiles of Different Widths

Figure 2.9 gives an example of several basis functions obtained by picking a tile extending from $\pi/4$ to $3\pi/8$, and considering various heights ranging from very short to very tall. In these figures, we again tried to keep the base frequency as low as possible, so that we explore the shape of the envelope rather than the oscillations within the envelope.

Continuing with this proposal, we consider comparable tilings, only we implement them using the Radon plane with wavelet packet bases. Figure 2.10 gives an example

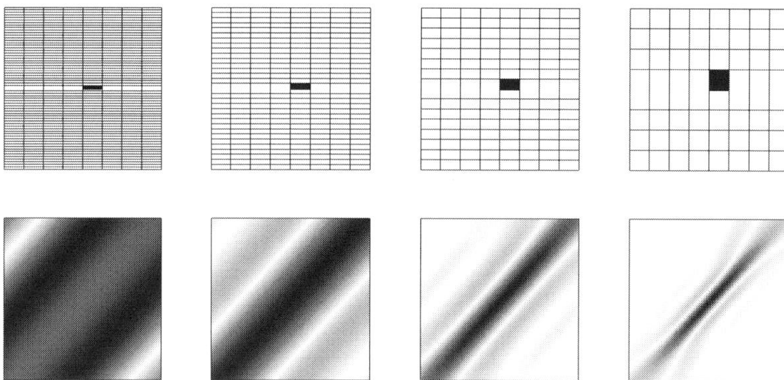

Figure 2.9. Basis Functions Derived from Tiles of Varying Heights

of several basis functions obtained from picking a Radon-domain waveform made of the tensor product of a wavelet packet in the ridge direction and a wavelet in the angular direction. The figure explores the results obtained by varying the scale of the wavelet packet in the ridge direction.

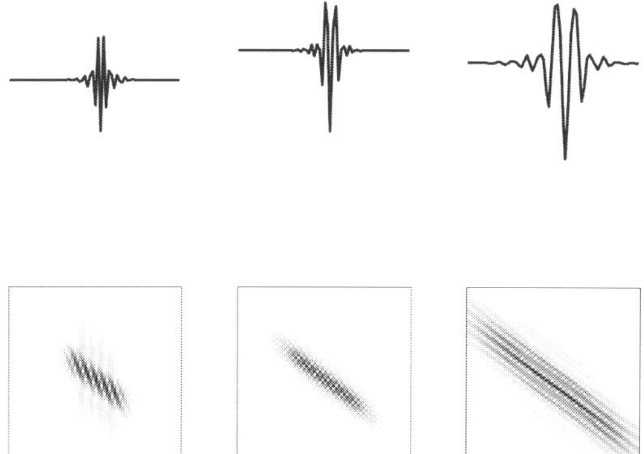

Figure 2.10. Basis functions derived from Radon-domain approach. Backprojections of Wavelet Packets at Different Widths in the ridge direction give different spatial waveforms

2.6.9 Analysis

We now give an example of using the Ridgelet Packet systems to analyze images. One basic purpose of image analysis is *sparse representation:* to use as few coefficients as possible to represent the object accurately. We will take some specific images, and analyze them in several different bases, and compare the representations for sparsity.

We consider an image with oriented linear features, depicted in Figure 2.11

We then analyze the object in the system of anisotropic cosine packets in the frequency domain. The figure 2.12 below illustrates the results, by showing the pseudopolar

Figure 2.11. Image with Oriented Linear Features

Fourier transform with the basis tiling overlaid upon it. The various panels indicate the underlying number of coefficients exceeding a fixed threshold.

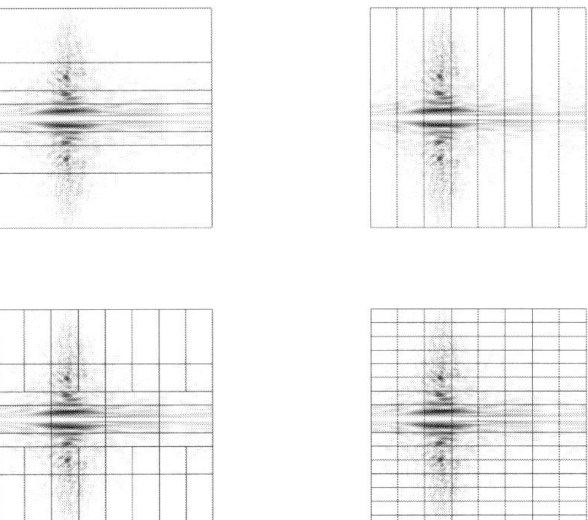

Figure 2.12. The pseudopolar Fourier transform overlaid with tilings defining 4 different ridgelet packet bases; the labels give the number of coefficients above threshold. Apparently, a decomposition splitting only on directions gives the sparsest representation

To solidify the reader's understanding, Figure 2.13 presents the present the associated graphic for the image data in original (Cartesian-product) Fourier space. The same four tilings are presented overlaid on the Fourier transform of the original image. The strong directionality of the image is evident from the concentration of energy along a line through the origin.

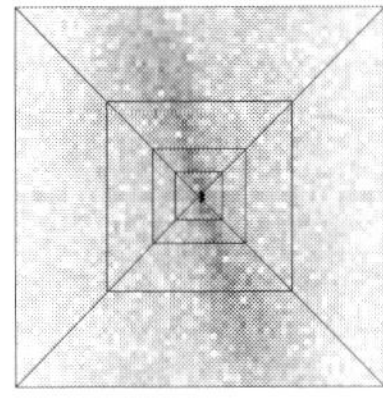
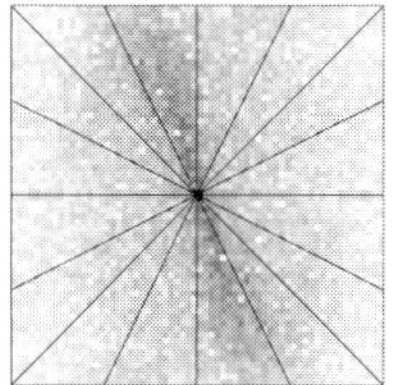
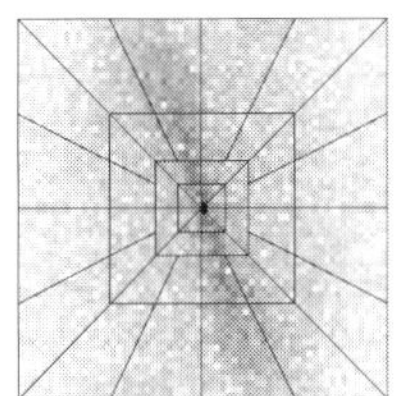
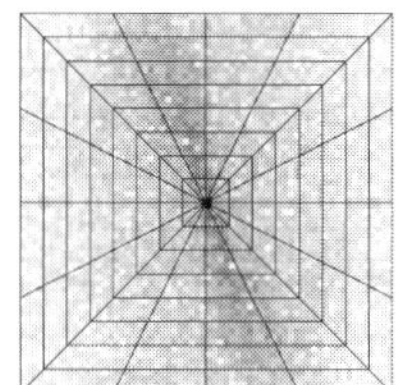

Figure 2.13. The two-dimensional Fourier transform overlaid with tilings defining 4 different ridgelet packet bases; the labels give the number of coefficients above threshold. Apparently, a decomposition splitting only on directions gives a larger number of coefficients exceeding threshold; while a decomposition splitting into more angular bins at higher frequencies gives fewer coefficients exceeding threshold.

2.7 ADAPTATION

We have defined, in the continuum case, a rather large collection of orthonormal bases; and in the digital data, a rather large collection of frames. Which of these is 'best' for a given dataset? In this section we describing an efficient computational method for finding an optimizing frame among the ridgelet packet bases.

2.7.1 Background on Best Basis

This question of adaptively choosing a basis has already been thoroughly studied in the context of time-frequency decompositions. The Cosine Packets Library and the Wavelet Packets Library define rather large collections of bases for representing signals – functions of time. Coifman and Wickerhauser [10] considered a library of bases $L = \{B\}$ with special properties (which were exhibited by both the cosine packets and wavelet packets libraries) and showed how to rapidly compute a 'best basis', one optimizing an expression of the form

$$\max_{B \in L} E(B)$$

where $E(B)$ (the 'entropy') is a measure of the quality of the basis. The same algorithm works equally well to find the minimizing basis.

Their algorithm works under these assumptions:

- Each basis is a particular subset from a dictionary of ϕ_γ, $\gamma \in \Gamma$;
- The bases in the library have a natural tree structure, i.e. there is a certain correspondence between subtrees of a complete tree and bases in a library;
- The 'entropy' is defined by an additive measure

$$E(B) = \sum_{\gamma \in B} e_\gamma.$$

The underlying principle that makes the Coifman-Wickerhauser algorithm work is the principle of dynamic programming, which allows to prove that bottom-up pruning of the complete tree is optimal.

The Coifman-Wickerhauser algorithm was developed originally in the context of time-frequency analysis, in connection with the Cosine Packet or Wavelet Packet bases. There the tree-ordering relationship of the bases was based on the tree-ordering property of the corresponding recursive dyadic partitions of the time and/or frequency domains, depending on whether we are considering the Cosine Packet or Wavelet Packet systems. The same general principle works also in two-dimensions, and has been used for example in constructing 2-d cosine packets bases for fingerprint image analysis [36] and in constructing 2-d brushlet bases [12]. For example, if one considers direct products of univariate 'time-frequency' bases, these correspond to direct products of pairs of univariate partitions, and to a natural tree order where one allows either vertical or horizontal splits. Other space-frequency bases can be constructed by recursive dyadic partitioning based on quadtree-splits – simultaneous splitting in both variables. Either family has the required tree property [4].

There are two general ways in which the Coifman-Wickerhauser algorithm can be applied, depending on the way in which the e_γ are specified.

First, one could be interested in finding a best-basis for an individual signal f, and then it makes sense to choose e_γ as a function of the coefficients of the signal f in basis B. If θ_γ is the γ-th coefficient of f in basis B, then it would be appropriate to set $e_\gamma = e(|\theta_\gamma|)$, where $e(t)$ is a concave function of t^2 – examples being $e(t) = \min(t^2, \lambda^2)$ and $e(t) = |t|^p$, $0 < p < 2$. Such are measures of the sparsity of the coefficients, and, subject to a fixed budget of coefficient energy $\sum_{\gamma \in B} |\theta_\gamma|^2$ they are small when the object has sparse coefficients – a few large coefficients and many small ones. (Note that in this setting our goal would be to minimize the $\sum_{\gamma \in B} e_\gamma$ rather than maximize it.)

Second, one could be interested in finding a best-basis for an ensemble of signals $\{f_m, m = 1, \ldots, M\}$. Then if θ_γ^m denotes the γ-th coefficient of f^m in basis B, it could make sense to let $e_\gamma = Ave\{e(\theta_\gamma^m)\}$, where e is one of the functions mentioned above in the single signal case, and then to minimize $E(B)$. Alternatively, we could take a statistical viewpoint, and look for a basis capturing the most squared variance on the diagonal; we would then let $e_\gamma = (Var_m\{\theta_\gamma^m\})^2$, and seek a basis maximizing $\sum_{\gamma \in B} e_\gamma$. The basis then has an interpretation as providing a best near-diagonalization within the cosine packets system [25]. In this case it also makes sense to consider maximizing the entropy derived from averaging $e(t) = t^4 - 3 \cdot t^2$ across the dataset; then $e_\gamma = Ave\{(\theta_\gamma^m)^4 - 3 \cdot (\theta_\gamma^m)^2\}$ is a measure of kurtosis, and the maximizing basis is the most kurtic basis; such bases are interesting as providing the bases which best expose the non-Gaussianity of the signal in a certain sense; compare [5, 6].

2.7.2 Application to Ridgelet Packets

We can adapt these existing best basis ideas to the Ridgelet Packet system in a natural way. In essence, Ridgelet Packet bases are all based on taking the data into the pseudopolar Fourier domain, viewing the pseudopolar FT as a complex valued $2n$ by $2n$ image, and then looking for the best anisotropic cosine packets basis for that image data. Bennett's thesis [4] has already explored the properties of best anisotropic cosine packets bases for real valued data; he studied the natural 2-D analog of the Coifman-

Wickerhauser algorithm in which anisotropic rectangles are allowed in the tiling. The extension to complex-valued data involves no new issues. Hence, applying the Bennett algorithm to the pseudopolar FFT involves nothing substantially new.

Figure 2.14 below shows the results of applying best anisotropic basis algorithm to the texture image of Figure 2.11.

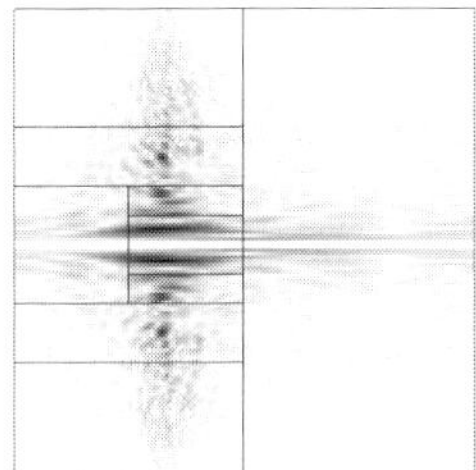
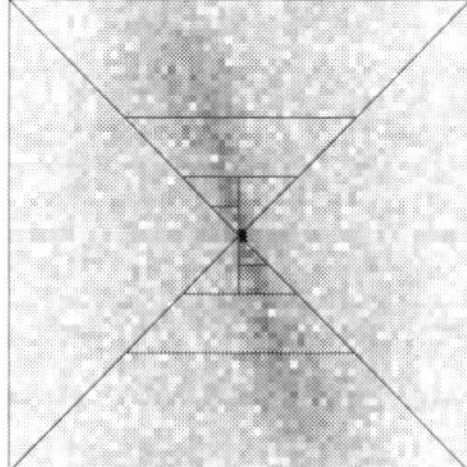

Figure 2.14. Left panel: Best rectangular partition overlaid on pseudopolar Fourier Plane. Right panel: Corresponding tiling of Cartesian Fourier Plane

2.8 DISCUSSION

This is obviously preliminary work; we can expect our efforts to be rapidly surpassed in various ways. We mention here some of the issues that come up in attempting to improve on the scheme developed here.

2.8.1 Improvements in the Digital Implementation

- *True Polar FFT.* In this article, we have based our efforts on an implementations using the pseudo Polar FFT and the Slant Stack. These are analogs, for digital data, of the continuum Fourier and Radon transforms. However, they are not precise analogs, since they do not fully represent the polar Fourier domain – using concentric squares rather than concentric circles. The discrepancy between squares and circles is particularly evident in analyzing radially symmetric objects such as a disk, which have Fourier transforms organized in a ring pattern (see Figure 4).

  In separate work, Averbuch et al. have developed a Polar FFT algorithm which starts from the pseudopolar FFT and then converts to a true polar form. While this algorithm does not offer a near isometry, the improved geometric fidelity might be important to have. A valuable next step would be to apply the Polar FFT algorithm in the present context.
- *Improved Coefficient Decay.* In the companion article [17], it was pointed out that the digital ridgelet transform based on ideas similar to those used in this paper does not have the same qualities of coefficient decay that were available in theory for the ortho ridgelet transform of continuous functions. One can expect the same statement to be true for any of the frames constructed using methods of this article. Donoho and Flesia speculate that the slower decay is owing to an implicit 'image mutilation' that occurs at the heart of the Slant Stack algorithm. This interesting possibility should be pursued further; perhaps it would lead to improved sparsity of not only ridgelet but all ridgelet packet representations.

- *True Space-Frequency Atoms.* As pointed out recently by Lars Villemoes, the usual setting for application of the Coifman-Wickerhauser algorithm with Cosine produces bases where the different elements do not have uniform Hesienberg constants. Roughly speaking, in the usual application, one has to make a choice of a "window transition length", and then all windows in the whole system make the transition from 1 to 0 in that same length, no matter whether the window is short or long. The result is that the basis elements corresponding to the long intervals have poor time-frequency localization.

  To respond to this problem, Villemoes [34] has developed an algorithm that provides uniformly good time-frequency localization. If Villemoes' algorithm were used in the ridgelet packet setting, it might substantially improve the coefficient decay of the best ridgelet packet basis. This would be a valuable direction to explore.

2.8.2 Limitations on the Ridgelet Packet Scheme

While the Ridgelet Packet system provides an interesting collection of new bases, all these bases are missing a key ingredient: a *translation parameter.* At best, the scheme offers a ridge translation parameter, associated with displacements vertically up or down in ridge space. From wavelets and Gabor systems, we are accustomed to having a translation-like parameter.

To understand this issue better, look at a typical member of a wavelet system in Fourier space;

$$\hat{\psi}_{j,k_1,k_2}(\xi) = \exp\{i\xi'(k_1,k_2)/2^j\} \cdot \hat{\psi}(\xi/2^j)/2^{j/2}.$$

The oscillating factor, $\exp\{i\xi'(k_1,k_2)/2^j\}$ is responsible for the translation effect. By comparison, a ridgelet packet, viewed in frequency space, is made up of several terms of the form

$$\exp\{ik_1(r-a)/b\} \cdot \exp\{ik_2(\theta-c)/d\} w_r((r-a)/b) \cdot w_\theta((\theta-c)/d).$$

Now qualitatively, the product of two windows $w_r((r-a)/b) \cdot w_\theta((\theta-c)/d)$ is simply a nice smooth function. So the oscillating factor

$$\exp\{ik_1(r-a)/b\} \cdot \exp\{ik_2(\theta-c)/d\}$$

would have to be able to create a translation or at least a pseudo translation effect. However, it is evident by inspection that, for large k_1, k_2 there is no such effect. Thus for example, under various assumptions on the behavior of a, b, c, d, the family can be shown to behave very differently than a translation family.

In response to this, it would seem worthwhile to search for a system of bases and associated algorithms with a more direct connection to translations.

An obvious way to do this would be to borrow ideas which were useful in the construction of curvelets [9]. Thus, we would separate the image into passbands based on bandpass filtering, spatially localize the bandpass images into square subimages via smooth windowing, and then apply ridgelet packet analysis to the square subimages.

ACKNOWLEDGEMENT

This research was supported by National Science Foundation grant DMS 95–05151, DMS 00–7726 and DMS 98–72890 (KDI), by AFOSR MURI 95–P49620–96–1–0028, and by DARPA

BAA-99-07. The authors would like to thank Nick Bennett and Yacov Hel-Or for helpful discussions. AGF would like to thank the Statistics Department at UC Berkeley for its hospitality.

REFERENCES

[1] P. Auscher, G. Weiss and M.V. Wickerhauser. *Local sine and cosine bases of Coifman and Meyer and the construction of smooth wavelets*, Wavelets: A tutorial in Theory and Applications, Academic Press, Boston, 237-256.

[2] E.H. Adelson and E.P. Simoncelli, *Orthogonal pyramid transforms for image coding*, Pro. SPIE, **845** (October 1987), Cambridge, MA.

[3] A. Averbuch, R.R. Coifman, D.L. Donoho, M. Israeli and J. Waldén, *Fast Slant Stack: A notion of Radon Transform for Data in a Cartesian Grid which is Rapidly Computible, Algebraically Exact, Geometrically Faithful and Invertible*, to appear: SIAM J. Sci. Comp.

[4] N. Bennett, *Fast algorithm for best anisotropic Walsh bases and relatives*, Appl. Comput. Harmon. Anal., **8** (2000), no. 1, 86-103.

[5] J.B. Buckheit, *Adaptive Wavelet Methods in Signal Processing*, Ph. D. Thesis, Department of Statistics, Stanford University,1996.

[6] J.B. Buckheit and D.L. Donoho, *WaveLab and Reproducible Research*, Wavelets in Statistics, A. Antoniadis and G. Oppenheim Ed., Springer-Verlag,New York, 1995, 55-82.

[7] E. Candès, *Harmonic Analysis of Neural Networks*, Appl. Comput. Harmon. Anal., **6** (1999), no. 2, 197-218.

[8] E. Candès. *Ridgelets: Theory and Applications*, Ph.D. Thesis, Department of Statistics, Stanford University,1998.

[9] Candès, E.J. and Donoho, D.L. (2000) Curvelets: a surprisingly effective nonadaptive representation of objects with edges. in *Curve and Surface Fitting: Saint-Malo 1999* Albert Cohen, Christophe Rabut, and Larry L. Schumaker (eds.) Vanderbilt University Press, Nashville, TN. ISBN 0-8265-1357-3

[10] R.R. Coifman and M.V. Wickerhauser, *Entropy-based algorithms for best basis selection*, IEEE Transactions on Information Theory, **38** (No.2 pt.2) (March 1992), 713-18.

[11] R.R. Coifman and Y. Meyer, *Remarques sur l'Analyse de Fourier à fenêtre*, C.R. Acad Sci. Paris,**312** (1991), 259-261.

[12] R.R. Coifman and F. Meyer, *Brushlets: a tool for directional image analysis and image compression*. Applied and Computational Harmonic Analysis, **4** (No.2)(April 1997),147-87.

[13] I. Daubechies, *The wavelet transform, time-frequency localization, and signal analysis.*, IEEE Trans. IT. ,**36** (1990), 961-1005.

[14] I. Daubechies, S. Jaffard and J.L. Journé, *A simple Wilson orthonormal basis with exponential decay*. SIAM J. Math. Anal., **24** (1990), 520-527.

[15] D.L. Donoho, *Orthonormal ridgelets and linear singularities*, SIAM J. Math. Anal. **31** (2000), no. 5, 1062-1099.

[16] D.L.Donoho,*Ridge functions and orthonormal ridgelets*, J. Approx. Theory., **111** (2001), no. 2, 143-179.

[17] D.L. Donoho and A.G. Flesia, *Digital Ridgelet Transform Based on True Ridge Functions*, To appear, Beyond Wavelets, J. Schmeidler and G.V. Welland, Eds., Academic Press, 2002.

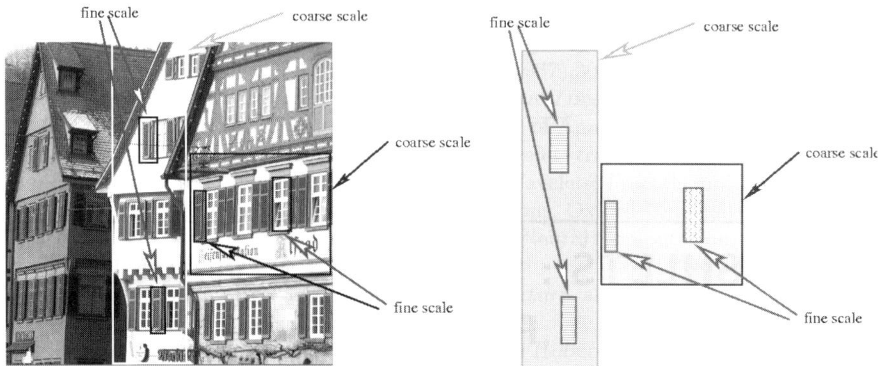

Figure 3.1. Small scale blocks that describe the shutters or the windows need to overlap on large scale blocks that describe the buildings. One cannot perform this type of analysis with local Fourier bases. As shown on the right one needs to have basis functions that lay over each others

of constant size, and then compute a Fourier expansion within each block. While this approach is easy to implement, it suffers from a number of drawbacks :

1 The size of the blocks should be adapted to the content inside the image : a large geometric feature should not belong to several small blocks, but should belong to a single large block.
2 The size of the blocks should be adapted to the frequencies of the complex exponential : short blocks for high frequency, large blocks for low frequencies.
3 The segmentation into blocks creates "blocking" artifacts when the transform is used for image coding.
4 One cannot have blocks of different sizes superimposed onto each other. As shown in figure 3.1, one should be able to describe the background with a large scale window, and the small foreground objects with small scale windows.

These issues can be solved if we replace the Fourier analysis with a multiresolution analysis such as the wavelet transform [6]. Two dimensional wavelet bases that are used in practice are tensor products of one dimensional bases. Let φ be a scaling function, and let ψ be the corresponding wavelet, we define four wavelet functions as follows :

$$\psi_k(x,y) = \begin{cases} \varphi(x)\varphi(y) \text{ if } k=0 \\ \varphi(x)\psi(y) \text{ if } k=1 \\ \psi(x)\varphi(y) \text{ if } k=2 \\ \psi(x)\psi(y) \text{ if } k=3 \end{cases} \tag{1.2}$$

The associated filter banks $m_k(\xi,\eta), k = 1,2,3$ can resolve 2.5 directions : horizontal, vertical and an undecided diagonal/anti-diagonal direction (see Figure 3.2). Replacing the octave band decomposition with a more general splitting of the Fourier domain allows us to resolve more orientations. Wavelet packets make it possible to adaptively construct an optimal tiling of the Fourier plane, and they have been used for image compression [10, 13]. However the tensor product of two real valued wavelet packets is always associated with four symmetric peaks in the frequency plane. The geometric interpretation of a large wavelet packet coefficient becomes then problematic : the intensity in the image is either oscillating as planar wave $e^{i(\omega_x x+\omega_y y)}$, or it is oscillating with the conjugate frequency $e^{i(\omega_x x-\omega_y y)}$. In order to remove the ghost in the conjugate direction, one needs to use filters that are zero in the upper left and lower right quadrants, or in the upper right and lower left quadrants. For instance, we could have (see Figure 3.3) :

$$m_3(\xi,\eta)=0 \quad \text{if} \quad \xi>0 \quad \text{and} \quad \eta<0, \quad \text{or if} \quad \xi<0 \quad \text{and} \quad \eta>0 \tag{1.3}$$

In order to construct such filters one could use two wavelets ψ_g and ψ_h that form an (approximate) Hilbert pair :

$$\psi_g(\xi)=\begin{cases} -i\psi_h(\xi), \text{ if } & \xi>0 \\ \ \ i\psi_h(\xi), \text{ if } & \xi<0 \end{cases} \tag{1.4}$$

with φ_h, φ_g being the corresponding scaling functions. Unfortunately, any tensor products of the form $\psi_h(x)\psi_g(y)$ will have its Fourier transform localized in only one quadrant. One way to solve this problem is to consider the tensor products of the wavelet ψ_h :

$$\begin{aligned} \psi_{h,1}(x,y) &= \varphi_h(x)\psi_h(y) \\ \psi_{h,2}(x,y) &= \psi_h(x)\varphi_h(y) \\ \psi_{h,2}(x,y) &= \psi_h(x)\psi_h(y) \end{aligned} \tag{1.5}$$

Similar tensor products $\psi_{g,k}$, $k=1,2,3$ can be defined for the wavelet ψ_g. One can then compute sums and differences of the wavelets $\psi_{h,k}$ and $\psi_{g,k}$, as was suggested by Selesnick in [14] :

$$\begin{aligned} \psi_i(x,y) &= \psi_{h,i}(x,y)+\psi_{g,i}(x,y) \\ \psi_{i+3}(x,y) &= \psi_{h,i}(x,y)-\psi_{g,i}(x,y) \qquad i=0,1,2 \end{aligned} \tag{1.6}$$

The filters m_i associated to the functions ψ_i are shown in Figure 3.4. A discrete implementation is obtained by combining two wavelet transforms associated with ψ_h and ψ_g respectively. The sums and differences of the coefficients are computed according to (1.6). This 2 times redundant wavelet transform can resolve 6 directions : $\frac{\pi}{12}, \frac{\pi}{4}, \frac{5pi}{12}$ (see Figure 3.4). Kingsbury described in [5] a similar construction using complex wavelets. Kingsbury's solution is 4 times redundant. Another construction, that does not result in orthogonal or even biorthogonal wavelets, was obtained using overcomplete steerable filters [4]. Other directionally oriented filter banks were constructed in [2]. While these

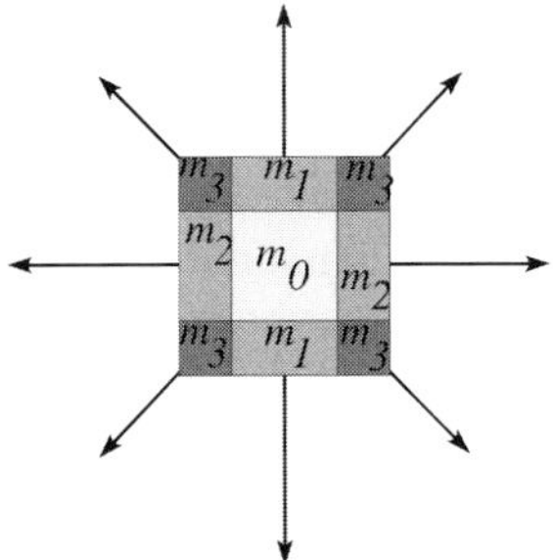

Figure 3.2. The wavelet filter banks m_1, m_2, m_3 can resolve 2.5 directions : horizontal, vertical and an undecided diagonal/anti-diagonal direction

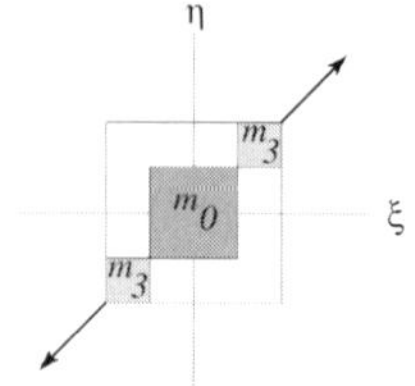

Figure 3.3. The filter m_3 can resolve the diagonal direction

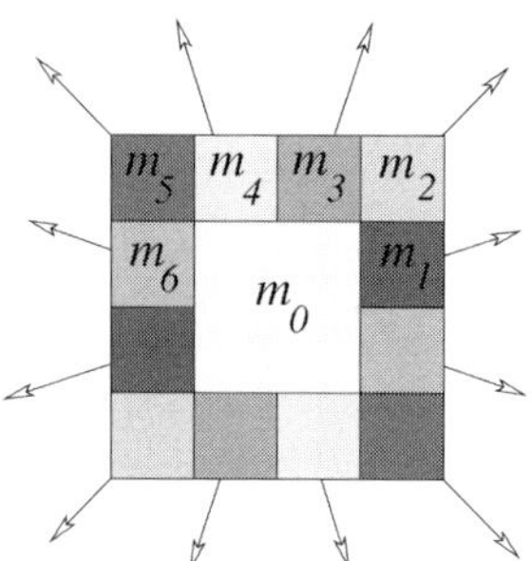

Figure 3.4. A 2 times redundant wavelet transform can resolve 6 directions : $\frac{\pi}{12}, \frac{\pi}{4}, \frac{5pi}{12}$

redundant wavelet transforms can be useful for image analysis, they suffer from the drawback of increasing the number of data by a factor 2 or 4. This becomes a major hurdle to any attempt at computing a sparse representation of an image.

In this work we propose to construct steerable wavelet packets. Inspired by the duality between local trigonometric bases and wavelet packets for a function of one variable [12], we propose to construct wavelet packets of two variables in the Fourier domain using trigonometric bases. We replace the local cosine bases by local Fourier bases. As a result our wavelet packets are complex valued functions with a phase. The construction is performed in the Fourier domain : the Fourier transform of the image is expanded into adjacent local Fourier bases. The method results in an expansion of the image into a set of steerable wavelet packets that we call *brushlets*.

A key ingredient of the construction of the brushlets is the window used for the local Fourier analysis. Indeed, the Fourier transform of the window constitutes the envelope of a brushlet. We therefore need to use windows that have a very fast decay in the Fourier domain, in order for the envelope to be well localized in space. We will present several choices of windows and study their properties. Finally, we can adaptively select the sizes and locations of the brushlets in order to obtain the most concise and precise representation of an image in terms of oriented textures with all possible directions, frequencies, and locations.

This paper is organized as follows. In the next section we review the construction of orthonormal windowed Fourier bases. This is followed in section 3 by a description of several optimized bells. The biorthogonal brushlet basis is described in section 4.

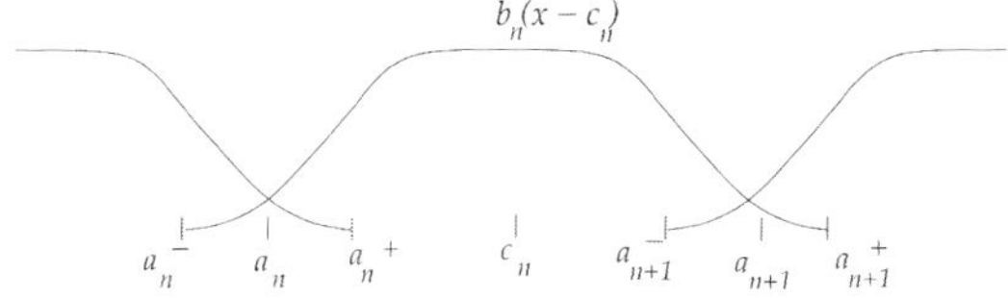

Figure 3.5. The bell $b_n(x - c_n)$ lives over the interval $[a_n^-, a_{n+1}^+]$

3.2 BIORTHOGONAL WINDOWED FOURIER BASES

As was explained in [12], wavelet packets can be constructed by expanding the Fourier transform of a signal into local cosine bases. We follow this approach and construct two dimensional wavelet packets in the Fourier domain. We replace the cosine bases with Fourier bases that will generate a unique two dimensional frequency. Our decomposition will correspond to a multiscale analysis in the spatial domain. The key ingredient of our construction is a bell used for the local Fourier bases. The design of a bell in the frequency domain is equivalent to the design of wavelet filters.

We review here the construction of smooth biorthogonal windowed Fourier bases in the one dimensional (1-D) case (references on the topic include for instance [1, 3, 16]). In two dimensions (2-D) we use tensor products of 1-D bases. We consider the more general setting where we have two biorthogonal bases [3, 8, 11]. Let $\bigcup_{n=-\infty}^{n=+\infty}[a_n, a_{n+1}[$ be a cover of $\mathbb{R}$. We define a neighborhood around each point a_n: $[a_n^-, a_n^+]$, where a_n^-, a_n, a_n^+ are such that (see Figure 3.5)

$$a_n^- = a_n - \delta < a_n < a_n^+ = a_n + \delta \tag{2.1}$$

Let $l_n = a_{n+1} - a_n$ be the length of the interval $[a_n, a_{n+1}]$, and let $c_n = (a_{n+1} + a_n)/2$ be the center of the interval $[a_n, a_{n+1}]$. Let b_n be a bell function supported on $[-l_n/2 - \delta, l_n/2 + \delta]$ (see Figure 3.6) such that

$$\forall x \in [-l_n - \delta, l_n + \delta], \qquad b_n^2(x) + b_n^2(2l_n - x) \neq 0 \tag{2.2}$$

and

$$\forall x \in [-l_n + \delta, l_n - \delta], \quad b_n(x) \neq 0 \tag{2.3}$$

We also define the "hump" function h (see Figure 3.6) :

$$\forall x \in [-\delta, \delta], \qquad h(x) = b_n(x - l_n) b_n(x + l_n) \tag{2.4}$$

We only consider the two overlapping case : the bell $b_n(x - c_N)$ only talks to the bells $b_{n-1}(x - c_{n-1})$ and $b_{n+1}(x - c_{n+1})$ (see Figure 3.5). Let

$$\theta_n(x) = \frac{1}{b_n^2(x) + b_n^2(-2l_n - x) + b_n^2(2l_n - x)} \tag{2.5}$$

then the dual bell $\tilde{b}_n$ is defined as follows:

$$\tilde{b}_n(x) = \begin{cases} \theta_n(x)\, b_n(x) \text{ if } & -l_n - \delta \le x \le -l_n + \delta \\ \dfrac{1}{b_n(x)} & \text{if } \quad -l_n + \delta \le x \le l_n - \delta \\ \theta_n(x)\, b_n(x) & \text{if } \quad l_n - \delta \le x \le l_n + \delta \\ 0 & \text{otherwise} \end{cases} \tag{2.6}$$

Let $E_{n,k}(x)$ be the family of complex exponentials on the interval $[a_n, a_{n+1})$:

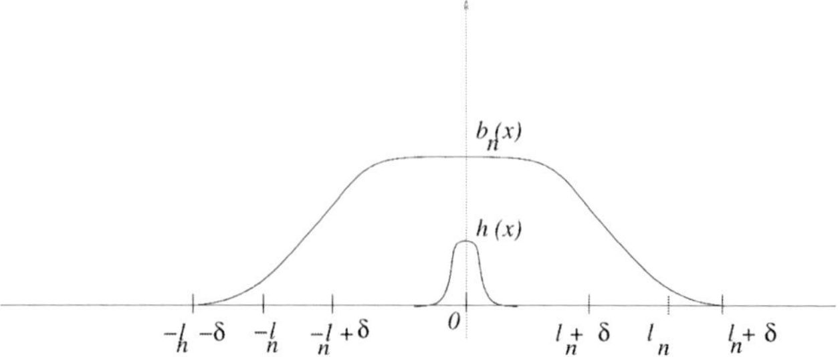

Figure 3.6. The bell b_n and the hump function h

$$E_{n,k}(x) = \frac{1}{\sqrt{l_n}} e^{-2i\pi k(\frac{x-a_n}{l_n})}. \tag{2.7}$$

We define the local Fourier basis functions as

$$u_{n,k}(x) = b_n(x-c_n)E_{n,k}(x)+h(x-a_n)E_{n,k}(2a_n-x)-h(x-a_{n+1})E_{n,k}(2a_{n+1}-x) \tag{2.8}$$

and the dual basis functions are defined as

$$\widetilde{u}_{n,k}(x) = \tilde{b}_n(x-c_n)E_{n,k}(x)+\tilde{h}(x-a_n)E_{n,k}(2a_n-x)-\tilde{h}(x-a_{n+1})E_{n,k}(2a_{n+1}-x) \tag{2.9}$$

One can use either one of the two bases to perform the analysis, and compute the coefficients, of a function f. One can then use the other basis for the synthesis, or reconstruction, of the function.

Lemma 3.2.1 $u_{n,k}$ *and* $\widetilde{u}_{n,k}$ *are Riesz biorthogonal bases:*

$$\int u_{n,k}(x)\, \widetilde{u}_{m,j}(x)\, dx = \delta_{j,k}\delta_{n,m} \tag{2.10}$$

$\forall f \in L^2(\mathbb{R})$,

$$\begin{aligned} f(x) &= \textstyle\sum_{n,k} \tilde{f}_{n,k}\, u_{n,k}(x) \qquad & \tilde{f}_{n,k} &= \int f(x)\, \widetilde{u}_{n,k}(x)\, dx \\ f(x) &= \textstyle\sum_{n,k} f_{n,k}\, \widetilde{u}_{n,k}(x) \qquad & f_{n,k} &= \int f(x)\, u_{n,k}(x)\, dx \end{aligned} \tag{2.11}$$

Furthermore : $\exists B > A > 0$ *such that,*

$$A \sum_{n,k} |f_{n,k}|^2 \leq \| \sum_{n,k} f_{n,k} u_{n,k} \|^2 \leq B \sum_{n,k} |f_{n,k}|^2 \tag{2.12}$$

The constants A and B are called the Riesz bounds. If $u_{n,k}$ is an orthonormal sequence, then $A = B = 1$. If f has unit norm, then

$$\frac{1}{B} \leq \sum_{n,k} |f_{n,k}|^2 \leq \frac{1}{A} \tag{2.13}$$

If A is much smaller than 1, then the coefficients $f_{n,k}$ in (2.13) can be very large. Conversely, if B is very large, the coefficients $f_{n,k}$ can become extremely small. In order to obtain decompositions that are numerically stable, and coefficients that neither explode nor vanish, one would like to have Riesz bounds close to 1.

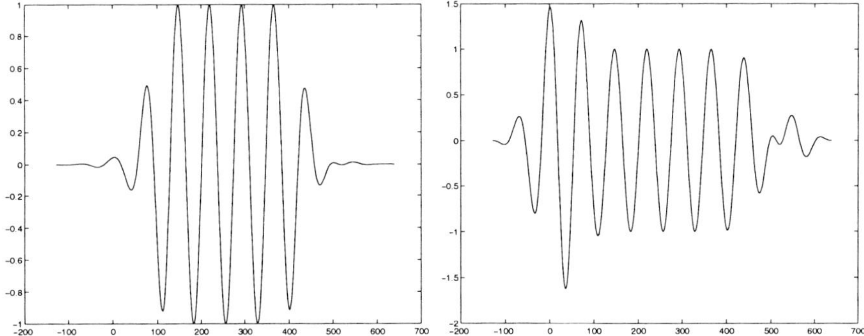

Figure 3.7. Left : real part of $u_{n,k}$, with $a_n = 0$, $a_{n+1} = 512$, and $\delta = 128$. Right : real part of the dual basis $\tilde{u}_{n,k}$

3.2.1 Implementation by folding

In practice, in order to expand a function f into the basis $u_{n,k}$ we do not calculate the correlation between f and the basis $\{u_{n,k}\}$. Instead we transform f restricted to $[a_n - \delta, a_{n+1} + \delta]$ into a smooth periodic function onto $[a_n, a_{n+1}]$, and expand it into the basis $\{E_{n,k}\}$. To do this we fold the overlapping parts of the window b_n and of the hump function h back into the interval, across the endpoints of the interval, with some folding and unfolding operators. The advantage of the procedure is that we can preprocess the data with the folding operators and then use a conventional FFT to calculate the expansion into the basis $\{E_{n,k}\}$. We will follow the construction of Wickerhauser in [1].

Unitary folding and unfolding. We define the unitary folding operator U_{a_n} and its adjoint the unfolding operator $U^*_{a_n}$ as follows:

$$U_{a_n} f(t) = \begin{cases} r(\frac{a_n - t}{\delta}) f(t) - r(\frac{t - a_n}{\delta}) f(2a_n - t), & \text{if } \ a_n - \delta < t < a_n, \\ r(\frac{t - a_n}{\delta}) f(t) + r(\frac{a_n - t}{\delta}) f(2a_n - t), & \text{if } \ a_n < t < a_n + \delta, \\ f(t), & \text{otherwise;} \end{cases} \tag{2.14}$$

$$U^*_{a_n} f(t) = \begin{cases} r(\frac{a_n - t}{\delta}) f(t) + r(\frac{t - a_n}{\delta}) f(2a_n - t), & \text{if } \ a_n - \delta < t < a_n, \\ r(\frac{t - a_n}{\delta}) f(t) - r(\frac{a_n - t}{\delta}) f(2a_n - t), & \text{if } \ a_n < t < a_n + \delta, \\ f(t), & \text{otherwise.} \end{cases} \tag{2.15}$$

We can then define the periodized folding and unfolding operators. These operators fold and unfold the right end of the segment with the left end. The periodized folding operator $W_{a_n, a_{n+1}}$ and its adjoint $W^*_{a_n, a_{n+1}}$ are defined as follows:

$$W_{a_n, a_{n+1}} f(t) =$$

$$\begin{cases} r(\frac{t - a_n}{\delta}) f(t) + r(\frac{a_n - t}{\delta}) f(a_n + a_{n+1} - t), & \text{if } a_n < t < a_n + \delta, \\ r(\frac{a_{n+1} - t}{\delta}) f(t) - r(\frac{t - a_{n+1}}{\delta}) f(a_n + a_{n+1} - t), & \text{if } a_{n+1} - \delta < t < a_{n+1}, \\ f(t), & \text{otherwise;} \end{cases} \tag{2.16}$$

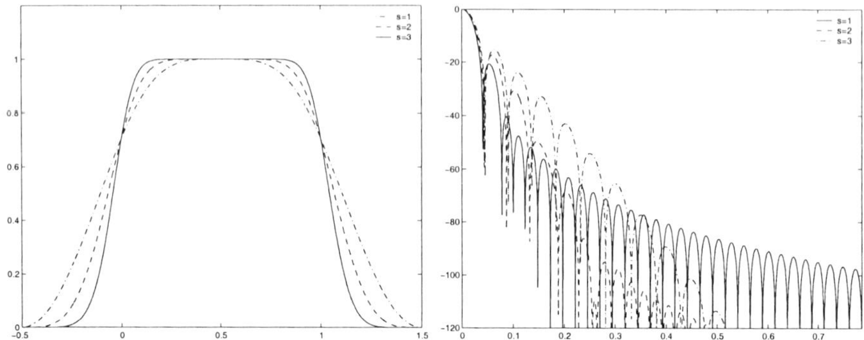

Figure 3.10. Left : orthonormal bells b^s, $s = 1, 2, 3$. Right : magnitude of the Fourier transform of b^s

$$\sum_{k=0}^{K-1} p^0_{n,k} \widetilde{\psi}_{n,k} \qquad \text{with} \quad p^0_{n,k} = \int p^0(x)\; \psi_{n,k}(x)\, dx \tag{3.5}$$

The norm of the residual error is

$$\| \sum_{k=K}^{\infty} p^0_{n,k}\; \widetilde{\psi}_{n,k} \|_2^2 \tag{3.6}$$

Matviyenko argues in [8] that this family of bells should yield a sparse representation of oscillatory signals of the form $c(x) = \cos(\omega x + \varphi)$.

Instead of reproducing exactly p^0 with one coefficient, Matviyenko designed a family of bells that minimize the residual error (3.6). Matviyenko shows that minimizing the sum

$$\sum_{k=K}^{\infty} |p^0_{n,k}|^2 \tag{3.7}$$

is related to minimizing the residual error (3.6). He then finds the bell b that minimizes (3.7) under the constraint :

$$b(x) + b(-x) = 1 \quad \text{for all } x \in [0, 1/2]. \tag{3.8}$$

The solution of the optimization problem is a bell $b^K(x)$ given by :

$$b^K(x) = \begin{cases} \frac{1}{2}(1 + \sum_{k=0}^{K-1} g_k \sin(k + 1/2)\pi x) & \text{if} \quad -0.5 \le x \le 0.5 \\ \frac{1}{2}(1 + \sum_{k=0}^{K-1} (-1)^k g_k \cos(k + 1/2)\pi x) & \text{if} \quad 0.5 \le x \le 1.5 \\ 0 & \text{otherwise} \end{cases} \tag{3.9}$$

The g_k are calculated numerically in [8]. K influences the steepness of the bell. All bells b_K are bounded by 1, and the dual bells $\tilde{b}_K$ are bounded by $(\sqrt{2} + 1)/2$. These bounds guarantee that the Riesz bounds will be $A = 1$ and $B = 2$ for all K. Figure 3.11 shows the bell b^K and the dual bell $\tilde{b}^K$ for $K = 1$ and 3. The magnitude of the Fourier transform of the bells b^K and $\tilde{b}^K$ (for an interval of N=512 samples) is shown in Figure 3.12. As K increases the side lobes become much smaller. This observation, and equation (3.7) seems to indicate that a large K should provide a smaller error, and a better frequency resolution. In practice, as shown in [9], small values of K often provide

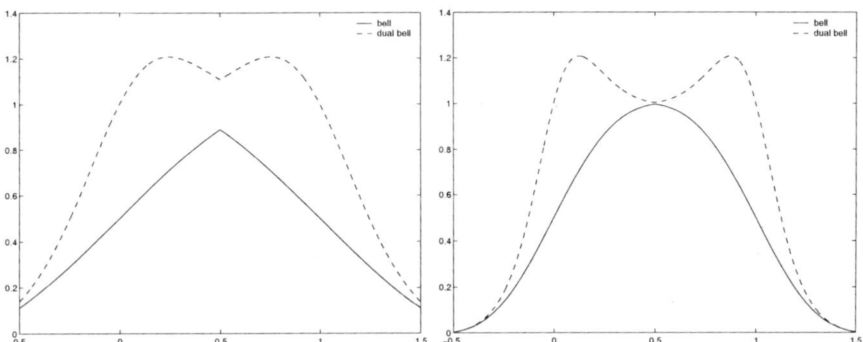

Figure 3.11. Matviyenko's optimized bell b^K, and dual bell $\tilde{b}^K$, for $K = 1$ (left), $K = 3$ (right)

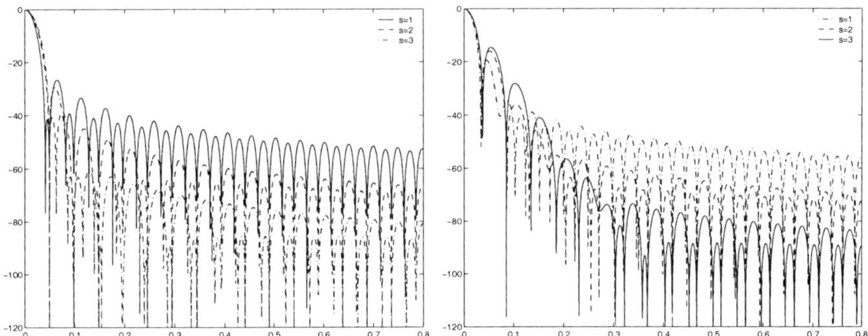

Figure 3.12. Left : Matviyenko's bell : magnitude of the Fourier transform of b^K. Right : magnitude of the Fourier transform of $\widetilde{b^K}$

better performances. One can either choose to use b^K or $\widetilde{b}^K$ to compute the coefficients. Because b^K is optimized to minimize the residual error, one should use b^K for the analysis and $\widetilde{b}^K$ for the reconstruction.

3.3.3 Modulated Lapped Biorthogonal Transform (MLBT)

A simple way to smooth $\sin \frac{\pi}{2}(x + 1/2)$ at both end-points is to take the square of the bell. As a result the following bell is in $C^1(\mathbb{R})$:

$$b(x) = \begin{cases} \sin^2 \left[\frac{\pi}{2}(x + 1/2)\right] = \dfrac{1 - \cos \pi(x + 1/2)}{2} & \text{if } x \in [-1/2, 1/2] \\ b(x) = b(1 - x) & \text{if } x \in [1/2, 3/2] \end{cases} \tag{3.10}$$

Malvar proposed in [7] the following dual bell :

$$\tilde{b}(x) = \begin{cases} \dfrac{1 - \cos[\pi(x + 1/2)^\alpha] + \beta}{2 + \beta} & \text{if } x \in [-1/2, 1/2] \\ b(x) = b(1 - x) & \text{if } x \in [1/2, 3/2] \end{cases} \tag{3.11}$$

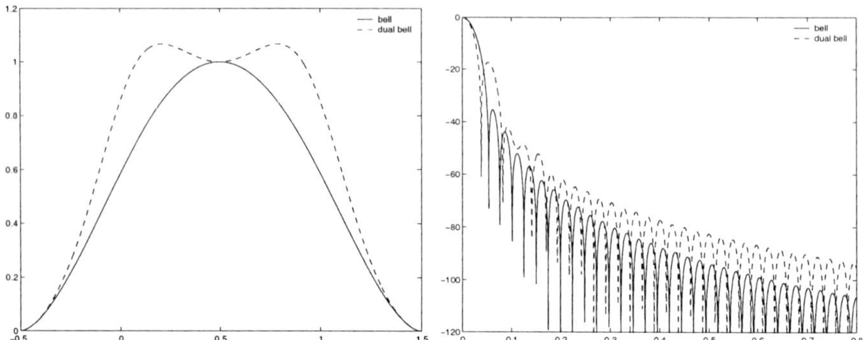

Figure 3.13. MLBT's bell b and the associated dual bell $\tilde{b}$. Right : magnitude of the Fourier transform of b and $\tilde{b}$

For $\alpha = 1$ and $\beta = 0$ we clearly find the square of $\sin \frac{\pi}{2}(x + 1/2)$. The bell is $C^1(\mathbb{R})$ if $\alpha \geq 1$. Because the bell is smoother, it will have a faster decay in the Fourier domain, and a better frequency selectivity of the associated basis functions [7].

The analysis bell b is derived from the dual bell $\tilde{b}$ using (2.6). In this paper we use the following values of the parameters : $\alpha = 0.85$, and $\beta = 0$. Figure 3.13 shows the graphs of b and $\tilde{b}$. The two bells are similar to the optimized bells of Matviyenko. The Riesz bounds are $A = 1$ and $B = 1.458$. The magnitude of the Fourier transform of the analysis bell and the dual bell (for an interval of N=512 samples) is shown in Figure 3.13. The Fourier transform of b has a wider main lobe than the Fourier transform of $\tilde{b}$ but a better stop-band attenuation (smaller side lobes).

3.4 BIORTHOGONAL BRUSHLET BASES

Inspired by the duality between local trigonometric bases and wavelet packets for a function of one variable [12], we propose to construct wavelet packets of two variables in the Fourier domain using trigonometric bases. We replace the local cosine bases by local Fourier bases.

3.4.1 One dimensional case

Let $f \in L^2(\mathbb{R})$, and let $\hat{f}$ be the Fourier transform of f. We define a cover of the frequency axis :

$$\bigcup_{n=-\infty}^{n=+\infty} [\omega_n - \frac{l_n}{2}, \omega_n + \frac{l_n}{2}]. \tag{4.1}$$

ω_n is the center of each interval of size l_n [1]. Let $u_{n,k}$ be the local Fourier basis associated with this cover. We expand $\hat{f}$ into the basis $u_{n,k}$

$$\hat{f} = \sum \hat{f}_{n,k} u_{n,k} \tag{4.2}$$

We then take the inverse Fourier transform . Let $\psi_{n,k}$ the inverse Fourier transform of $u_{n,k}$.

[1] the center of the interval was called c_n when the analysis was performed in the original time domain, since we work in the frequency domain we prefer to use ω_n

$$f = \sum \hat{f}_{n,k} \psi_{n,k} \tag{4.3}$$

Since the Fourier transform is a unitary operator, we obtain a new pair of biorthogonal bases by applying the inverse Fourier transform on $u_{n,k}$ and $\tilde{u}_{n,k}$.

Lemma 3.4.1 *$\{\psi_{m,j}, \tilde{\psi}_{n,k} \quad j,k,m,n \in \mathbb{Z}\}$ are biorthogonal bases for $L^2(\mathbb{R})$.* □

We call $\{\psi_{n,k}\}$ and $\left\{\tilde{\psi}_{n,k}\right\}$ the biorthogonal brushlet basis. From (2.8) we have

$$\psi_{n,k}(x) = \frac{1}{\sqrt{l_n}} e^{2i\pi\omega_n x} \left\{(-1)^k \widehat{b}_n(x - \frac{k}{l_n}) - 2i \sin(\pi l_n x) \hat{h}(x + \frac{k}{l_n})\right\} \tag{4.4}$$

We can introduce the "steepness factor" of the window b_n,

$$\sigma = \frac{\delta}{l_n} \tag{4.5}$$

we also introduce the window b_σ supported on $[-1/2 - \sigma, 1/2 + \sigma]$ such that

$$b_n(x) = b_\sigma(\frac{x}{l_n}) \tag{4.6}$$

and the hump function h_σ supported on $[-\sigma, \sigma]$ such that

$$h(x) = h_\sigma(\frac{x}{l_n}) \tag{4.7}$$

We have

$$\widehat{b}_n(x) = l_n \hat{b}_\sigma(l_n x) \tag{4.8}$$

then (4.4) can be rewritten as follows

$$\psi_{n,k}(x) = \frac{1}{\sqrt{l_n}} e^{2i\pi\omega_n x} \left\{(-1)^k l_n \hat{b}_\sigma(l_n x - k) - 2i \sin(\pi l_n x) l_n \hat{h}_\sigma(l_n x + k)\right\} \tag{4.9}$$

then we have

$$\psi_{n,k}(x) = \sqrt{l_n} e^{2i\pi a_n x} e^{i\pi l_n x} \left\{(-1)^k \hat{b}_\sigma(l_n x - k) - 2i \sin(\pi l_n x) \hat{h}_\sigma(l_n x + k)\right\} \tag{4.10}$$

We note in (4.10) that l_n appears as a scaling factor of the analysis, and k is the translation index of the brushlet. $\psi_{n,k}$ has an expression similar to a wavelet, however as opposed to a real valued wavelet, $\psi_{n,k}$ is a complex valued function with a phase. The phase encodes the orientation of the brushlet pattern in the two-dimensional case. b_σ and h_σ are even real valued functions, thus $\hat{b}_\sigma$ and $\hat{h}_\sigma$ are also even real valued functions.

The function $\psi_{n,k}$ is composed of two terms, localized around k/l_n, and around $-k/l_n$, that are oscillating with the frequency ω_n. The main term is an exponential multiplied by $\hat{b}_\sigma$. Because b_σ is compactly supported, $\hat{b}_\sigma$ has an infinite support. As explained in the previous section, a careful choice of b allows us to have a Fourier transform $\hat{b}_\sigma$ with a fast decay.

The envelope of the second term is the Fourier transform of the hump function h. If one chooses $\hat{b}_\sigma$ with a fast decay, then $\hat{h}_\sigma$ will also have a fast decay. We can also control the magnitude of $\hat{b}_\sigma$ using σ : since $|\hat{h}_\sigma(x)| \leq \sigma$, the second term can be made as small as possible. However, when σ tends to zero the first term is not localized anymore. There is a tradeoff between the localization of $\hat{b}_\sigma$ and the magnitude of the second term.

3.4.2 Discrete implementation of the brushlet expansion

We assume that the original signal f has been sampled at N equally spaced mesh nodes

$$F_n = f(n\Delta x) \quad n = 0, 1, \dots, N-1 \tag{4.11}$$

We calculate the discrete Fourier transform of the sequence F_n using an FFT. We obtain N samples

$$\widehat{F}_k = \hat{f}(\frac{k}{\Delta x}) \quad , \qquad k = -\frac{N}{2}, -\frac{N}{2}+1, \dots, 0, \dots, \frac{N}{2}-1 \tag{4.12}$$

We then divide the set of integers

$$\left\{-\frac{N}{2}, \dots, \frac{N}{2}-1\right\} \tag{4.13}$$

into $\frac{N}{l}$ intervals of equal size l. For each interval

$$[k\frac{1}{\Delta x}, (k+l-1)\frac{1}{\Delta x}]$$

we expand $\widehat{F}$ into the discrete orthonormal windowed Fourier basis. First we calculate $T_{k,k+l-1}(\widehat{F})$ using the discrete smooth periodic restriction operator. We then expand $T_{k,k+l-1}(\widehat{F})$ into the basis $E_{n,k}$ using an FFT of size l. On $[k\frac{1}{\Delta x}, (k+l-1)\frac{1}{\Delta x}]$, $\widehat{F}$, is uniquely characterized by the samples at the mesh points

$$\frac{k}{\Delta x}, \dots, \frac{k+l-1}{\Delta x}$$

Since $T_{k,k+l-1}$ is a unitary isomorphism $T_{k,k+l-1}(\widehat{F})$ is also uniquely characterized by the samples

$$\frac{k}{\Delta x}, \dots, \frac{k+l-1}{\Delta x}$$

Finally the discrete Fourier transform of $T_{k,k+l-1}(\widehat{F})$ is characterized by the samples at the mesh points

$$0, \frac{1}{l}\Delta x, \dots, \frac{l-1}{l}\Delta x$$

Let $\Psi_{n,k}$ be the discrete version of the basis function $\psi_{n,k}$. We note that even though the support of $\Psi_{n,k}$ is larger than $[0, \Delta x\frac{l-1}{l}]$, $\Psi_{n,k}$ is entirely characterized by the samples

$$0, \frac{1}{l}\Delta x, \dots, \frac{l-1}{l}\Delta x$$

This result is similar to the subsampling operation in multiresolution analysis [6, 12]. Here the subsampling is performed by selecting subinterval of the global Fourier transform and expanding them into a local Fourier basis. In (4.10) we can replace x by the sample $k\frac{\Delta x}{l}$, $k = 0, \dots, l-1$, and l_n by $\frac{l}{\Delta x}$ and we obtain the discrete version of $\psi_{n,k}$:

$$\Psi_{n,k}(k) = \sqrt{l}\, e^{2i\pi\frac{jk}{l}} e^{i\pi k}\left\{(-1)^n \hat{b}_\sigma(k-n) - 2i\,\sin(\pi k)\hat{h}_\sigma(k+n)\right\} \tag{4.14}$$

with

$$0 \le n \le l-1 \quad \text{and} \quad k = 0, \dots, l-1.$$

Figures 3.14 and 3.15 show the basis functions $\Psi_{n,k}$ and the dual function $\widetilde{\Psi}_{n,k}$ for two

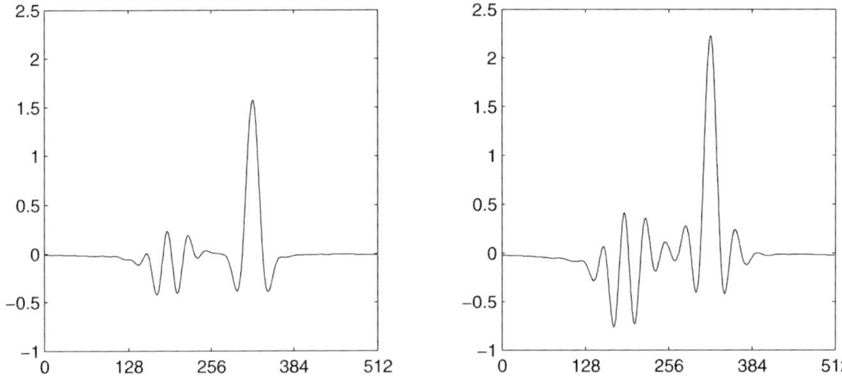

Figure 3.14. Left : the basis function $\Psi_{n,k}$. Right : the dual function $\widetilde{\Psi}_{n,k}$. We used the MLBT window with $\alpha = 0.85$. $N = 512, l = 16, \delta = 8, \omega_n = 8$ and $k = 320$

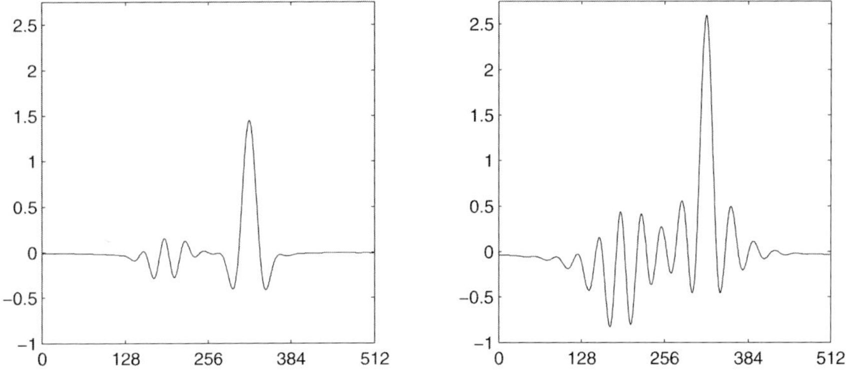

Figure 3.15. Left : the basis function $\Psi_{n,k}$. Right : the dual function $\widetilde{\Psi}_{n,k}$. We used Matviyenko's window with $K = 3$. $N = 512, l = 16, \delta = 8, \omega_n = 8$ and $k = 320$

different windows. All the graphs were obtained with $N = 512, l = 16, \delta = 8, \omega_n = 8$ and $k = 320$. In figure 3.14 we used the MLBT window with $\alpha = 0.85$. In figure 3.15 we used Matviyenko's window with $K = 3$. The main part of the function $\Psi_{n,0}$ is similar to the scaling function of a wavelet. This should be the case, since $\Psi_{n,0}$ corresponds to the window that is nearest to 0 in the Fourier domain. As expected we observe the Fourier transform of the hump function, $\hat{h}$, on the left of main part. We can lower the amplitude of $\hat{h}$ in $\Psi_{n,k}$ by having a window b_n that is steeper, or by decreasing δ. Unfortunately, decreasing δ has the effect of spreading the main part of $\Psi_{n,k}$. As we make the window b_n steeper, the Fourier transform of the dual of the hump function, $\widehat{\tilde{h}}$, becomes larger.

3.4.3 Two-dimensional case

We are now in the position of constructing two dimensional brushlets. We define a partition of the frequency plane obtained by the lattice cubes :

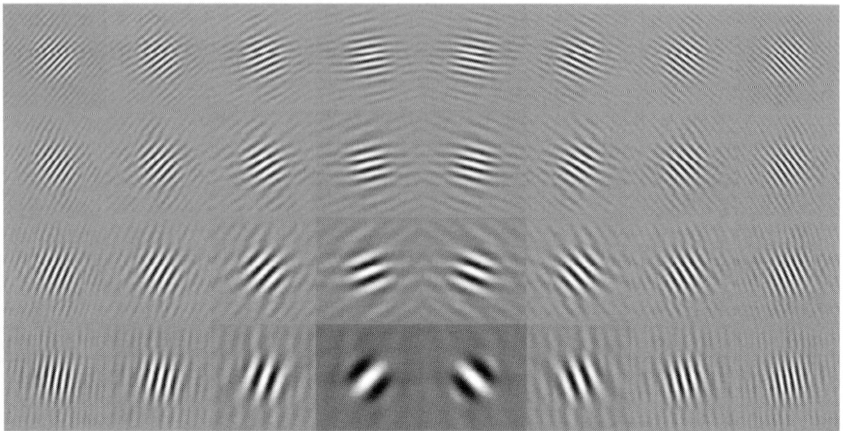

Figure 3.16. Basis functions $\psi_{m,j} \otimes \psi_{n,k}$ for the frequencies $(m\frac{2\pi}{512}, m\frac{2\pi}{512})$, with $(m,n) \in \{-48,-32,-16,0,16,32,48\}^2$. $\psi_{n,k}$ is represented as an image, where a large positive value is coded with white, and a large negative value is coded with black. We have $h_m = l_n = 16, \delta = 8$

$$\bigcup_{n=-\infty}^{\infty} \bigcup_{m=-\infty}^{\infty} [\xi_m - \frac{h_m}{2}, \xi_m + \frac{h_m}{2}] \otimes [\eta_n - \frac{l_n}{2}, \eta_n + \frac{l_n}{2}] \tag{4.15}$$

of the frequency axis. (ξ_m, η_n) is the center of each rectangle of size $h_m \times l_n$. We consider the separable tensor products of bases $\psi_{m,j}$, and $\psi_{n,k}$. We have

Lemma 3.4.2 *The sequence $\psi_{m,j} \otimes \psi_{n,k}$ is an orthonormal basis for $L^2(\mathbb{R}^2)$.* □

We have :

$$\begin{aligned}\psi_{m,j}(x) \otimes \psi_{n,k}(y) = \sqrt{h_m\ l_n} e^{2i\pi(\xi_m x + \eta_n y)} & \\ & \left[(-1)^j \hat{b}_\sigma(h_m x - j) - 2i \sin(\pi h_m\, x) \hat{h}_\sigma(h_m x + j)\right] \\ & \left[(-1)^k \hat{b}_\sigma(l_n y - k) - 2i\ \sin(\pi l_n\ y) \hat{h}_\sigma(l_n y + k)\right]\end{aligned} \tag{4.16}$$

The tensor product $\psi_{m,j}(x) \otimes \psi_{n,k}(y)$ is an oriented pattern oscillating with the frequency (ξ_m, η_n) and localized at $(j/h_m, k/l_n)$. The size of the pattern is inversely proportional to the size of the analyzing window: $h_m \times l_n$ in the Fourier space.

Directional image analysis. Figure 3.16 shows the basis functions $\psi_{m,j} \otimes \psi_{n,k}$ for several values of the frequency (ξ_m, η_n), at a fixed scale (i.e fixed h_m and l_n). The function is represented as an image, where a large positive value is coded with white, and a large negative value is coded with black. The image size was 512×512, and the analysis windows in the Fourier domain were defined by $h_m = l_n = 16, \delta = 8$. The frequencies shown are

$$(m\frac{2\pi}{512}, m\frac{2\pi}{512}) \quad (m,n) \in \{-48,-32,-16,0,16,32,48\}^2 \tag{4.17}$$

This figure illustrates the selective orientation analysis performed by the brushlets : at a given scale the brushlets can resolve many more orientations (10 for this scale) than standard wavelet packets. As the scale gets coarser (smaller h_m, l_n), one can resolve even more directions.

Clearly the advantage of the brushlet over a local Fourier analysis stems from the fact that one can perform a multiscale analysis of an image. We illustrate this property with the brushlet expansion of the image Barbara, shown in Figure 3.17. A first expansion was performed with a tiling of the Fourier plane into four quadrants. We have $x_0 = -256, x_1 = 0, x_2 = 255$, and similarly $y_0 = -256, y_1 = 0, y_2 = 255$. The four sets of brushlets have the orientations $\frac{\pi}{4} + k\frac{\pi}{2}, k = 0, \ldots, 3$. Figure 3.18 shows the imaginary part of the brushlets coefficients for each of the four quadrants of the Fourier plane. Since the signal is real, the coefficients are antisymmetric with respect to the origin. The upper right quadrant contains textures with patterns oriented along the direction $\frac{\pi}{4}$: the right leg, the mouth, the eyes, and the left arm. In the upper left window, textures with patterns oriented along the direction $\frac{3\pi}{4}$: the left leg, the nose, the right arm.

A second expansion has been performed using a finer grid. Each quadrant was further divided into four quadrants. The brushlet expansion was calculated for this finer tiling. The sixteen set of brushlets have twelve different orientations as shown in Figure 3.19. The orientations $\frac{\pi}{4} + k\frac{\pi}{2}$ are associated with two different frequencies. Figure 3.19 shows the imaginary part of the brushlet expansion. Again the coefficients are antisymmetric with respect to the origin. The four lattice squares around the origin characterize the DC terms of the expansion. The other squares correspond to higher frequency textures. We note that the texture of the legs, and on the scarf have been completely removed from the four DC regions, and are present in the regions that have the directions 1 and 12 as shown in Figure 3.19.

We note that the decomposition achieved by wavelet packets does not permit us to localize a unique frequency, for instance in the positive part of the Fourier space. Indeed two symmetric windows are always associated with a real wavelet. As a result a wavelet packet expansion will require many more coefficients to describe a pattern with an arbitrary orientation; whereas the same pattern can be coded with a single brushlet coefficient.

Directionally oriented filter banks (e.g. [2, 15]) have been used for image compression and image analysis. They do not allow however an arbitrary partitioning of the Fourier plane. Furthermore in our method the tiling can be adapted to the image content : we can adaptively select the size and location of the windows $[\xi_m - h_m/2, \xi_m + h_m/2] \otimes [\eta_n - l_n/2, \eta_n + l_n/2]$ with the best basis algorithm.

Figure 3.17. Original 512 x 512 Barbara image

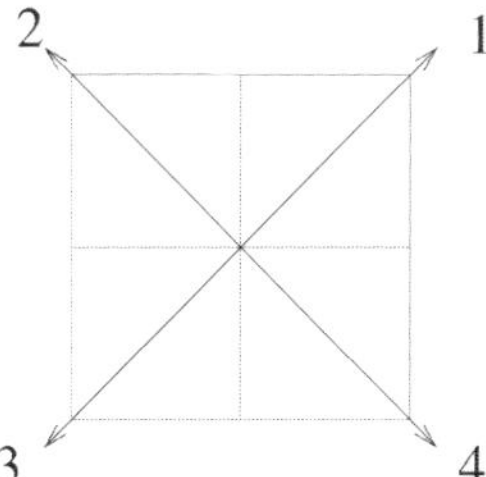

Figure 3.18. Imaginary part of the brushlets coefficients for each of the four quadrants of the Fourier plane. Since the signal is real, the coefficients are antisymmetric with respect to the origin. The upper right quadrant contains textures with patterns oriented along the direction $\frac{\pi}{4}$: the right leg, the mouth, the eyes, and the left arm. In the upper left window, textures with patterns oriented along the direction $\frac{3\pi}{4}$: the left leg, the nose, the right arm

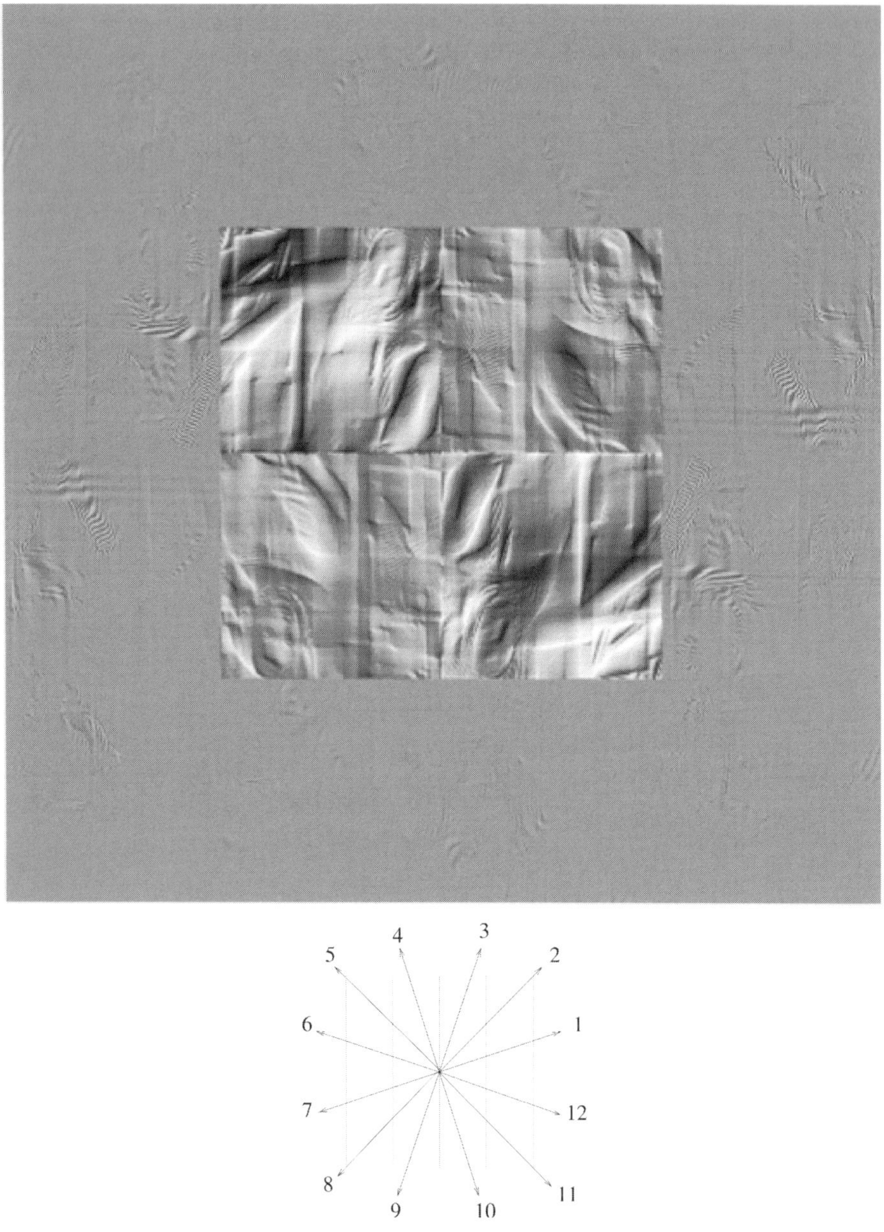

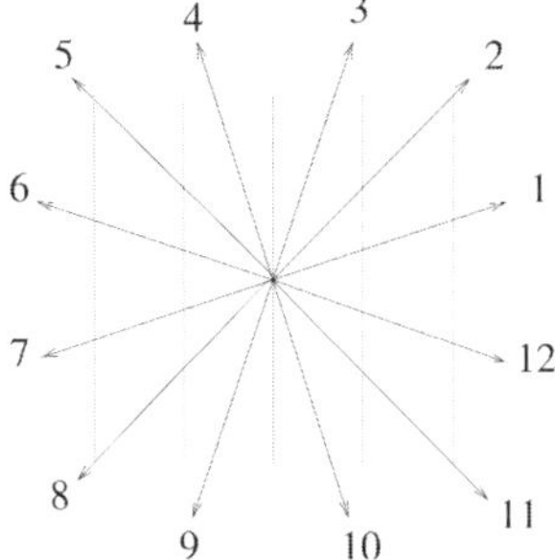

Figure 3.19. Imaginary part of the brushlet expansion. The four lattice squares around the origin characterize the DC terms of the expansion. The other squares correspond to higher frequency textures. The texture of the legs, and on the scarf have been completely removed from the four DC regions, and are present in the regions that have the directions 1 and 12.

Figure 3.20. These simple images cannot be coded efficiently with any existing transform : wavelets, Fourier, DCT, brushlets, etc.

3.5 CONCLUSION

We have addressed the problem of describing an image with a library of steerable wavelet packets. Inspired by the duality between local trigonometric bases and wavelet packets, we constructed wavelet packets of two variables in the Fourier domain using local Fourier bases. Our wavelet packets are complex valued functions with a phase. We have shown that the brushlets can resolve without any ambiguity many more orientations than standard wavelet packets. In theory our brushlets have infinite support. However we have shown that a careful design of the window of the local Fourier basis yields a brushlet with a very fast decay.

This construction demonstrates that there are many other ways to analyze and represent images that go beyond the standard bi-dimensional wavelet transform. We believe that the following questions are areas for future studies, and will give rise to more "natural" basis functions for images.

1 Our brushlets have a square support. One should be able to construct elongated brushlets with several possible aspect ratio. One approach to this problem consists in resampling the Fourier transform $\hat{f}$ in polar coordinates (ρ, θ) and expanding $\hat{f}(\rho, \theta)$ with local Fourier bases.

2 We have only investigated the efficient representation of periodic patterns. Many other type of texture exist in images. The texture formed by the coffee beans or the flowers in the images shown in Figure 3.20 cannot be efficiently coded by brushlets. In fact, these deceptively simple images cannot be efficiently coded with any transform that is available to us today : wavelet transform, Fourier transform, etc.

REFERENCES

[1] P. Auscher, G. Weiss, and M.V. Wickerhauser, *Local sine and cosine bases of Coifman and Meyer*, Wavelets-A Tutorial, Academic Press, 1992, pp. 237–256.

[2] R.H. Bamberger and M.J.T. Smith, *A filter bank for the directional decomposition of images: theory and design*, IEEE Trans. on Signal Processing (1992), 882–893.

[3] C.K. Chui and X. Shi, *Characterization of biorthogonal cosine wavelets*, J. Fourier Anal. Appl. **3** (1997), no. 5, 560–575.

[4] W.T. Freeman and E.H. Adelson, *The design and use of steerable filters*, IEEE Trans. PAMI **13, No9** (1991), 891–906.

[5] N. Kingsbury, *Image processing with complex wavelets*, Phil. Trans. R. Soc. Lond. A **357** (1999), 2543–2560.

[6] S. Mallat, *A wavelet tour of signal processing*, Academic Press, 1999.

[7] H.S. Malvar, *Biorthogonal and nonuniform lapped transforms for transform coding with reduced blocking and ringing artifacts*, IEEE Transactions on Signal Processing **46(4)** (1998), 1043–1053.

[8] G. Matviyenko, *Optimized local trigonometric bases*, Applied and Computational Harmonic Analysis **3** (1996), 301–323.

[9] F.G. Meyer, *Image compression with adaptive local cosines : A comparative study*, International Conference on Image Processing, ICIP'01, Thessaloniki, Greece, Oct. 2001, IEEE Press, 2001.

[10] F.G. Meyer, A.Z. Averbuch, and J-O. Strömberg, *Fast adaptive wavelet packet image compression*, IEEE Trans. on Image Processing (2000), 792–800.

[11] F.G. Meyer and R.R. Coifman, *Brushlets: a tool for directional image analysis and image compression*, Applied and Computational Harmonic Analysis (1997), 147–187.

[12] Y. Meyer, *Wavelets and operators*, Cambridge University Press, 1993.

[13] K. Ramchandran and M. Vetterli, *Best wavelet packet bases in a rate-distortion sense*, IEEE Trans. on Image Processing **2** (1993), no. 2, 160–175.

[14] I. Selesnick, *The design of Hilbert transform pairs of wavelet bases*, *to appear in* IEEE Trans. on Signal Processing, 2001.

[15] E.P. Simoncelli and E.H. Adelson, *Nonseperable extensions of quadrature miror filters to multiple dimensions*, Proc. of the IEEE (1990), 652–664.

[16] M.V. Wickerhauser, *Adapted wavelet analysis from theory to software*, A.K. Peters, 1995.

Beyond Wavelets
G. V. Welland (Editor)
© 2003 Elsevier Science (USA) All rights reserved

4

CONTOURLETS

M. N. DO AND M. VETTERLI

Department of Electrical and Computer Engineering, Beckman Institute
University of Illinois at Urbana-Champaign, Urbana, IL 61801, USA
minhdo@uiuc.edu

Department of Communication Systems, Swiss Federal Institute of Technology
1015 Lausanne, Switzerland
and
Department of Electrical Engineering and Computer Science
University of California at Berkeley, Berkeley, CA 94720, USA
Martin.Vetterli@epfl.ch

Abstract

This chapter focuses on the development of a new "true" two-dimensional representation for images that can capture the intrinsic geometrical structure of pictorial information. Our emphasis is on the discrete framework that can lead to algorithmic implementations. We propose a double filter bank structure, named the only *pyramidal directional filter bank*, by combining the Laplacian pyramid with a directional filter bank. The result is called the only *contourlet transform*, which provides a flexible multiresolution, local and directional expansion for images. The contourlet transform can be designed to satisfy the anisotropy scaling relation for curves, and thus offers a fast and structured curvelet-like decomposition sampled signals. As a result, the proposed transform provides a sparse representation for two-dimensional piecewise smooth signals that resemble images. The link between the developed filter banks and the continuous-space constructions is set up precisely in a newly defined directional multiresolution analysis. Finally, we show some numerical experiments demonstrating the potential of the new transform in several image processing tasks.

4.1 INTRODUCTION AND MOTIVATION

We are interested in the construction of *efficient linear expansion* for *two-dimensional signals, which are smooth away from discontinuities across smooth curves.* Such signals resemble natural images where discontinuities are generated by *edges* – points in the image where there is a sharp contrast in intensity, whereas edges are often gathered along smooth *contours*, which are created by typically smooth boundaries of physical objects. Efficiency of a linear expansion means that the coefficients for signals belonging to the class of interest are *sparse*, and thus it implies efficient representations for such functions, using a non-linear approximation (NLA) scheme.

Over the last decade, wavelets have had a growing impact on signal processing, mainly due to their good NLA performance for piecewise smooth functions in one dimension [1–3]. Unfortunately, this is not the case in two dimensions. In essence, wavelets are good at catching point or zero-dimensional discontinuities, but as mentioned above, two-dimensional piecewise smooth functions resembling images have one-dimensional discontinuities. Intuitively, wavelets in 2-D obtained by a tensor-product of one dimensional wavelets will be good at isolating the discontinuities at edge points, but will not see the smoothness along the contours. This indicates that more powerful representations are needed in higher dimensions.

Recently, Candès and Donoho [5, 6] pioneered a new system of representation, named *curvelet*, that was shown to achieve optimal approximation behavior in a certain sense for 2-D piecewise smooth functions in $\mathbb{R}^2$ where the discontinuity curve is a C^2 function.[1] More specifically, an M-term non-linear approximation for such piecewise smooth functions using curvelets has L^2 square error decaying like $O(M^{-2})$, and this is the best rate that can be achieved by a large class of approximation processes [7]. An attractive property of the curvelet system is that such correct approximation behavior is simply obtained via thresholding a *fixed* transform. The key features of the curvelet elements is that they exhibit very high directionality and anisotropy.

The original construction of the curvelet transform [5] was intended for functions defined in the *continuum* space $\mathbb{R}^2$. The development of *discrete* transforms for sampled images that has all the features promised by curvelets in the continuous domain remains a challenge, especially when critical sampling is desirable. Furthermore, as the curvelet transform was defined in the frequency domain, it is not clear how curvelets are sampled in the spatial domain. In fact, in [8], one of the fundamental research challenges for curvelets was stated as: "is there a spatial domain scheme for refinement which, at each generation doubles the spatial resolution as well as the angular resolution?". This is what we will try to explore in the following.

First, we will identify the key features that make curvelets an efficient representation for 2-D piecewise smooth functions with smooth discontinuity curves. Based on this, we propose a filter bank structure that can deal effectively with piecewise smooth images with smooth contours. The resulting image expansion is a frame composed of contour segments, and thus is named *contourlet*. We then derive an analysis framework that connects the proposed discrete transform to the frames in the continuous-domain, which can be particularized to a curvelet-like expansion. Thus our scheme provides an effective method to implement the discrete curvelet transform. Furthermore, the resulting transform has very small redundancy, being almost critically sampled. Finally, we will show some numerical experiments demonstrating the potential of the contourlet transform in several image processing tasks.

---

[1] C^p is the space of functions that are bounded and p-times continuously differentiable.

4.2 REPRESENTING 2-D PIECEWISE SMOOTH FUNCTIONS

4.2.1 Curvelet construction

In a nutshell, the curvelet transform [5] is obtained by filtering and then applying a windowed ridgelet transform [4] to each bandpass image. In $\mathbb{R}^2$, ridgelets are constant along ridge lines $x_1 \cos(\theta) + x_2 \sin(\theta) = const$ and are wavelets (with a scale s) along the orthogonal direction. In frequency domain, such a ridgelet function is essentially localized in the corona $|\omega| \in [2^s, 2^{s+1}]$ and around the angle θ. The ridgelet transform provides a sparse representation for smooth objects with straight edges. In summary, the curvelet decomposition is composed of the following steps [5] (also see Figure 4.2(a)):

1 Subband decomposition of the object into a sequence of subbands.
2 Windowing each subband into blocks of appropriate size, depending on its center frequency.
3 Applying the ridgelet transform to these blocks.

The motivation behind the curvelet transform is that by smooth windowing, segments of smooth curves would look straight in sub-images, hence they can be captured efficiently by a local ridgelet transform. Subband decomposition is used to keep the number of ridgelets at multiple scales under control by the fact that ridgelets of a given scale live in a certain subband. The window's size and subband frequency are coordinated such that curvelets have support obeying the key *anisotropy scaling relation* for curves [5, 6]:

$$width \propto length^2. \tag{2.1}$$

4.2.2 Non-linear approximation behaviors

We next sketch illustrations on the non-linear approximation behaviors for 2-D piecewise smooth functions using different expansions. Rather than being rigorous, the following discussion aims at providing an intuition that can serve as a guideline for our construction of the pyramidal directional filter banks and contourlets ladder. For a complete and rigorous discussion, we refer to [7].

Consider a simple "Horizon" model of piecewise smooth functions $f(x_1, x_2)$ defined on the unit square $[0, 1]^2$:

$$f(x_1, x_2) = 1_{\{x_2 \geq c(x_1)\}} \qquad 0 \leq x_1, x_2 \leq 1,$$

where the boundary of two pieces (or the contour) $c(x_1)$ is in C^p and has finite length inside the unit square. Clearly, such a 2-D function has complexity equivalent to a 1-D function, namely its contour $c(x_1)$. The reason for studying this model is that the approximation rates for 2-D piecewise smooth functions resembling images are typically dominated by the discontinuity curves.

Let's first consider how a wavelet system performs for such function. Assume that the orthonormal wavelet transform with the separable Haar wavelet is employed. At the level j, wavelet basis functions have support on dyadic squares of size 2^{-j} (see Figure 4.1(a)). Let n_j be the number of dyadic squares at level j that intersect with the contour on the unit square. Since the contour has finite length, it follows that

$$n_j \sim O(2^j). \tag{2.2}$$

Thus, there are $O(2^j)$ nonzero wavelet coefficients at the scale 2^{-j}. This is the problem of the separable wavelet transform for 2-D piecewise smooth functions. For the 1-D piecewise smooth function, the number of significant wavelet coefficients at each scale is bounded by a constant; in the 2-D case this number grows exponentially as the scale gets finer. The total number of nonzero wavelet coefficients up to the level J is

$$N_J = \sum_{j=0}^{J} n_j \sim O(2^J). \tag{2.3}$$

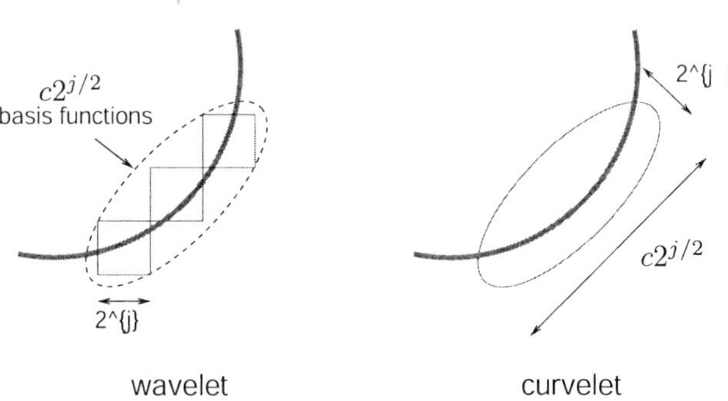

Figure 4.1. Non-linear approximation of a 2-D piecewise smooth function using wavelets and curvelets. Curvelet basis functions can be viewed as a local grouping of wavelet basis functions into linear structures so that they can capture the smooth discontinuity curve more efficiently.

Along the discontinuity curve c, it is easy to see that these nonzero wavelet coefficients decay like $O(2^{-j})$ at the j-th level. Next, suppose we keep only $M = N_J$ nonzero coefficients up to the level J in the wavelet expansion. Then the error due to truncation of the wavelet series is

$$\begin{aligned} ||f - \hat{f}_M^{(wavelet)}||^2 &\sim \sum_{j=J+1}^{\infty} 2^j (2^{-j})^2 \\ &\sim O(2^{-J}). \end{aligned} \tag{2.4}$$

Combining (2.3) and (2.4) we obtain the following non-linear approximation rate of the wavelet expansion for the "Horizon" model

$$||f - \hat{f}_M^{(wavelet)}||^2 \sim O(M^{-1}). \tag{2.5}$$

Therefore, when the discontinuity curve c is sufficiently smooth, $c \in C^p$ with $p > 1$, wavelet approximation is suboptimal. It is important to note that the smoothness of the discontinuity curve is irrelevant to the performance of the wavelet approximation.

How can we improve the performance of the wavelet representation when the discontinuity curve is known to be smooth? Simply looking at the wavelet scheme in Figure 4.1(a) suggests that rather than treating each significant wavelet coefficient along the discontinuity curve independently, one should group the nearby coefficients since their locations are locally correlated. Recall that at the level j, the essential support of the wavelet basis functions has size 2^{-j}. The curve scaling relation (2.1) suggests that we can group about $c2^{j/2}$ nearby wavelet basis functions into one basis function with a linear structure so that its width is proportional to its length squared (see Figure 4.1). This grouping operation reduces the number of significant coefficients at the level j from $O(2^j)$ to $O(2^{j/2})$. Consequently, this new representation provides the same approximation error as wavelets in (2.4) with only $M' \sim \sum_{j=0}^{J} 2^{j/2}$ or $O(2^{J/2})$ coefficients. In other words, the M-term non-linear approximation using this improved wavelet representation decays like

$$||f - \hat{f}_M^{(improved-wavelet)}||^2 \sim O(M^{-2}). \tag{2.6}$$

Comparing with (2.5), we see that for C^2 discontinuity curves, the new representation is superior compared to wavelets and in fact achieves the optimal rate. The curvelet system achieves this optimality using a similar argument. In the original curvelet construction [5], the linear structure of the basis function comes from the ridgelet basis while the curve scaling relation is ensured by suitable combination of subband filtering and windowing.

4.2.3 A filter bank approach for sparse image expansions

The original definition of the curvelet transform as described in Section 4.2.1 poses several problems when one translates it into the discrete world. First, since it is a block-based transform, either the approximated images have blocking effects or one has to use overlapping windows and thus increase the redundancy. Secondly, the use of ridgelet transform, which is defined on a polar coordinate, makes the implementation of the curvelet transform for discrete images on rectangular coordinates very challenging. In [8–10], different interpolation approaches were proposed to solve the polar versus rectangular coordinate transform problem, all required overcomplete systems. Consequently, the version of the discrete curvelet transform in [9] for example has a redundancy factor equal to $16J + 1$ where J is the number of multiscale levels.

Comparing the wavelet scheme with the curvelet scheme in Figure 4.1, we see that the improvement of curvelets can be loosely interpreted as a grouping of nearby wavelet coefficients, since their locations are locally correlated due to the smoothness of the discontinuity curve. Therefore, we can obtain a sparse image expansion by first applying a multiscale transform and then applying a local directional transform to gather the nearby basis functions at the same scale into linear structures. In essence, we first use a wavelet-like transform for *edge* detection, and then a local directional transform for *contour segment* detection. Interestingly, this approach is similar to the popular Hough transform [11] for line detection in computer vision.

With this insight, we proposed a *double filter bank* approach for obtaining sparse expansions for typical images with smooth contours (Figure 4.2(b)). In our newly constructed *pyramidal directional filter bank* [12], the Laplacian pyramid [13] is first used to capture the point discontinuities, then followed by a directional filter bank [14] to link point discontinuities into linear structures. The overall result is an image expansion using elementary images like contour segments, and thus it is named the *contourlet transform*.

The contourlet transform offers a flexible multiresolution and directional decomposition for images, since it allows for a different number of directions at each scale. For the contourlet transform to satisfy the *anisotropy scaling law*, as in the curvelet transform, we simply need to impose that the number of directions is doubled at every *other* finer scale of the pyramid [12].

The contourlet transform is almost *critically sampled*, with a small redundancy factor of up to 1.33. Comparing this with a much larger redundancy ratio of the discrete implementation of the curvelet transform [9] mentioned above, the contourlet transform

4.3.2 Directional decomposition

In 1992, Bamberger and Smith [14] introduced a 2-D directional filter bank (DFB) that can be maximally decimated while achieving perfect reconstruction. The DFB is efficiently implemented via a l-level tree-structured decomposition that leads to 2^l subbands with wedge-shaped frequency partition as shown in Figure 4.4.

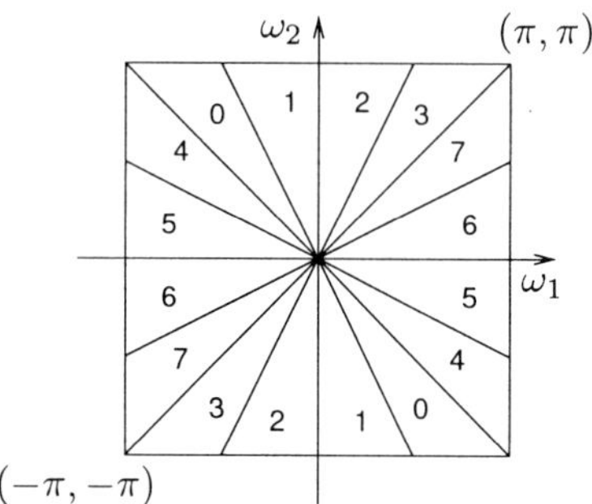

Figure 4.4. Directional filter bank frequency partitioning where $l = 3$ and there are $2^3 = 8$ real wedge-shaped frequency bands.

The original construction of the DFB in [14] involves modulating the input signal and using diamond-shaped filters. Furthermore, to obtain the desired frequency partition, an involved tree expanding rule has to be followed (see [16, 17] for details). As a result, the frequency regions for the resulting subbands do not follow a simple ordering as shown in Figure 4.4 based on the channel indices.

In [27], we propose a new formulation for the DFB that is based only on the QFB's with fan filters. The new DFB avoids the modulation of the input image and has a simpler rule for expanding the decomposition tree. Intuitively, the wedge-shaped frequency partition of the DFB is realized by an appropriate combination of directional frequency splitting by the fan QFB's and the "rotation" operations done by resampling, which are illustrated in Figure 4.5 and Figure 4.6, respectively.

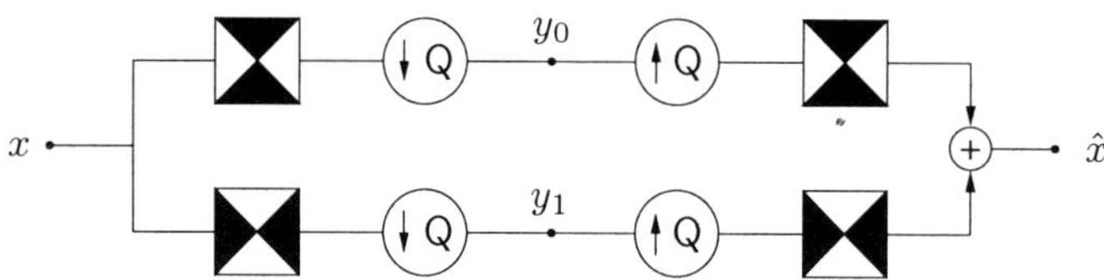

Figure 4.5. Two-dimensional spectrum splitting using the quincunx filter banks with fan filters. The black regions represent the ideal frequency supports of each filter.

Using the multirate identities, we can transform a l-level tree-structured DFB into a parallel structure of 2^l channels with equivalent filters and overall sampling matrices. Denote these equivalent synthesis filters as $G_k^{(l)}$, $0 \leq k < 2^l$, which correspond to the subbands indexed as in Figure 4.4. The oversampling matrices have diagonal form as:

(a)

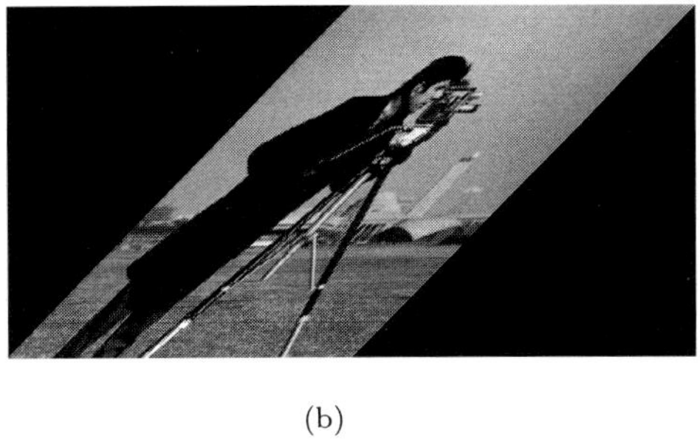

(b)

Figure 4.6. Example of a resampling operation that is used effectively as a rotation operation for the DFB decomposition. (a) The "cameraman" image. (b) The "cameraman" image after being resampled.

$$\mathsf{S}_k^{(l)} = \begin{cases} \operatorname{diag}(2^{l-1}, 2) & \text{for } 0 \leq k < 2^{l-1} \\ \operatorname{diag}(2, 2^{l-1}) & \text{for } 2^{l-1} \leq k < 2^l, \end{cases} \tag{3.1}$$

which correspond to the basically horizontal and basically vertical subbands, respectively.

With this, it is easy to see that the family

$$\left\{ g_k^{(l)}[n - \mathsf{S}_k^{(l)} m] \right\}_{0 \leq k < 2^l,\ m \in \mathbb{Z}^2}, \tag{3.2}$$

obtained by translating the impulse responses of the synthesis filters $G_k^{(l)}$ over the sampling lattices $\mathsf{S}_k^{(l)}$, is a *basis* for discrete signals in $l^2(\mathbb{Z}^2)$. This basis exhibits both directional and localization properties. Figure 4.7 demonstrates this fact by showing the impulse responses of equivalent filters from an example DFB. These basis functions have linear supports in space and span all directions. Therefore (3.2) resembles a local Radon transform and the basis functions are referred to as *Radonlets*.

4.3.3 Multiscale and directional decomposition

The directional filter bank (DFB) is designed to capture the high frequency components (representing directionality) of images. Therefore, low frequency components are handled poorly by the DFB. In fact, with the frequency partition shown in Figure 4.4, low frequencies would "leak" into several directional subbands, hence DFB does *not* provide a sparse representation for images. To improve the situation, low frequencies should be removed before using the DFB. This provides another reason to combine the DFB with a multiresolution scheme.

Therefore, the LP permits further subband decomposition to be applied on its bandpass images. Those bandpass images can be fed into a DFB so that directional information can be captured efficiently. The scheme can be iterated repeatedly on the coarse image (see Figure 4.8). The end result is a double iterated filter bank structure, named

pyramidal directional filter bank (PDFB), which decomposes images into directional subbands at multiple scales. The scheme is flexible since it allows for a different number of directions at each scale.

With perfect reconstruction LP and DFB, the PDFB is obviously perfect reconstruction, and thus it is a frame operator for 2-D signals. The PDFB has the same redundancy

Figure 4.7. Impulse responses of 32 equivalent filters for the first half channels of a 6-levels DFB that use the Haar filters. Black and gray squares correspond to $+1$ and -1, respectively. Because the basis functions resemble "local lines", we call them *Radonlets*.

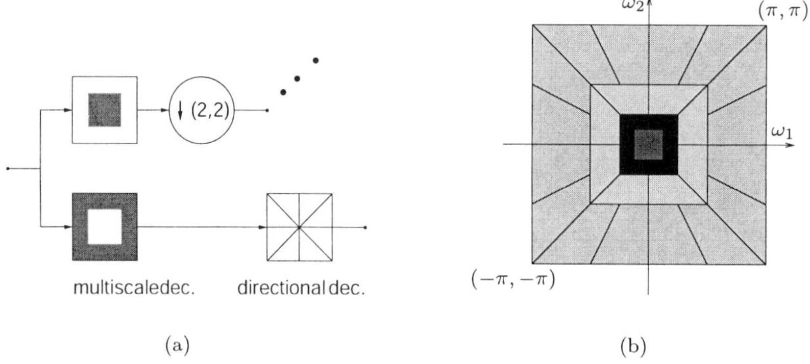

Figure 4.8. Pyramidal directional filter bank. (a) Block diagram. First, a standard multiscale decomposition into octave bands is computed, where the lowpass channel is subsampled while the highpass is not. Then, a directional decomposition with a DFB is applied to each highpass channel. (b) Resulting frequency division, where the number of directions is increased with frequency.

as the LP: up to 33% when subsampling by two in each dimension. Combining the tight frame and orthogonal conditions for the LP and DFB, respectively, it is easy to obtain the following result for the PDFB [12].

Proposition 4.3.1 *The PDFB is a tight frame with frame bounds equal to* 1 *when orthogonal filters are used in both the LP and the DFB.*

Let us point out that there are other multiscale and directional decompositions such as the cortex transform [18] and the steerable pyramid [19]. Our PDFB differs from those in that it allows different number of directions at each scale while nearly achieving critical sampling. In addition, we make the link to continuous-domain construction in Section 4.4

4.3.4 PDFB for curvelets

Next we will demonstrate that a PDFB where *the number of directions is doubled at every* ***other*** *finer scale in the pyramid* satisfies the key properties of curvelets discussed in Section 4.2.1. That is, we apply a DFB with $\lfloor n_0 - j/2 \rfloor$ levels or $2^{\lfloor n_0 - j/2 \rfloor}$ directions to the bandpass image b_j of the LP. Thus, the PDFB provides an efficient discrete implementation for the curvelet transform.

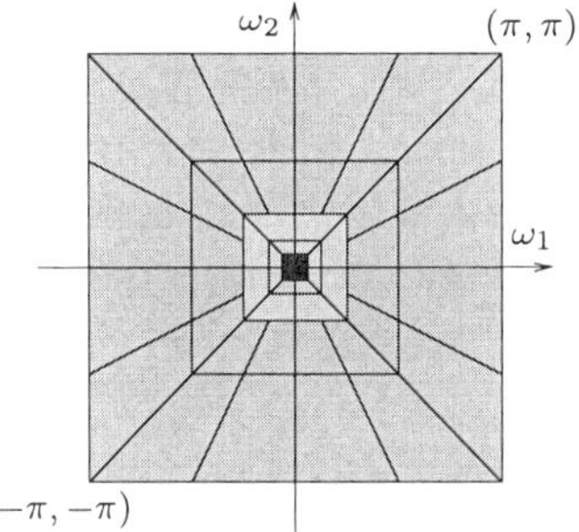

Figure 4.9. Resulting frequency division by a pyramidal directional filter bank for the curvelet transform. As the scale is refined from coarse to fine, the number of directions is doubled at every **other** octave band.

A LP, with downsampling by two in each direction, is taken at every level, providing an octave-band decomposition: the LP bandpass image b_j at the level j creates a subband with a corona support based on the interval $[\pi 2^{-j}, \pi 2^{-j+1}]$, for $j = 1, 2, \ldots, J$. Combining this with a directional decomposition by a DFB, we obtain the frequency tiling for curvelets as shown in Figure 4.9.

In terms of basis functions, a coefficient in the LP subband b_j corresponds to a basis function that has local support in a square of size about 2^j. Then, a basis function from a DFB with $\lfloor n_0 - j/2 \rfloor$ iterated levels has support in a rectangle of length about $2^{n_0 - j/2}$ and width about 1. Therefore, in the PDFB, a basis function at the pyramid level j has support as:

$$width \approx 2^j \quad \text{and} \quad length \approx 2^j.2^{n_0 - j/2} = 2^{n_0} 2^{j/2}, \tag{3.3}$$

which clearly satisfies the anisotropy scaling relation (2.1) of curvelets.

Figure 4.10 graphically depicts this property of a PDFB implementing a curvelet transform. As can be seen from the two pyramidal levels shown below, the support size of the LP is reduced by four times while the number of directions of the DFB is doubled. With this, the support size of the PDFB basis images are changed from one level to next in accordance with the curve scaling relation. Also note that in this representation, as the scale is getting finer, there are more directions.

4.4 MULTIRESOLUTION ANALYSIS

As for the wavelet filter bank, the iterated PDFB can be associated with a continuous-domain system, which we call *contourlet*. This connection will be made precise by studying

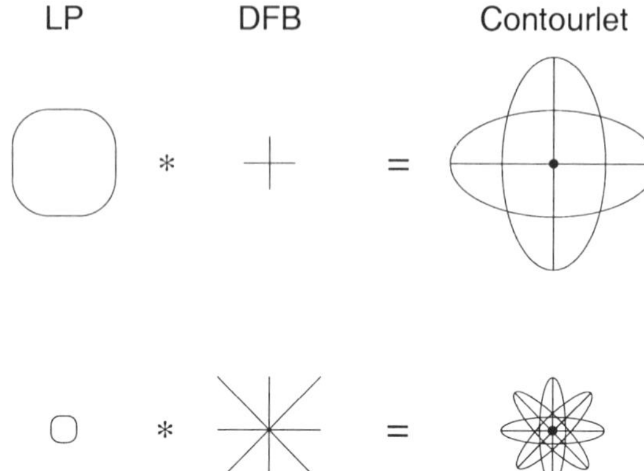

Figure 4.10. Illustration of the contourlet basis images that satisfy the curve scaling relation. From the upper line to the lower line, the scale is reduced by four while the number of directions is doubled.

the embedded grids of approximation as in the multiresolution analysis for wavelets [20, 21]. The new elements are multiple directions and the combination with multiscale.

4.4.1 Multiscale

Suppose that the LP in the PDFB uses orthogonal filters and downsampling by two is taken in each dimension. Under certain conditions, the lowpass filter G in the LP uniquely defines an orthogonal scaling function $\phi(t) \in L^2(\mathbb{R}^2)$ via the two-scale equation [22, 3]

$$\phi(t) = 2 \sum_{n \in \mathbb{Z}^2} g[n]\phi(2t - n)$$

Denote

$$\phi_{j,n} = 2^{-j}\phi\left(\frac{t - 2^j n}{2^j}\right), \qquad j \in \mathbb{Z}, n \in \mathbb{Z}^2. \tag{4.1}$$

Then the family $\{\phi_{j,n}\}_{n \in \mathbb{Z}^2}$ is an orthonormal basis of V_j for all $j \in \mathbb{Z}$. The sequence of nested subspaces $\{V_j\}_{j \in \mathbb{Z}}$ satisfies the following invariance properties:

$$\begin{aligned} &\text{Shift invariance:} \quad f(t) \in V_j \Leftrightarrow f(t - 2^j k) \in V_j, \quad \forall j \in \mathbb{Z}, k \in \mathbb{Z}^2 \\ &\text{Scale invariance:} \quad f(t) \in V_j \Leftrightarrow f(2^{-1}t) \in V_{j+1}, \quad \forall j \in \mathbb{Z}. \end{aligned}$$

In other words, V_j is a subspace defined on a uniform grid with intervals $2^j \times 2^j$, which characterize the image approximation at the resolution 2^{-j}. The difference image in the LP carries the details necessary to increase the resolution of an image approximation. Let W_j be the orthogonal complement of V_j in V_{j-1} (also see Figure 4.11)

$$V_{j-1} = V_j \oplus W_j$$

The LP can be considered as an oversampled filter bank where each polyphase component of the difference signal comes from a separate filter bank channel like the coarse signal [15]. Let $F_i(z), 0 \leq i \leq 3$ be the synthesis filters for these polyphase components.

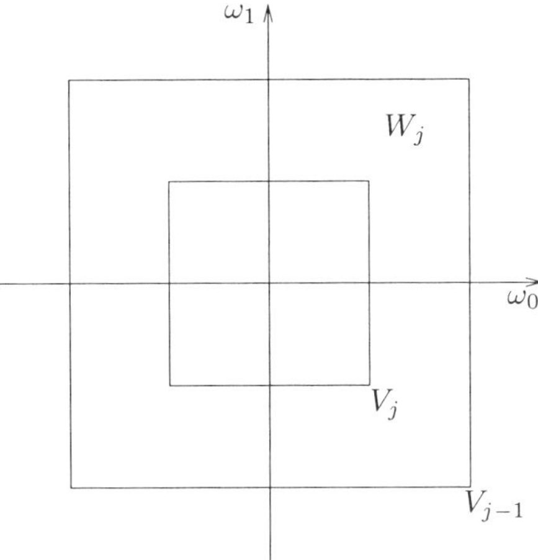

Figure 4.11. Multiscale subspaces generated by the Laplacian pyramid

Note that $F_i(z)$ are highpass filters. As in the wavelet filter bank, we associate with each of these filters a continuous function $\psi^{(i)}(t)$ where

$$\psi^{(i)}(t) = 2 \sum_{n \in \mathbb{Z}^2} f_i[n]\phi(2t - n).$$

Proposition 4.4.1 *([15]) Suppose that the LP with orthogonal filter generates an MRA. Then for a scale j,* $\{\psi^{(i)}_{j,n}\}_{0 \le i \le |\mathsf{M}|-1,\, n \in \mathbb{Z}^d}$ *is a tight frame of* W_j. *For all scales,* $\{\psi^{(i)}_{j,n}\}_{0 \le i \le |\mathsf{M}|-1,\, j \in \mathbb{Z},\, n \in \mathbb{Z}^2}$ *is a tight frame of* $L^2(\mathbb{R}^d)$. *In all cases, the frame bounds are equal to 1.*

Since W_{j+1} is generated by four prototype functions, in general it is *not* a shift invariant subspace, unless $F_i(z)$ are shifted versions of a filter, or

$$F_i(z) = z^{-k_i} F(z) \tag{4.2}$$

where k_i are the coset representatives of the downsampling lattice $(2, 2)$

$$k_0 = (0,0)^T,\ k_1 = (1,0)^T,\ k_2 = (0,1)^T, \text{ and } k_3 = (1,1)^T. \tag{4.3}$$

Nevertheless, based on this, we can mimic W_{j+1} to be a shift invariant subspace by denoting

$$\mu_{j,2n+k_i}(t) = \psi^{(i)}_{j+1,n} = \sum_{m \in \mathbb{Z}^2} f_i[m]\phi_{j,n+m}(t). \tag{4.4}$$

With this notation, the family $\{\mu_{j,n}\}_{n \in \mathbb{Z}^2}$ is a tight frame of W_{j+1} and it resembles a uniform grid on $\mathbb{R}^2$ of intervals $2^j \times 2^j$.

4.4.2 Multiple Directions

Suppose that the DFB's in the PDFB use orthogonal filters. In the PDFB, the discrete basis (3.2) of the DFB can be regarded as a change of basis for the continuous subspaces

$$\rho_{j,k,n}^{(l)}(t) = \rho_{j,k}^{(l)}(t - 2^j \mathsf{S}_k^{(l)} n) \tag{4.12}$$

Proof: By direct substitution and a change of variable. □

Consequently, the subspaces $W_{j+1,k}^{(l)}$ satisfy the following shift invariant property:

$$f(t) \in W_{j+1,k}^{(l)} \quad \Leftrightarrow \quad f(t - 2^j S_k^{(l)} n) \in W_{j+1,k}^{(l)}, \quad \forall n \in \mathbb{Z}^2. \tag{4.13}$$

This says that the directional multiscale subspaces $W_{j+1,k}^{(l)}$ are defined on a rectangular grid with intervals $2^{j+l-1} \times 2^{j+1}$ (or $2^{j+1} \times 2^{j+l-1}$, depending on whether it is basically horizontal or vertical). By substituting (4.4) into (4.11), we can write the prototype function $\rho_{j,k}^{(l)}(t)$ directly as a linear combination of the scaling function $\phi_{j,m}(t)$ as

$$\begin{aligned}
\rho_{j,k}^{(l)}(t) &= \sum_{i=0}^{3} \sum_{n} g_k^{(l)}[2n + k_i] \left(\sum_{m \in \mathbb{Z}^2} f_i[m] \phi_{j,n+m} \right) \\
&= \sum_{m \in \mathbb{Z}^2} \underbrace{\left(\sum_{i=0}^{3} \sum_{n \in \mathbb{Z}^2} g_k^{(l)}[2n + k_i] f_i[m - n] \right)}_{c_k^{(l)}[m]} \phi_{j,m}(t).
\end{aligned} \tag{4.14}$$

The sequence $c_k^{(l)}[m]$ resembles a summation of convolutions between $g_k^{(l)}[m]$ and $f_i[m]$, thus it is a highpass and directional filter. Equation (4.14) reveals the "contourlet-like" behavior of the prototype function $\rho_{j,k}^{(l)}(t)$ where it is seen as a grouping of "edge-detection" elements at a scale j and along a direction k.

4.4.3 Multiscale and multidirection

Finally, integrating over scales we have the following result for the *contourlet frames* on the space $L^2(\mathbb{R}^2)$.

Theorem 4.4.1 *For a sequence of finite positive integers* $\{l_j\}_{j \leq j_0}$ *the family*

$$\{\phi_{j_0,n}(t),\ \rho_{j,k,n}^{(l_j)}(t)\}_{j \leq j_0,\ 0 \leq k \leq 2^{l_j}-1,\ n \in \mathbb{Z}^2} \tag{4.15}$$

is a tight frame of $L^2(\mathbb{R}^2)$. *For a sequence of finite positive integers* $\{l_j\}_{j \in \mathbb{Z}}$, *the family*

$$\{\rho_{j,k,n}^{(l_j)}(t)\}_{j \in \mathbb{Z},\ 0 \leq k \leq 2^{l_j}-1,\ n \in \mathbb{Z}^2} \tag{4.16}$$

is a directional wavelet tight frame of $L^2(\mathbb{R}^2)$. *In each case, the frame bounds are equal to 1.*

Proof: This result is obtained by applying Proposition 4.4.3 to the following decompositions of $L^2(\mathbb{R}^2)$ into mutual orthogonal subspaces:

$$L^2(\mathbb{R}^2) = V_{j_0} \oplus \left(\bigoplus_{j \le j_0} W_j \right), \quad \text{and}$$

$$L^2(\mathbb{R}^2) = \bigoplus_{j \in \mathbb{Z}} W_j.$$

□

As discussed in Section 4.3.4, the tight frame in (4.15) provides a curvelet-like expansion when the number of directions is doubled at every other finer scale. This means that if at the scale 2^{j_0} we start with an l_{j_0}-level DFB (which has $2^{l_{j_0}}$ directions) then at finer scales 2^j, $j < j_0$, the number of decomposition levels by the DFB should be:

$$l_j = \lfloor l_{j_0} - (j - j_0)/2 \rfloor, \quad \text{for } j \le j_0. \tag{4.17}$$

Thus the embedded grid of approximation for the curvelet PDFB expansion at the scale 2^j is $2^{\lfloor n_0 + j/2 \rfloor} \times 2^j$ for basically horizontal directions and $2^j \times 2^{\lfloor n_0 + j/2 \rfloor}$ for near vertical directions, where $n_0 = l_{j_0} - j_0/2 + 2$. Figure 4.14 illustrates this sampling pattern at different scales and directions. The main point to note here is that in the refinement process, one spatial dimension is refined at twice the speed as the other spatial dimension.

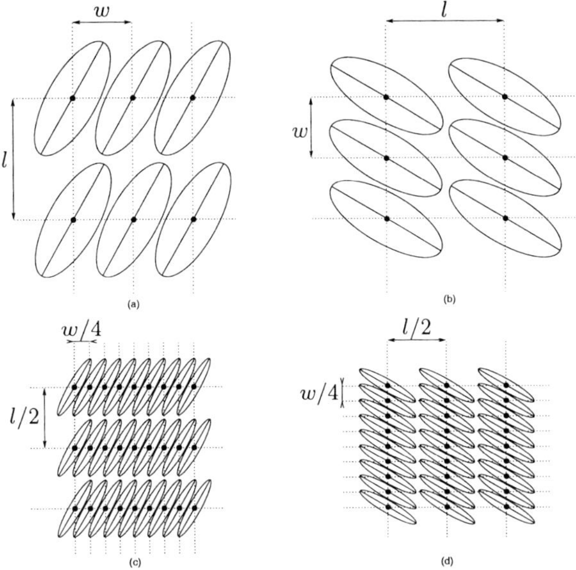

Figure 4.14. Embedded grids of approximation in spatial domain for a PDFB that implements the curvelet transform. These are four illustrative subspaces $W_{j,k}$ representing coarser vs. finer scales and basically horizontal vs. basically vertical directions. Each subspace is spanned by the shifts of a curvelet prototype function given in (4.11). The sampling intervals match with the supports of the prototype function, for example width w and length l, so that the shifts would tile the $\mathbb{R}^2$ plane.

Figure 4.10 and Figure 4.14 give a complete view of the multiresolution approximation of our curvelet construction based on the PDFB. They clearly show a refinement scheme

and then computing the decomposition as well as the reconstruction process. However, Harten's method cannot be directly applied to the more interesting and generally used pyramidal filtering algorithms in which the standard wavelet transforms are implemented. This is because we have to work only with fixed size and fixed value filters in this context, and these rigid filters can not be directly used to compute the adaptive divided difference tables at each grid point.

The ENO method retains a fixed wavelet transform and locally modifies the function near discontinuities so that wavelet filters are applied to smooth data. By recording how the changes are made, the original discontinuous function can be exactly recovered by using the original inverse filters. Indeed, by applying the idea of using one-sided information near the discontinuities, we directly extend the functions from both sides of the discontinuities, thus we can apply the standard wavelet transforms on these extended values such that there are no large coefficients generated in the high frequencies and the low frequency approximations are essentially non-oscillatory, and therefore Gibbs' phenomenon can be completely avoided. The extension idea in wavelet methods, such as extension obtained by spline wavelet methods, has been used in constructing wavelets for closed intervals [3], [16] and [17]. However, those approaches usually modify the wavelet basis at the boundary of the interval rather than the function.

In addition, in this modified wavelet transform, the low frequency part preserves the piecewise smoothness of the original function. In particular, the jumps in the low frequency part is not spread widely as in the standard transform. Therefore, the same ENO idea can be recursively used for the coarser levels of the low pass coefficients. By doing so, the multiresolution framework can be kept too.

The resulting wavelet transform retains all the desirable properties of the standard transform: it is stable and can have uniformly arbitrarily high order of approximation (with a rigorous uniform order of the error bound), it concentrates the large coefficients to the low frequencies, it preserves the multiresolution framework and fast transform algorithms, and it is easy to implement. Furthermore, since we do not fully adopt the ENO schemes, in particular, we do not build the divided difference table and compare the smoothness of all possible stencils at every point, the extra cost (in floating point operations) required by the modified ENO-wavelet transforms is insignificant. In fact, it is of the order $O(dl)$ where d is the number of discontinuities and $l+1$ the stencil length. Compared to the cost of the standard wavelet transform, which is of the order $O(nl)$ where n is the size of the data, the ratio of the extra cost over that of the standard transform is of the order $O(\frac{d}{n})$ which is independent of l and negligible when n is large.

Besides, since the designed ENO-wavelet transforms play the same role as the standard wavelet transforms in the applications, it is natural and even more beneficial to use them in conjunction with the standard adaptive nonlinear techniques such as hard and soft thresholding in many applications such as image compression and denoising. We will discuss those applications briefly at the last part of this chapter.

The arrangement of the chapter is as follows. In section 5.2, we start by reviewing the standard wavelet transforms. Then we give a general idea to construct the ENO-wavelet transforms. In section 5.3, we state the stability results and an error bound for the ENO-wavelet approximation which shows that the error in the ENO-wavelet approximation depends only on the size of the derivative of the function *away* from the discontinuities. Finally, in section 5.4, we discuss some possible applications of the ENO-wavelet transforms including function approximation, image compression and signal denoising, and we give some numerical examples.

5.2 THE ENO-WAVELET ALGORITHM

In this section, we give the general idea to construct ENO-wavelet transforms for piecewise smooth functions.

5.2.1 ENO-wavelet at Discontinuities

Before we present the adaptive ENO-wavelet transforms, we want to briefly recall some basic knowledge in the standard wavelet transforms. In this section, we do not intend to cover all fundamentals in wavelet theory, we just want to use this opportunity to introduce some notations used in the standard wavelet transforms so that they can be used in our ENO-wavelet transforms. For readers who are interested in the standard wavelet theory, please see [13], [21], [38], [42] and many other relevant references that we do not list here. We also want to point out that in this chapter, we only discuss the design of ENO-wavelet transforms using Daubechies orthonormal wavelet orthonormal wavelet frameworks. The idea can be easily extended to other types of wavelets such as biorthogonal wavelets , but that is not our focus here.

To simplify the discussion, we assume zeros have been padded to the data at the boundaries.

The standard wavelet transforms are based on translation and dilation. Suppose $\phi(x)$ and $\psi(x)$ are the scaling function and the corresponding wavelet respectively with finite support $[0, l]$ where l is a positive integer. It's well known that $\phi(x)$ satisfies the basic dilation equation :

$$\phi(x) = \sqrt{2}\sum_{s=0}^{l} c_s\phi(2x - s); \tag{2.1}$$

and $\psi(x)$ satisfies the corresponding wavelet equation :

$$\psi(x) = \sqrt{2}\sum_{s=0}^{l} h_s\phi(2x - s); \tag{2.2}$$

where the c_s's and h_s's are constants called low pass and high pass filter coefficients respectively.

Wavelet $\psi(x)$ having p vanishing moments means:

$$\int \psi(x)x^j dx = 0, \qquad for \quad j = 0, 1, \cdots, p-1. \tag{2.3}$$

We will use the following standard notations:

$$\phi_{j,i}(x) = 2^{\frac{j}{2}}\phi(2^j x - i), \tag{2.4}$$

and

$$\psi_{j,i}(x) = 2^{\frac{j}{2}}\phi(2^j x - i). \tag{2.5}$$

Consider the subspace V_j of L^2 defined by:

$$V_j = Span\{\phi_{j,i}(x), i \in Z\},$$

and the subspace W_j of L^2 defined by:

$$W_j = Span\{\psi_{j,i}(x), i \in Z\}.$$

The subspaces V_j's, $-\infty < j < \infty$, form a multiresolution of L^2 with the subspace W_j being the difference between V_j and V_{j+1}. In fact, the L^2 space has an orthonormal decomposition as:

$$L^2 = V_J \oplus \sum_{j=J}^{\infty} W_j.$$

The projection of a L^2 function $f(x)$ onto the subspace V_j is defined by:

$$f_j(x) = \sum_i \alpha_{j,i}\phi_{j,i}(x), \tag{2.6}$$

where

$$\alpha_{j,i} = \int f(x)\phi_{j,i}(x)dx, \quad i = \cdots, -1, 0, 1, \cdots, \tag{2.7}$$

which we call low frequency wavelet coefficients (they are often called scaling coefficients in the literature). Similarly, we can project $f(x)$ onto W_j by:

$$w_j(x) = \sum_i \beta_{j,i}\psi_{j,i}(x), \tag{2.8}$$

where

$$\beta_{j,i} = \int f(x)\psi_{j,i}(x)dx. \quad i = \cdots, -1, 0, 1, \cdots, \tag{2.9}$$

which we call high frequency wavelet coefficients (often called wavelet coefficients in the literature). In this paper, we use the term wavelet coefficients to denote both low and high frequency coefficients. Therefore, the function $f(x)$ can be decomposed by:

$$f(x) = f_j(x) + \sum_{t=j}^{\infty} w_t(x).$$

The projection $f_j(x)$ is called the linear approximation of the function $f(x)$ in the subspace V_j.

From (2.4) and (2.5), the projection coefficients $\alpha_{j,i}$ and $\beta_{j,i}$ of $f(x)$ in the subspaces V_j and W_j can be easily computed from the coefficients $\alpha_{j+1,i}$ by the so called fast wavelet transform :

$$\alpha_{j,i} = \sum_{s=0}^{l} c_s \alpha_{j+1,2i+s}; \tag{2.10}$$

and

$$\beta_{j,i} = \sum_{s=0}^{l} h_s \alpha_{j+1,2i+s}. \tag{2.11}$$

The standard linear wavelet approximation can achieve arbitrary high accuracy away from discontinuities, but it oscillates near the jumps. The intuitive reason for the oscillations is that some stencils cross jumps and cause the corresponding high frequency coefficients to becoming large and therefore, more information is lost when the high frequency coefficients are discarded.

In Figure 5.1, we display a piecewise continuous function (left) and its DB-6 wavelet coefficients (right) with low frequencies at the left end and high frequencies at the right end. From the right picture, we see that most of the high frequency coefficients are zeros, except for a few large coefficients which are computed near jumps. Figure 5.2 displays the linear approximation (dash-dotted line) compared to the initial function (dotted line).

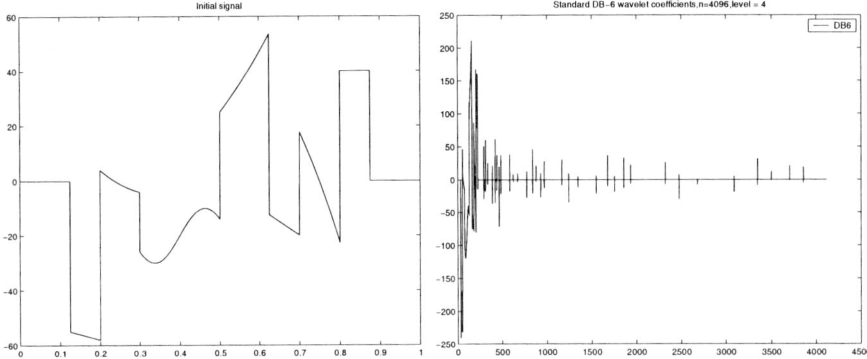

Figure 5.1. The initial function (left) and its DB6 coefficients (right). Most of the high frequency coefficients (right part) are zero except for a few large coefficients computed near the jumps.

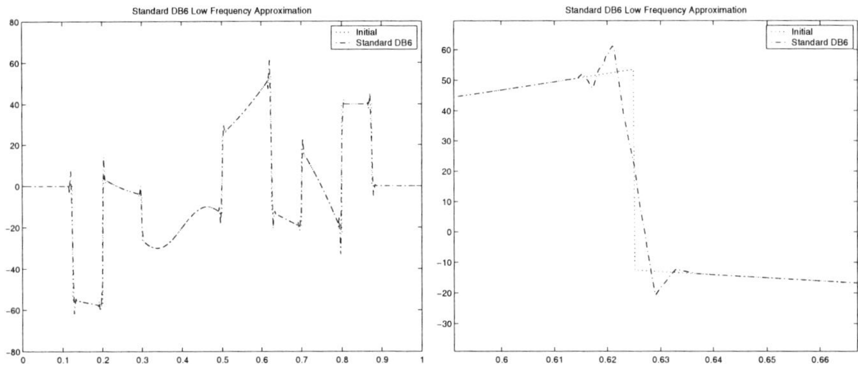

Figure 5.2. The approximation function (left) and its zoom in (right), Oscillations are generated near the discontinuities in the linear approximation.

The right picture is the zoom-in to show the approximation behavior near a jump. In this figure, we clearly see oscillations (also known as Gibbs' phenomenon) near discontinuities.

Since the oscillations are generated by discarding large high frequency coefficients which are computed on the stencils crossing discontinuities, to get rid of the oscillations, we want to avoid stencils crossing discontinuities. This motivates us to apply the ENO idea to avoid stencils crossing jumps.

In addition to the standard wavelet transforms, our ENO-wavelet transforms are composed of two phases: locating the jumps and forming the approximations at the discontinuities. Firstly, to better explain the algorithm, we assume that the location of the jumps are known, and we give the ENO-wavelet approximations at the discontinuities by using one-sided information to avoid oscillations. Then, we give some methods to detect the location of the discontinuities.

We want to modify the standard wavelet transforms near the jumps such that oscillations can be avoided in the approximation. From ENO schemes, we borrow the idea of using one-sided information to form the approximation and avoid applying the wavelet filters crossing the discontinuities. In order to simplify the explanation, we also assume that the discontinuities are well separated so that the modification we will make at one jump will not interact the modification at another jump. Therefore, we can just consider the local modification near one jump. The main tool which we use to modify the standard wavelet transforms at the discontinuities is function extrapolation in the function spaces or in the wavelet spaces.

The first way is to extend the function directly at the discontinuity by extrapolation from both sides. Then we can apply the standard wavelet transforms on the extended functions and avoid computing wavelet coefficients using information from both sides.

To maintain the same approximation accuracy near the discontinuity as that for away from the discontinuity, the extrapolation has to be $(p-1)$-th order accurate if the wavelet functions have p vanishing moments. For instance, we use constant extrapolation for Haar wavelet, $(p-1)$-th order extrapolation for Daubechies-$2p$ orthogonal wavelets which have p vanishing moments.

We use the diagram in Figure 5.3 to show how to extend the function and compute the ENO-wavelet coefficients.

As shown in Figure 5.3, the discontinuity is located between $\{x(2i+l-2), x(2i+l-1)\}$. We extend the function from both sides of the discontinuity using $(p-1)$-th order extrapolation, i.e. we use the information from the left side of the jump to extrapolate the function over $\hat{x}(2i+l-1), \cdots, \hat{x}(2i+2l-2)$; use the information from the right side to extrapolate the function over $\bar{x}(2i), \cdots, \bar{x}(2i+l-2)$. And then for $i \leq m \leq i+k-2$, where $l = 2k-1$, we can compute the wavelet coefficients $\hat{\alpha}_{j,m}$ and $\hat{\beta}_{j,m}$ from the left side, and compute $\bar{\alpha}_{j,m}$ and $\bar{\beta}_{j,m}$ from the right side by using the standard wavelet transforms respectively.

In general, we have the low frequency wavelet coefficients on the finer levels instead of knowing the function values themselves near the discontinuities. We extrapolate these finer level coefficients from both sides of the discontinuities to obtain the values of $\hat{\alpha}_{j+1,m}$ and $\bar{\alpha}_{j+1,m}$, and use the fast wavelet transforms (2.10) and (2.11) to compute the coarser level coefficients.

There are many methods to extrapolate the extended values. For example, a straightforward way is to use p-point polynomial extrapolation such as Lagrange polynomials or Taylor expansion polynomials. In our numerical experiments in this chapter, we use Lagrange polynomial extrapolation for noise free data, and least square extrapolation [45] for noisy data.

There is a storage problem for this direct function extrapolation or the extrapolation of the finer coefficients. Indeed, it doubles the number of the wavelet coefficients near every discontinuity. To retain the perfect invertible property, using the notation in Figure 5.3, we need to store the ENO-wavelet coefficients $\hat{\alpha}_{j,m}$ and $\hat{\beta}_{j,m}$ from the left side, also $\bar{\alpha}_{j,m}$ and $\bar{\beta}_{j,m}$ from the right side. Thus, the output sequences are no longer the same size as the input sequences. In many applications, such as image compression, this extra storage requirement definitely needs to be avoided.

Facing this challenge, we have proposed a better way, which we called coarse level extrapolation , to accomplish our goals. The idea is to extrapolate the coarser level wavelet coefficients near the discontinuities instead of the function values or the finer level wavelet coefficients.

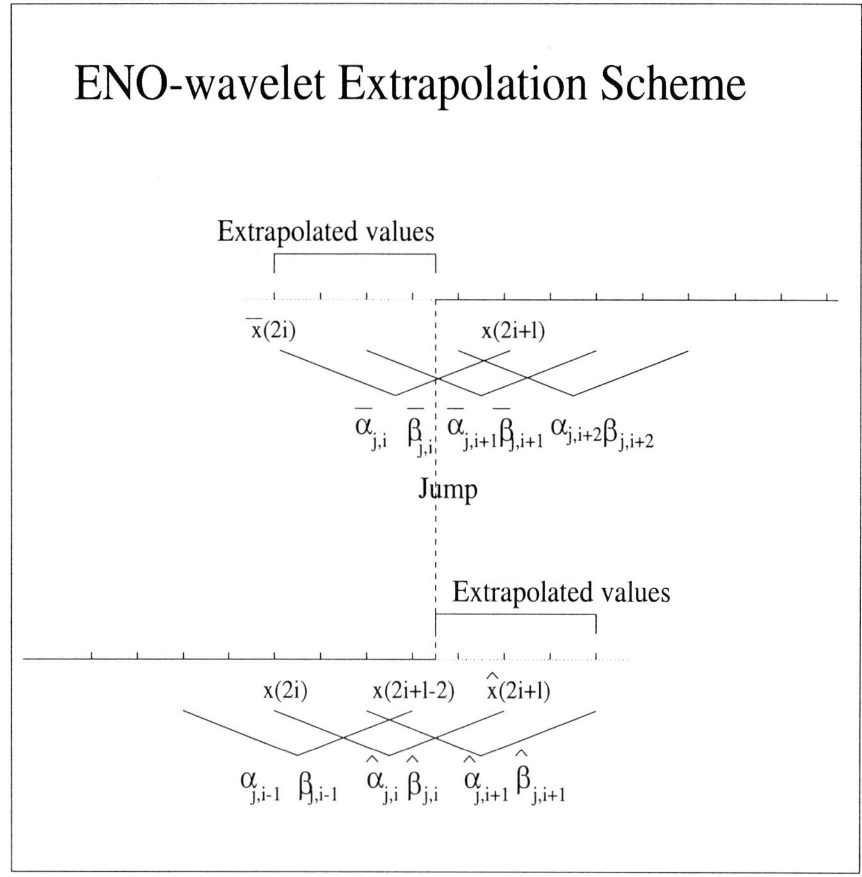

Figure 5.3. Coarse Level Extrapolation Illustration. From the left side of the discontinuity, we extrapolate the low frequency coefficients $\hat{\alpha}_{j,m}$ to determine corresponding high frequency coefficients $\hat{\beta}_{j,m}$ and store them. From the right side of the discontinuity, we extend the high frequency coefficients $\bar{\beta}_{j,m}$ to determine and store the low frequency coefficients $\bar{\alpha}_{j,m}$.

We still use Figure 5.3 to illustrate these schemes. We consider the left side of the jump first.

In the direct function extrapolation case, the computation process is to directly extrapolate the finer level wavelet coefficients, and then compute the extended coarser level wavelet coefficients $\hat{\alpha}_{j,m}$ and $\hat{\beta}_{j,m}$, $i \leq m \leq (i+k-2)$ using the standard filters. We reverse the order of this process in our coarse level extrapolation. More precisely, we extrapolate the coarser level low frequency coefficients $\hat{\alpha}_{j,m}$ using the known low frequency coefficients from the left, and extend the coarser level high frequency coefficients $\hat{\beta}_{j,m}$ to zero (or some pre-defined values), then determine the extended finer level wavelet coefficients.

However, in Daubechies' orthonormal wavelet transforms, we cannot arbitrarily prescribe both $\hat{\alpha}_{j,m}$ and $\hat{\beta}_{j,m}$ simultaneously. This is because they are not linearly independent. Let's take $m = i$ as an example. Assume that we have prescribe both $\hat{\alpha}_{j,i}$ and $\hat{\beta}_{j,i}$ as given values, this means that we have implicitly extended the finer level values $\hat{\alpha}_{j+1,2i+l-1}$ and $\hat{\alpha}_{j+1,2i+l}$ satisfying:

$$\begin{pmatrix} \hat{\alpha}_{j,i} \\ \hat{\beta}_{j,i} \end{pmatrix} = \begin{pmatrix} \sum_{s=0}^{l-2} c_s \alpha_{j+1,2i+s} + c_{l-1}\hat{\alpha}_{j+1,2i+l-1} + c_l \hat{\alpha}_{j+1,2i+l} \\ \sum_{s=0}^{l-2} h_s \alpha_{j+1,2i+s} + h_{l-1}\hat{\alpha}_{j+1,2i+l-1} + h_l \hat{\alpha}_{j+1,2i+l} \end{pmatrix}.$$

As we have $\frac{h_{l-1}}{c_{l-1}} = \frac{h_l}{c_l}$, this implies that we can only prescribe one of the coarse level coefficients $\hat{\alpha}_{j,i}$ and $\hat{\beta}_{j,i}$ and determine the other one by the above relationship. Thus we have two choices:

(1) We can extrapolate the low frequency coefficients $\hat{\alpha}_{j,m}$ first, then determine the corresponding high frequency coefficients $\hat{\beta}_{j,m}$
(2) Or we can extend $\hat{\beta}_{j,m}$ to zero first, then determine the corresponding $\hat{\alpha}_{j,m}$.

Again by symmetry, we have two analogous choices for the right side of the jump.

Using this coarse level extrapolation technique, we can easily solve the storage problem which we have in the direct function extrapolation. In fact, we just need to store the high frequency coefficients $\hat{\beta}_{j,m}$ for choice (1) and the low frequency coefficients $\hat{\alpha}_{j,m}$ for choice (2). In our implementation, we use choice (1) for the left side of the jumps and choice (2) for the right side of the jumps, therefore we store $\hat{\beta}_{j,m}$ and $\bar{\alpha}_{j,m}$ for every m. This satisfies the standard wavelet storage scheme, i.e. storing one low frequency and one high frequency coefficients for every stencil.

Since we know the way we extend the data at the discontinuities, we can easily extrapolate the low frequency coefficients $\hat{\alpha}_{j,m}$ from the left sides of the discontinuities. Using them together with the stored high frequency coefficients $\hat{\beta}_{j,m}$, we can exactly recover data at the left sides by applying the standard inverse filters. Similarly, the data at right sides of the discontinuities can also be exactly restored.

Of course, in the ENO-wavelet transforms, to retain the perfect invertibility property, we need to store all adaptive information, i.e. the locations of the discontinuities. In our implementation in this chapter, we just use one extra bit for each stencil near the discontinuities to indicate it contains a discontinuity. In the application of compression, which aims to reduce the total storage of representing an image, these extra bits need to be taken into account carefully, we will discuss it in the last section of this chapter.

For each stencil crossing a jump, an extra cost (in floating point operation) is required in the extrapolation low frequency coefficients, which is of the order $O(1)$ per stencil, and in the computation of the corresponding high and low frequency coefficients, which is of the order $O(l)$ per stencil. Overall, the extra cost over the standard wavelet transform is of the order $O(dl)$ where d is the number of discontinuities. Compared to the cost of the standard wavelet transform, which is of the order $O(nl)$ where n is the size of data, the ratio of the extra cost over that of the standard transform is $O(\frac{d}{n})$, which is independent of l and negligible when n is large.

5.2.2 Locating the Discontinuities

In the previous subsection, we showed how to modify the standard wavelet transforms at the discontinuities to avoid oscillations if we know the exact location of the jumps. In this subsection, we introduce the methods to detect the exact location of the discontinuities for piecewise smooth functions with and without noise. First we give a method for smooth data.

Our purpose is to avoid wavelet stencils crossing discontinuities. Theoretically, a discontinuity can be characterized by comparing the left and right limit of the derivatives $f^{(m)}(x)$ at the given point x, i.e. we call a point x a discontinuity if for some $m < p$, we have:

$$f^{(m)}(x-) \neq f^{(m)}(x+).$$

We define the intensity of a jump in the m-th derivative at x as

$$[f^{(m)}(x)] = |f^{(m)}(x+) - f^{(m)}(x-)|.$$

It is well known that the high pass filters in wavelet transforms measure the smoothness of functions: they produce smaller values at smoother regions, and larger values at rougher regions. In fact, it has been shown in [6], [34] and [44] that if a function $f(x)$ is Lipschitz $\mu \leq p$ at x, i.e. $|f(x+\delta) - f(x)| \leq \delta^{\mu}$ for any small δ, the corresponding high frequency wavelet coefficients are of the order of $O(\Delta x^{\mu})$. From this, it is easy to obtain that at smooth regions, the magnitudes of high frequency coefficients $|\beta_{j,i}|$ have the order of $|f^{(p)}(x)|O(\Delta x^{p})$. By Taylor expansion, if both $|\beta_{j,i-1}|$ and $|\beta_{j,i}|$ are in a smooth region, we have

$$|\beta_{j,i}| = (1 + O(\Delta x))|\beta_{j,i-1}|,$$

where the constant in the term $O(\Delta x)$ depends on the high order (larger than p) derivatives of $f(x)$. On the other hand, if a stencil contains a discontinuity, no matter a discontinuity in function value ($m = 0$) or in its m-th derivative, the magnitude of the corresponding high frequency coefficient $|\beta_{j,i}|$ is of the order of $O(\Delta x^{(m)})$, i.e.

$$|\beta_{j,i}| = |[f^{(m)}(x_0)]|O(\Delta x^{m}),$$

which is at least one order lower than that at the smooth regions.

Therefore, we can design a method to detect the discontinuities as follows: For each standard stencil, suppose we know that the previous standard stencil does not contain any discontinuities, if we have $|\beta_{j,i}| \leq \tau|\beta_{j,i-1}|$, where $\tau > 1$ is a given constant, then we treat the current stencil as a smooth stencil. Otherwise, we conclude that there are discontinuities contained in it.

The choice of constant τ depends on the grid size Δx, and also the intensity of the jumps. In fact, the ratio between a high frequency coefficient at the rough regions and that at the smooth regions is of the order of $|[f^{(m)}(x)]|O(\Delta x^{(m-p)})$. When Δx becomes small, this ratio is large. We can choose τ as any number such that

$$(1 + O(\Delta x)) \leq \tau \leq \min_{x}\{|[f^{(m)}(x)]|O(\Delta x^{(m-p)})\}, \tag{2.12}$$

provided the above minimal number is larger than $1 + O(\Delta x)$. This is always true for piecewise smooth functions with small enough grid size Δx. Obviously, the jump detection procedure can capture all jumps in the m-th derivative with intensity larger than $O(\Delta x^{(p-m)})$. On the other hand, when a jump in the m-th derivative has small intensity, which is even less than $O(\Delta x^{(p-m)})$, this jump can not be detected by the above described method. However, the error caused by missing this jump is also very small, which is at the same order of the error bound we will give in section 5.3. In practice, especially when we just care about the jumps in function values, we have a large range to select τ.

The extra cost introduced by this comparison jump identification method over the standard wavelet transforms is just the comparison $|\beta_{j,i}| > \tau|\beta_{j,i-1}|$ for each stencil.

The above described detection method may not be reliable if the function is polluted by noise, especially when the noise is "large". This is because the high frequency coefficients β's may not be able to measure the correct order of smoothness of the functions. Indeed, the high frequency coefficients have the order $\|f^{(p)}(x)+\sigma n^{(p)}(x)\|O(\Delta x^p)$, where $n(x)$ is the random noise and σ a positive number indicating the noise level. In general, the derivatives of the noise $n^{(p)}(x)$ have large values. The noise term $\sigma n^{(p)}(x)$ can dominate the function term $f^{(p)}(x)$ if the noise level σ is large and thus, the high frequency coefficients β's may not be able to detect certain discontinuities, e.g. if the jump is small or the discontinuity is in the higher derivatives. In this situation, we need to use heuristics to locate the exact position of the essential discontinuities [11].

5.2.3 A Simple Example

In the last part of this section, to better illustrate the idea of the construction of ENO-wavelet transforms, we give a simple example in the ENO-Haar case. We consider computing the transform coefficients of the following initial data containing two discontinuities at [0, 1] and [2, 10] respectively:

$$\left(0\ 0\ 0\ 1\ 1\ 1\ 2\ 10\ 11\ 12\right).$$

The standard Haar produces the low and high frequency coefficients:

$$\alpha=\left(0\ \tfrac{1}{\sqrt{2}}\ \tfrac{2}{\sqrt{2}}\ \tfrac{12}{\sqrt{2}}\ \tfrac{23}{\sqrt{2}}\right), \beta=\left(0\ -\tfrac{1}{\sqrt{2}}\ 0\ -\tfrac{8}{\sqrt{2}}\ -\tfrac{1}{\sqrt{2}}\right).$$

We notice that comparing to their neighbors, there are two relatively large high frequency coefficients corresponding to the two jumps. The corresponding linear approximation by setting $\beta=0$ is:

$$\left(0\ 0\ 0.5\ 0.5\ 1\ 1\ 6\ 6\ 11.5\ 11.5\right),$$

which does not recover the discontinuities correctly.

Using the ENO-Haar wavelet, we break the initial data sequence into three smooth pieces as shown in the following two rows:

$$\begin{pmatrix} & y\ 1\ 1\ 1\ 1\ w & \\ 0\ 0\ 0\ x & & z\ 10\ 11\ 12\end{pmatrix},$$

where x, y, z and w are some smooth extensions of the corresponding pieces. In fact, we extend x in a way such that the low frequency coefficient $\hat{\alpha}_2$ (boxed in (2.13)) based on the stencil $(0,x)$ is the same as the previous α_1, which is based on the stencil $(0,0)$ giving $x=0$. Similarly, we extend y in a way such that the high frequency coefficient $\bar{\beta}_2$ (boxed in (2.13)) is zero giving $y=1$. Therefore we compute the high frequency coefficients $\hat{\beta}_2$ based on stencil $(0,x)$ and the low frequency coefficients $\bar{\alpha}_2$ based on stencil $(y,1)$ by using the corresponding standard filters giving $\hat{\beta}_2=0$ and $\bar{\alpha}_2=\frac{2}{\sqrt{2}}$. Similarly, we determine $w=0$ according to $\hat{\alpha}_4=\alpha_3$ (boxed in (2.13)), then compute $\hat{\beta}_4=\frac{2}{\sqrt{2}}$, and $z=10$ by $\bar{\beta}_4=0$ and then $\bar{\alpha}_4=\frac{20}{\sqrt{2}}$. Thus we have the coefficients:

$$\alpha=\begin{pmatrix} & \frac{2}{\sqrt{2}}\ \frac{2}{\sqrt{2}} & \boxed{\frac{2}{\sqrt{2}}} & \\ 0\ \boxed{0} & & \frac{20}{\sqrt{2}} & \frac{23}{\sqrt{2}}\end{pmatrix}, \beta=\begin{pmatrix} & \boxed{0}\ 0 & \frac{2}{\sqrt{2}} & \\ 0\ 0 & & \boxed{0} & -\frac{1}{\sqrt{2}}\end{pmatrix}. \tag{2.13}$$

Since we know how we extended $\hat{\alpha}_2$, $\bar{\beta}_2$, $\hat{\alpha}_4$ and $\bar{\beta}_4$, we do not need to store them. In fact, we just need to store the low and high frequency coefficients as:

$$\alpha = \left(0\ \tfrac{2}{\sqrt{2}}\ \tfrac{2}{\sqrt{2}}\ \tfrac{20}{\sqrt{2}}\ \tfrac{23}{\sqrt{2}}\right), \beta = \left(0\ 0\ 0\ \tfrac{2}{\sqrt{2}}\ -\tfrac{1}{\sqrt{2}}\right),$$

which have the same storage schemes as the standard Haar wavelet transform.

When we reconstruct the linear approximation, we can first recover $\hat{\alpha}_2$, $\bar{\beta}_2$, $\hat{\alpha}_4$ and $\bar{\beta}_4$ by the same way as in the forward transform, and then apply the standard inverse filters to the smooth data to build the approximation. In fact, in this case the linear approximation is

$$\left(0\ 0\ 0\ 1\ 1\ 1\ 1\ 10\ 11.5\ 11.5\right).$$

We notice that the first discontinuity is perfectly retained, and the second one is more accurate than that of the standard transform, although it is not exactly recovered. More importantly, this approximation preserves the discontinuities sharply in contrast to the standard Haar wavelet which takes the average at the discontinuity.

We would like to close this section by making the following two remarks,

(i) The ENO-wavelet transforms are just simple modifications of the standard wavelet transforms near discontinuities. The computational complexity of the algorithms remains $O(n)$ and they are relatively easy to implement.

(ii) Like other wavelet transforms, 2-dimensional or even higher dimensional transforms can be formed by tensor products. In the numerical example section, we will give a 2-dimensional example.

5.3 THEORY: ERROR BOUND AND STABILITY

In this section, we present the ENO-wavelets approximation error bound for piecewise continuous functions and the stability of the algorithm. We do not give proof. They can be found in [11] and [46].

Given a function $f(x)$ in L^2, in standard wavelet theory [38] [21] [42], it can be linearly approximated by its projection $f_j(x)$ in V_j as in (2.6) and (2.7). This linear approximation has a standard error estimate which we state in the following theorem, see also [42].

Theorem 5.3.1 *Suppose the wavelet $\psi(x)$ generated by scaling function $\phi(x)$ has p vanishing moments, $f_j(x)$ is the approximation of $f(x)$, which has bounded p-th order derivative, in V_j with basis $\phi_{j,k}(x)$, then,*

$$\|f(x) - f_j(x)\| \le C(\Delta x)^p \|f^{(p)}(x)\|, \tag{3.1}$$

where $\Delta x = 2^{-j}$ and C is a constant which is independent of j.

This theorem holds for the L^2 norm in general. Moreover, if the scaling function and its wavelet have finite support, then it also holds for the L^∞ norm.

In this theorem, we can see that the approximation error is controlled by two factors. One is the p-th power of the spatial step Δx; the other is the norm of the p-th derivative of the function. This error bound does not hold if the function does not have finite p-th derivative. This implies that the approximation could be poor for irregular functions even if the spatial step Δx is small. For piecewise continuous functions, especially functions with large jumps, the approximation error cannot be controlled as for smooth functions, In fact, in the standard approximation function $f_j(x)$, oscillations are generated near the discontinuous points and they will not disappear even if the spatial step size is reduced (Gibbs' phenomenon).

Table 5.1. The Comparison of maximum error of the standard DB4 and the ENO-DB4 approximations for the smooth function $f(x) = \exp[-(\frac{1}{x} + \frac{1}{1-x})], 0 < x < 1$. They have the same error and both achieve second order of accuracy which agrees with the results in Theorem 1 for the smooth functions.

level	DB4 E_∞	ENO-DB4 E_∞	$Order_\infty$
4	3.316e-5	3.316e-5	
3	7.650e-6	7.650e-6	2.104
2	1.590e-6	1.590e-6	2.232
1	2.972e-7	2.973e-7	2.406

frequencies $f_j(x)$ to approximate $f(x)$ directly. Here, we use some 1-D numerical examples to illustrate the approximation abilities of the ENO-wavelet transforms. We will demonstrate the error bound 3.2 given in section 5.3. In particular, we show results for the ENO-Haar, ENO-DB4 and ENO-DB6 wavelet transforms.

In all examples, for simplicity, we just consider functions with zero values at the boundary. For non-zero boundary functions, we can easily extend the function by zero and treat the boundaries as discontinuities.

To illustrate the performance of ENO-wavelet transforms, we show graphical comparisons of the standard wavelet approximations and corresponding ENO-wavelet approximations. In addition, we compare the L_∞ and L_2 errors of the standard wavelet approximations and the ENO-wavelet approximations at different levels by measuring $E_{\infty,j} = \inf_x \|f(x) - f_j(x)\|$, which is computed by finding the largest difference on the finest grid, and $E_{2,j} = \|f(x) - f_j(x)\|_2$. Using them, we compute the orders of accuracy defined by:

$$Order_\infty = \log_2 \frac{E_{\infty,i}}{E_{\infty,i-1}},$$

and

$$Order_2 = \log_2 \frac{E_{2,i}}{E_{2,i-1}},$$

which indicates the order of accuracy of the approximation in the L_∞ norm and L_2 norm respectively.

Since we consider noise free examples in this part, we use the method for noise free data described in section 5.2.2 to detect the positions of the discontinuities. And we select $a = 2$ (as used in the algorithms in section 5.2) for all 1-D examples.

Firstly, we compare the approximations for smooth functions. Table 5.1 shows the results of comparison of DB4 with ENO-DB4 approximations for the function $f(x) = \exp[-(\frac{1}{x} + \frac{1}{1-x})], 0 < x < 1$,

We see from the table that for smooth functions, the ENO-wavelet transforms have exactly the same approximation error as the standard wavelet transforms. Both of them maintain the approximation order 2, which agree with the results in Theorem 1. In fact, we notice that in this situation, no singularity is detected, the ENO-wavelet algorithms perform the standard transforms for completely smooth functions as we expected.

Next, we apply Haar and ENO-Haar, DB4 and ENO-DB4, and DB6 and ENO-DB6 transforms to a piecewise smooth function and compare the approximation error. Figure 5.4 shows the comparison of the order of accuracy in the L_∞ and L_2 norm. It is clear that both L_∞ and L_2 order of accuracy for ENO-wavelet transforms are of the order 1, 2 and 3 for ENO-Haar, ENO-DB4 and ENO-DB6 respectively. And they agree with the results in Theorem 2. In contrast, standard wavelet transforms do not retain the corresponding order of accuracy for piecewise smooth functions.

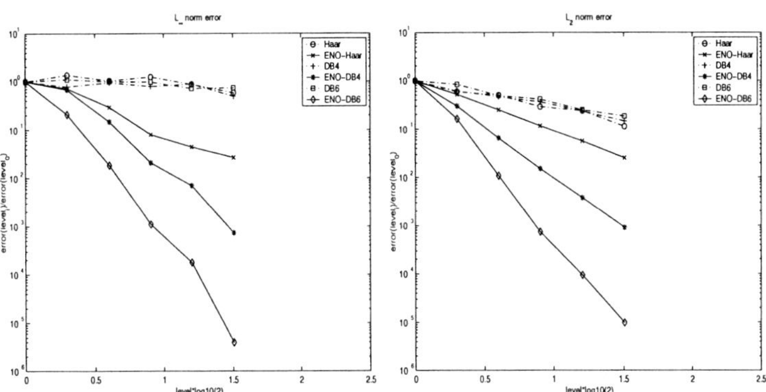

Figure 5.4. The approximation accuracy comparison of ENO-wavelet and wavelet transforms. Both L_∞ (left) and L_2 (right) order of accuracy show that ENO-wavelet transforms maintain the order 1, 2 and 3 for ENO-Haar, ENO-DB4 and ENO-DB6 respectively and they agree with the results of Theorem 2. In contrast, standard wavelet transforms do not retain the order of accuracy for piecewise smooth functions.

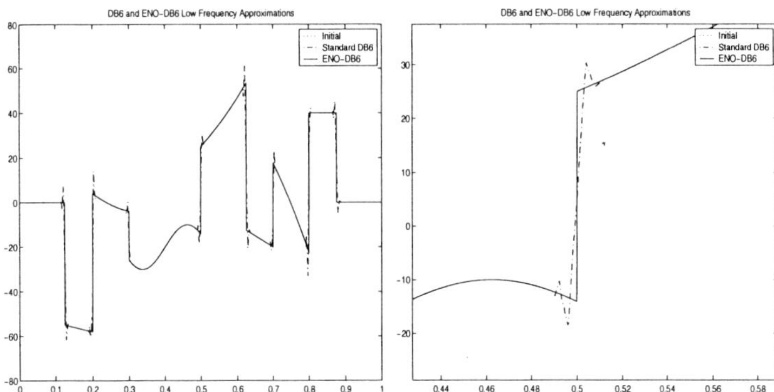

Figure 5.5. The 4-level ENO-DB6 (solid line) and the standard DB6 (dash-dotted line) Approximation. The standard DB6 generates oscillations near discontinuities, but the ENO-DB6 does not.

To see the Gibbs' oscillations, we display the 4-level ENO-DB6 and standard DB6 approximations to a piecewise smooth function in Figure 5.5. In the left picture, we show the original function (dotted line), the standard wavelet linear approximations (dash-dotted) and the ENO-wavelet approximations (solid line). The right pictures is the zoom-in of the left picture near a discontinuity. We clearly see the Gibbs' oscillations in the standard approximations; in contrast, the ENO-wavelet approximations preserve the jump accurately.

In Figure 5.8, we also present the standard DB6 wavelet coefficients (dotted line) and the ENO-DB6 wavelet coefficients (solid line) respectively. The left part corresponds to the low frequency coefficients and the right part the high frequency coefficients. We notice that there are some large standard high frequency coefficients near the discontinuities.

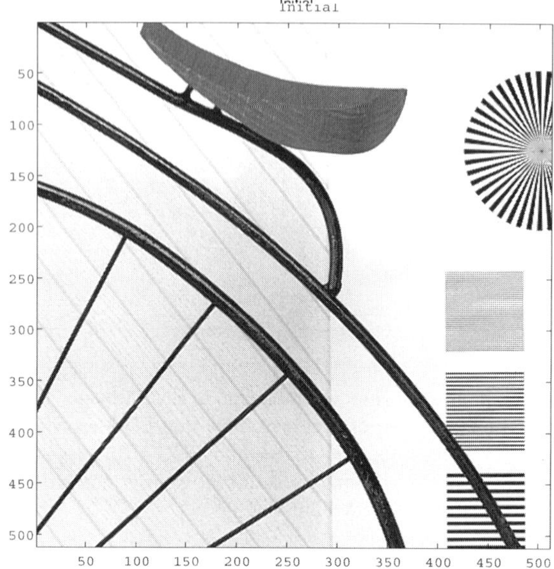

Figure 5.6. Original 2-d Image

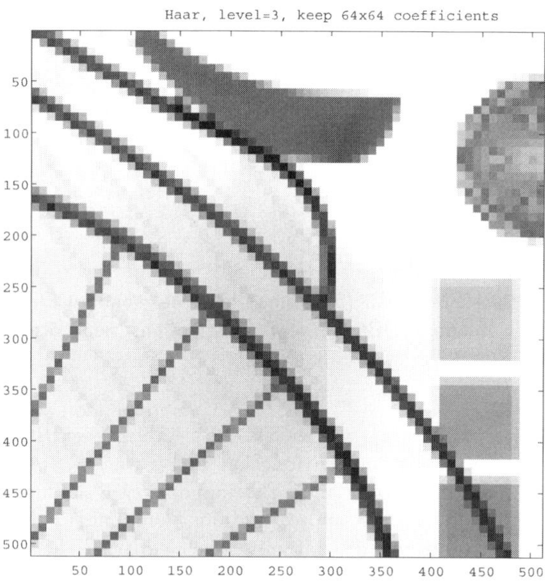

Figure 5.7. The 3-level standard Haar Approximation, the edges are fuzzier than that in the next picture. Most detail information is lost.

On the other hand, no large high frequency coefficients are present in the ENO-wavelet

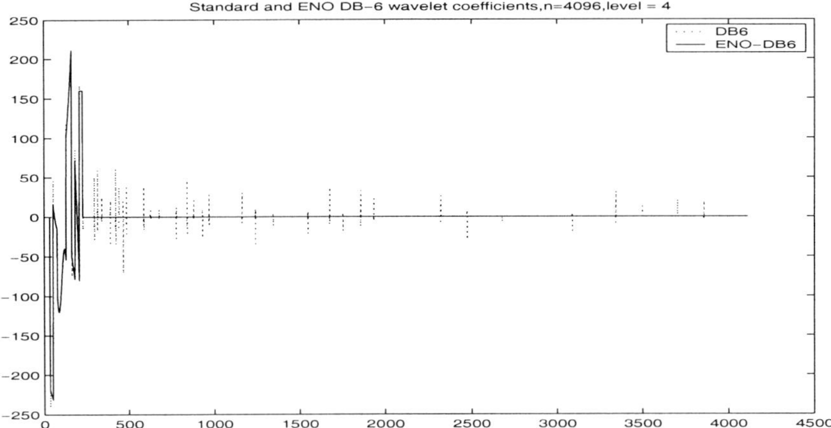

Figure 5.8. The 4-level ENO-DB6 coefficients (solid line) and the standard DB6 coefficients (dotted line). There are large high frequency coefficients near the discontinuities in the standard DB6 transform but not in the ENO-DB6 transform.

coefficients. This illustrates that the ENO-wavelet coefficients have better distribution than standard wavelet coefficients, i.e., no large coefficients in the high frequencies and the energy is concentrated in the low frequency end.

5.4.2 Image Compression

Digital image compression aims to reduce the storage requirement of digital images with (or without) losing information (they are called lossy (or lossless) compression). Wavelet based lossy image compression algorithms have been the leading methods in high ratio (the ratio of original file size over the compressed file size) compression. The most notable work in this area goes to Shapiro's EZW compression [40]. Many studies have been conducted along this direction, including the remarkable work of Said and Pearlman's coding algorithm based on set partitioning in hierarchical trees (SPIHT) [39].

A wavelet based image compression algorithm usually consists of three steps, namely transform, quantization and coding . Transform means that instead of working directly on the pixel values of the digital images themselves, one uses wavelets transforms to compute the wavelet coefficients so that the spatial correlations between pixels in the original images can be decoupled. In other words, in smooth regions where the pixel values are close, the generated high frequency coefficients have small magnitudes and eventually will not be retained. Quantization refers to truncating the real valued wavelet coefficients into a finite set of fixed values so that they can be used in the coding process. In this step, the small wavelet coefficients are usually quantized to zero. Therefore, the more small wavelet coefficients a transform generates the better compression it achieves. Obviously, this step is a lossy and non-invertible process. Coding converts finite quantized wavelet coefficients into binary bit streams for storage.

How to use ENO-wavelet transforms together with quantization and coding steps to form complete image compression algorithms is an ongoing research topic. Many open questions, especially those questions related to the quantization and coding steps, need to be answered. In this chapter, we do not intend to answer those questions. Instead, we

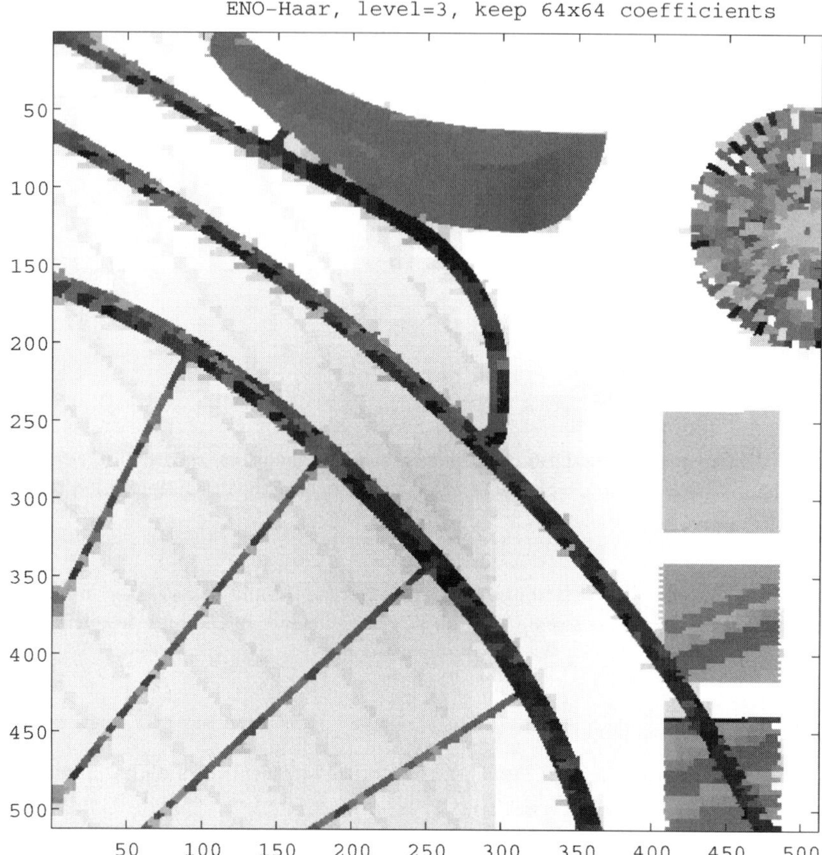

Figure 5.9. The 3-level ENO-Haar Approximation. Edges are preserved sharply, more detail information is retained.

focus on the transform step and discuss the potential of using ENO-wavelet transforms to obtain more efficient compression algorithms. We show pictures to illustrate such potential.

As we have explained, one of the most important reasons for the success of wavelets in image compression is their ability to approximate smooth functions efficiently. Many classes of digital images can be modeled as piecewise smooth functions connected by large discontinuities (edges and boundaries of objects). Using a small number of coefficients one can obtain very high accuracy approximations to the smooth regions. On the other hand, jumps (edges) indicate important features and should be retained. Standard wavelet transforms require many coefficients to represent jumps (edges) and tend to generate a Gibbs phenomenon called jump (edge) artifacts when compression techniques are applied.

In addition, in order to reconstruct the edges correctly, not only does one need to store those large jump related high frequency coefficients, but also to record their positions (coordinates). Otherwise, the decompression process will not be able to put them in the

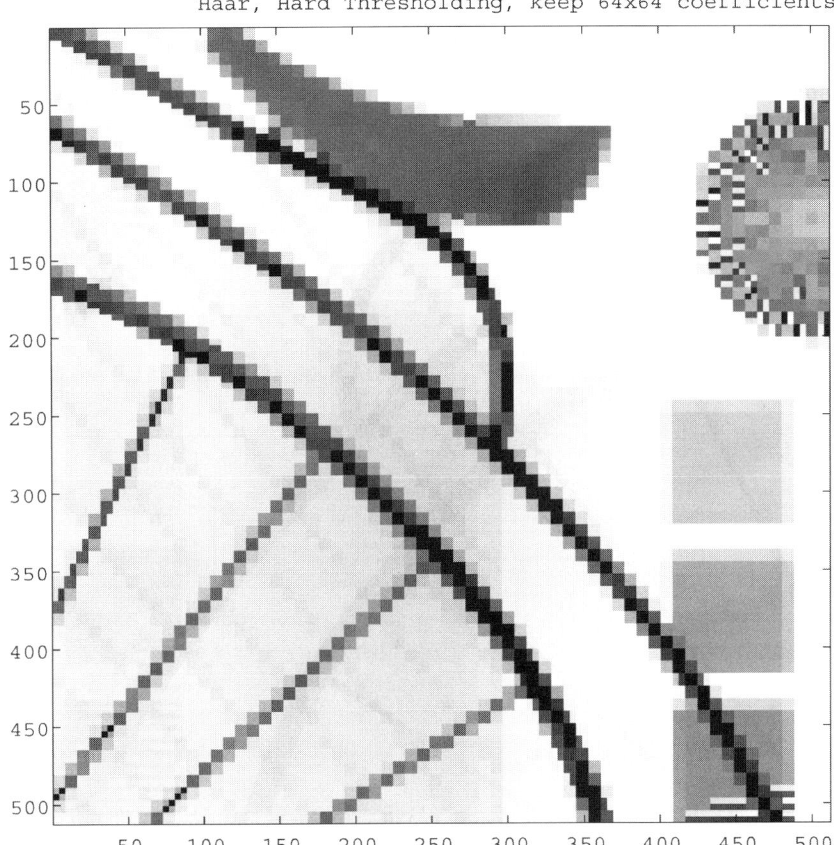

Figure 5.10. The 3-level standard Haar hard thresholding approximation. More details are preserved. Edges are still fuzzy.

correct places to rebuild the images. In fact, recording the position information often consumes more space than remembering those coefficient values themselves. How to code the positions of the large coefficients is the most important question answered by EZW and SPIHT algorithms, where the tree structures are used to code both the positions and coefficient values. Therefore, if one can reduce the number of large high frequency coefficients, a better compression may be obtained.

ENO-wavelet transforms give methods which can reduce the number of large high frequency coefficients. ENO-wavelet transforms maintain all advantages of the standard wavelet transforms, such as multiresolution data structures and concentrating energy to fewer large coefficients. Their performance in smooth regions is the same as that of the standard wavelet transforms. More importantly, ENO-wavelet transforms do not generate large high frequency coefficients near jumps because all filters are applied to smooth functions; thus more efficiently represent jumps (edges).

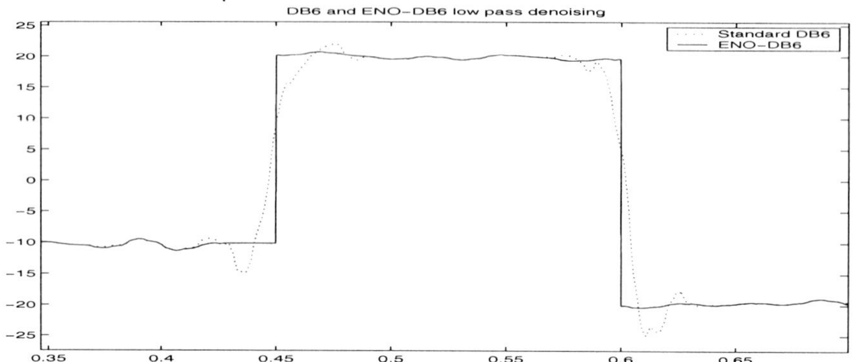

Figure 5.14. A zoom-in of the denoising example at a discontinuities. The ENO-DB6 reconstruction retains the sharp jumps but the standard DB6 one does not.

bits. However, the best way to accomplish this is under investigation.

5.4.3 Signal Denoising

Similar to the standard wavelet transforms, it is also very natural to use ENO-wavelet transforms in signal denoising . In this case, the extra bits used to indicate discontinuities in the signal are no longer a problem. Therefore, the design of the algorithms becomes more straight forward. As we said in the introduction, one can modify a signal denoising method based on the standard wavelet transforms to a method based on the ENO-wavelet transforms, simply by replacing the standard wavelet transforms by the ENO-wavelet transforms, for instance, the translation invariant wavelet denoising method [15] can be used to together with the ENO-wavelet transforms.

The major advantage of the ENO-wavelet based denoising over the standard wavelet denoising is that ENO-wavelet transform denoising can remove the oscillation without smearing edges, while the smearing problem is a drawback for standard wavelet denoising algorithms. ENO-wavelet transforms use filtered (less noisy) values to form extensions at jumps which makes the method more stable and more effectively denoises near discontinuities. However, as we explained in section 5.2.2, the presence noise makes it harder to detect jumps. One often needs to use some heuristic to find the correct locations. Edge detection methods can be used together with the ENO-wavelet transforms.

To illustrate the idea, here we just give an 1-D signal denoising example by using simple linear thresholding (truncating all high frequency coefficients). We apply the ENO-DB6 wavelet transform to a piecewise constant signal polluted by Gaussian random noise (see Figure 5.12). Here, since we have noise in the data, we do not use 2-nd order polynomial extrapolation as we did to the smooth data, instead, we use least square extrapolation at jumps. Despite the presence of noise in the initial data, the level-6 ENO-DB6 reconstruction (solid line in Figure 5.13) still retains the sharp edges (see zoom-in picture in Figure 5.14) compared to the standard DB6 reconstruction (dotted line in Figure 5.13) which produces oscillations at the discontinuities and also smears them.

ACKNOWLEDGEMENTS

This work is supported in part by grants ONR-N00017-96-1-0277, NSF DMS-9973341, NSF DMS-0073916 and ARO DAAD19-99-1-0141.

REFERENCES

[1] S. Amat, F. Arandiga, A. Cohen, R. Donat, G. garcia and M. von Oehsen, *Data Compresion with ENO Schemes: A Case Study*, ACHA 11, 273-288 (2001)

[2] S. Amat, F. Arandiga, A. Cohen and R. Donat, *Tensor product multiresolution analysis with error control for compact image representations*, to appear in Signal Processing, 2002.

[3] L. Anderson, N. Hall, B. Jawerth and G. Peters (1993). *Wavelets on closed subsets of the real line*, in Recent Advances in Wavelet Analysis, Schumaker L.L. and G. Webb (eds.), pp. 1-61, Academic Press, New York, 1993.

[4] F. Arandiga and R. Donat, *A Class of Nonlinear Multiscale Decompositions*, Preprint, 1999.

[5] F. Arandiga and R. Donat, *Nonlinear Multiscale Decompositions: The approach of A. Harten*, Numerical Algorithms 23 (2000) 175-216.

[6] A. Arneodo, *Wavelet Analysis of Fractals: From the Mathematical Concepts to Experimental Reality*, in Wavelets: Theory and Applications, Ed. G. Erlebacher, M. Hussaini, L. Jameson, Oxford Univ. Press, 1996.

[7] G. Beylkin, R. Coifman, and V. Rokhlin, *Fast Wavelet Transforms and Numerical Algorithms*, Comm. Pure Appl. Math, 44(1991), pp141-183.

[8] E. Candes and D. Donoho, *Ridgelets: a Key to Higher-dimensional Intermittency?*, Phil. Trans. R. Soc. Lond. A(1999).

[9] E. Candes and D. Donoho, *Curvelets - A Surprisingly Effective Nonadaptive Representation for Objects with Edges*, Curves and Surfaces, L. L. Schumaker et al. Eds, Vanderbilt University Press, Nashville, TN. 1999.

[10] A. Chambolle, R. DeVore, N. Lee and B. Lucier, *Nonlinear Wavelet Image Processing: Variational Problems, Compression, and Noise Removal Through Wavelet Shrinkage*, IEEE Tran. Image Proc., Vol. 7, No. 3, Mar. 1998, pp319-333.

[11] T. Chan and H. M. Zhou, *Adaptive ENO-wavelet transforms for Discontinuous Functions*, UCLA CAM Reports, 99-21, June, 1999 (also submitted to SIAM, Numer. Anal.).

[12] T. Chan and H. M. Zhou, *Optimal Construction of Wavelet Coefficients Using Total Variation Regularization in Image Compression*, UCLA CAM Reports, 00-27, July, 2000.

[13] C. K. Chui, *Wavelet: A Mathematical Tool for Signal Analysis*, SIAM, 1997.

[14] R. Coifman and A. Sowa, *New Type of Total Variation Diminishing Flows*, Preprint.

[15] R. Coifman and D. Donoho, *Translation Invariant De-Noising*, Wavelets and Statistics, A. Antoniadis and G. Oppenheim Eds, Springer-Verlag, 1995, pp125-150.

[16] A. Cohen, I. Daubechies, B. Jawerth, and P. Vial, *Multiresolution analysis, wavelets and fast algorithms on an interval*, Comptes Rendus Acad. Sci. Paris, 316 (Serie 1), pp. 417–421, 1993.

[17] A. Cohen, I. Daubechies, and P. Vial, *Wavelets on the interval and fast wavelet transforms*, Applied and Computational Harmonic Analysis 1, pp. 54–81, 1993.

[18] A. Cohen and B. Matei, *Compact representations of images by edge adapted multiscale transforms*, to appear in the proceedings of the IEEE ICIP conference, Tessaloniki, 2001.

[19] P. Claypoole, G. Davis, W. Sweldens and R. Baraniuk, *Nonlinear Wavelet Transforms for Image Coding*, Correspond. Author: Baraniuk, Dept. of Elec. and Comp. Sci., also Submitted to IEEE Tran. on Image Proc., Preprint, 1999.

[20] I. Daubechies, *Orthonormal Bases of Compactly Supported Wavelets*, Comm. Pure Appl. Math. 41(1988), pp909-996.

[21] I. Daubechies, *Ten Lectures on Wavelets*, SIAM 1992.

[22] S. Durand and J. Froment, *Artifacts Free Signal Denoising with Wavelets*, in the Proceedings of ICASSP 2001 VI, Salt Lake City, Utah, May 7-11, 2001, pp3685-3689.

[23] D. Donoho, *De-noising by Soft Thresholding*, IEEE Trans. Inf. Th. 41(1995), pp613-627.

[24] D. Donoho, *Wedgelets: Nearly-Minimax Estimation of Edges*, Tech. Report, Dept. of Stat., Stanford Univ., 1997.

[25] D. Donoho, *Orthonormal Ridgelets and Linear Singularities* Tech. Report, Dept. of Stat., Stanford Univ., 1998.

[26] D. Donoho, I. Daubechies, R. DeVore, M. Vetterli, *Data Compression and Harmonic Analysis*, Dept. of Stat., Stanford Univ., Preprint, 1998.

[27] D. Donoho, I. Johnstone, *Adapting to Unknown Smoothness via Wavelet Shrinkage*, J. Amer. Stat. Assoc., Vol. 90, 1995, pp1200-1224.

[28] A. Harten, *Discrete Multi-resolution Analysis and Generalized Wavelet*, Appl. Numer. Math., Vol. 12, 1993, pp153-192.

[29] A. Harten, *Multiresolution Representation of Data, II. General Framework*, Dept. of Math., UCLA, CAM Report 94-10, April 1994.

[30] A. Harten, *Multiresolution Representation of Cell-Averaged Data*, Dept. of Math., UCLA, CAM Report 94-21, July 1994.

[31] A. Harten, *Multiresolution Algorithms for the Numerical Solution of Hyperbolic Conservation Laws*, Comm. Pure Appl. Math. 48(1995), pp1305-1342.

[32] A. Harten, B. Engquist, S. Osher and S. Chakravarthy, *Uniformly High Order Essentially Non-Oscillatory Schemes, III*, Journal of Computational Physics, v71 (1987), pp.231-303.

[33] E. Hernandez and G. Weiss, *A First Course on Wavelets*, CRC Press, 1996.

[34] S. Jaffard, *C. R. Acad. Sci. Paris*, Serie I, 79, 1989, pp.308.

[35] E. Le Pennec and S. Mallat, *Image Compression with Geometrical Wavelets*, in IEEE Conference on Image Processing(ICIP), Vancouver, September, 2000.

[36] S. Mallat, *Multiresolution Approximation and Wavelet Orthonormal Bases of* $L^2(R)$, Tran. Amer. Math. Soc. 315(1989), pp.69-87.

[37] S. Mallat, *A Theory of Multiresolution Signal Decomposition: The Wavelet Representation*, IEEE Trans. PAMI 11 (1989), pp. 674-693.

[38] S. Mallat, *A Wavelet Tour of Signal Processing*, Academic Press, 1998.

[39] A. Said and W. Pearlman, *A New, Fast, and Efficient Image Codec Based on Set Partitioning in Hierarchical Trees*, IEEE Trans. Circ. and Sys. for Video Tech., 6(3), 1996, pp243-250.

[40] J. Shapiro, *Embedded Image Coding Using Zerotrees of Wavelet Coefficients*, IEEE Trans. on Signal Processing, 41(12): pp.3445-3462, 1993.

[41] C. W. Shu, *High Order ENO and WENO Schemes for Computational Fluid Dynamics*, Lecture Notes in Computational Science and Engineering 9: High-Order Methods for Computational Physics, T. Barth and H. Deconinck (Eds.), Springer, 1999, pp439-582.

[42] G. Strang and T. Nguyen, *Wavelets and Filter Banks*, Wellesley-Cambridge Press, 1996.

[43] W. Sweldens, *The Lifting Scheme: A Construction of Second Generation Wavelets*, SIAM J. Math. Anal., Vol. 29, No. 2, 1997, pp511-546.

[44] P. Tchamitchian, *Wavelets, Functions, and Operators*, in Wavelets: Theory and Applications, Ed. G. Erlebacher, M. Hussaini, L. Jameson, Oxford Univ. Press, 1996.

[45] J. R. Williams and K. Amaratunga, *A Discrete Wavelet Transform without Edge Effects* MIT, IESL Tech. Report No. 95-02, 1995.

[46] H. M. Zhou and T. Chan, *The Stability of Adaptive ENO-wavelet Transforms for Piecewise Contiunous Functions*, Preprint.

Tony F. Chan
Department of Mathematics
University of California, Los Angeles
Los Angeles, CA 90095-1555
chan@math.ucla.edu

Hao-Min Zhou
Department of Applied and Computational Mathematics
California Institute of Technology, Mail Code 217-50
Pasadena, CA 91125
hmzhou@acm.caltech.edu

Beyond Wavelets
G. V. Welland (Editor)
© 2003 Elsevier Science (USA) All rights reserved

6

A MECHANICAL IMAGE MODEL FOR BAYESIAN TOMOGRAPHIC RECONSTRUCTION

SHIYING ZHAO AND HAIYAN CAI

Department of Mathematics and Computer Science University of Missouri - St. Louis
8001 Natural Bridge Road, St. Louis, MO 63121
zhao@arch.umsl.edu and *cai@arch.umsl.edu*

Abstract

We present an explicit mechanical image model for Bayesian reconstruction from tomographic data. The image intensity of each pixel is modeled by a transverse motion of a so-called *"pixtron"* with unknown mass. A prior energy for Bayesian tomographic reconstruction is therefore interpreted as the total kinetic energy of a collection of pixtrons. With the log-likelihood as the potential energy restricting the motion of pixtrons, the minimization of a log-posterior is an analogue of the *principle of least action in the classical mechanics*. We show that the Gaussian Markov random field prior can be viewed as the kinetic energy of free motion of pixtrons.

With the framework of the mechanical image model, we propose a novel image prior for Bayesian tomographic reconstruction based on level-set evolution of an image driven by the mean curvature motion. As it has been studied in image processing with nonlinear diffusion, this prior encourages the stabilization of an edge while the reconstructed image is smoothed along both sides of the edge. A distinguished feature of our approach is that the curvature term itself appears in the image prior rather than in the resulting differential equation derived from the total variation method. An algorithm of iterated coordinate descent has been implemented with the proposed prior using Brent's method for one-dimensional optimization. Our simulation results demonstrate that our algorithm can outperform existing priors for preserving sharp edges during tomographic reconstruction without introducing additional artifacts.

6.1 INTRODUCTION AND BACKGROUND

6.1.1 Introduction

The Bayesian approach to tomographic reconstruction has now established itself as a powerful and practical tool in several clinical applications including positron emission tomography (PET)and single-photon emission computed tomography (SPECT) [1]. By incorporating data acquisition with image restoration models, Bayesian methods can significantly improve image quality for reconstruction from noise and incomplete tomographic data. While data acquisition models are determined by the physical effects, image restoration models are largely based on our knowledge on image processing.

An important problem in Bayesian reconstruction is the design of edge-preserving image models as *a priori* distributions. Various edge-preserving prior models have been proposed in recent years. Earlier models relied on modifications of a specific form of prior functions, such as the quadratic function [2]. Later models took more advantages of new technologies of image processing, including building line processes [3], adapting total variation approach [4], applying multiresoultion analysis in the wavelet domain [5], and incorporating nonlocal boundary information [6]. Although these proposed methods are successful to certain degrees, the search for more natural and efficient image models is far from over.

Recently, there has been a new movement towards nonlinear partial differential equation (PDE) based methods in image processing, which is motivated by a more systematic approach to restoring images with sharp edges, as well as for image segmentation [7]- [10]. In these PDE based approaches, the image is diffused (denoised) according to a nonlinear anisotropic diffusion PDE, designed to diffuse less near edges. Moreover, the PDEs are designed to possess certain desirable geometrical properties such as affine invariance and causality. The level set method, devised in 1987 by Osher and Sethian for computation of moving fronts in interfacial dynamics [7], can be viewed as a method in this family. It yields the so called morphological principle, which has a large significance in the theory of image processing [10].

In this paper, we first introduce an explicit mechanical image model for Bayesian tomographic reconstruction. The image intensity of each pixel is modeled by the transverse movement of a so-called *"pixtron"* with unknown mass. A prior energy for Bayesian tomographic reconstruction is therefore interpreted as the total kinetic energy of a collection of pixtrons. The log-likelihood, on the other hand, is an analogue of potential energy which restricts the motion of pixtrons. During tomographic reconstruction, the kinetic energy of each pixtron indicates the rate of change to the intensity at the pixel where the pixtron is located. By using this image model, we can easily adapt PDE-based image-processing transforms to the context of tomographic reconstruction.

Based on this mechanical image model, we apply the level-set method of Osher and Sethian [7] to obtain the kinetic energy of the motion of pixtrons induced by a level-set evolution of the image under reconstruction. We show that the Gaussian Markov random field (GMRF) prior can be viewed as the kinetic energy due to the level-set evolution with a constant speed. We then propose a novel image prior for Bayesian tomographic reconstruction, which is motivated by the studies of the mean curvature motion for image denoising. In contrast to GMRF, the kinetic energy of a pixtron driven by the mean curvature motion is much lower than the average when the pixtron is located at a sharp edge of the image, so that it is more likely to be trapped by its potential. Consequently, this prior encourages the stabilization of an edge while smoothing the image along both sides of the edge. A distinguished feature of this prior is that the curvature term itself appears in the image prior (see (2.20) below) rather than in the resulting differential equation derived from the total variation method.

The Bayesian maximum *a posteriori* (MAP) reconstruction with the proposed prior is implemented using the method of iterated coordinate descent (ICD) [11] with Brent's one-dimensional optimization algorithm [12]. Our simulation results demonstrated that the proposed algorithm is capable of (1) retaining important features of anatomy morphological principle for both noise and noise-free tomographic data; (2) preserving sharp edges while denoising; and (3) introducing fewer additional artifacts than existing image priors.

The outline of this paper is as follows. First, in Sections 1.2 and 1.3, we briefly describe some background of statistical reconstruction methods to PET emission data. The proposed mechanical image model is introduced in Section 2.1, and the new image prior for Bayesian tomographic reconstruction is derived in Section 2.2. Some issues on our numerical implementation of the ICD algorithm with the proposed image prior are discussed in Section 2.3. In Section 3.1, we present our simulation results. Finally, in Sections 3.2 and 3.3, we make some concluding remarks on further development of the proposed level-set method for tomographic reconstruction.

6.1.2 Positron Emission Tomography

Positron emission tomography or PET is a medical diagnostic technique that allows one to obtain images of metabolic processes in tissues and blood flow or neuroreceptors activities of the brain. It is applied in cancer detection of the brain, breast, heart and lung. It is also used to measure in detail the functioning of distinct areas of human brain and to study the chemical process involved in the working of human brains. Applications of PET imaging can also be found in other areas of cardiology, neurology, and oncology.

In PET, a metabolically active tracer is introduced into the human body. Such a tracer is a biological molecule that carries a positron-emitting isotope with it. The isotope decays in human body and emits positrons. These positrons will travel only for a short distance before they encounter nearby electrons and cause the emission of gamma photons. The photons travel in nearly opposite directions along a randomly determined line in the space. Detection of two such photons indicates the previous existence of a positron along this line. Therefore a large photon count in a given direction implies a higher density of positron emitting isotope. To detect the arriving photons, rings of detectors are set up outside the human body. The spatial distribution of the tracer within the human body can now be obtained indirectly in the form of photon counts from the rings, and the image can be reconstructed with reconstruction algorithms. In this way, for example, one can combine glucose with a positron-emitting tracer to show where glucose is being used in the brain, the heart muscle, or a growing tumor.

A classical reconstruction algorithm that has been applied to the computed tomography (CT) and magnetic resonant imaging (MRI) is called filtered back-projection. In this algorithm, the counts of photons obtained from a pair of detectors is approximated by a line integral of the image alone the straight line that passes through the detectors. The algorithm is based on the theory of integral geometry which asserts that the image can be uniquely reconstructed from the collection of line integrals through the Radon transform.

Since PET imaging inherently involves physical processes with random outcomes, the images obtained with the filtered back-projection method are always noisy. Another reason for the unsatisfactory performance of such a deterministic method is that PET is a seriously ill-posed inverse problem due to a relatively low resolution of tomographic data. Over the last two decays, a variety of statistical reconstruction methods have been introduced to improve the quality of the images. In these approaches, the problem of image reconstruction becomes the problem of parameter estimation. The maximum likelihood estimate (MLE) was the first method proposed [13] in tomographic reconstruction. This method became computationally feasible when it was shown in [14, 15] that the expectation-maximization (EM) iterative algorithm could be nicely adapted to solve the maximization problem.

Although MLE shows clear improvements over the filtered back-projection method, the images obtained are still noisy. Also, simulation studies have shown that the algorithm for MLE is unstable in the sense that the quality of the images deteriorates as the number of iterations in the EM algorithm increases. These problems from MLE methods promote the studies of Bayesian reconstruction methods. The most important feature in Baysian approaches is that by incorporating *a priori* information into the probability distribution of a image, one can model the values of image pixels or, equivalently, the intensity parameters of a joint Poisson distribution as *correlated* variables. Under the constraints of an *a priori* distribution, the model parameters can no longer be allowed to vary independently. The image value at one pixel depends on the values of other pixels in the same image. A good prior distribution should capture such correlations among the pixel values. These correlations should represent fundamental properties of a image.

6.1.3 Bayesian Tomographic Reconstruction Method

To reconstruct an emission PET image $f = \{f_1, \ldots, f_N\}$ from the tomographic data $y = \{y_1, \ldots, y_M\}$, the basic statistical PET model (see [16]) assumes that y are independent Poisson observations with mean $\bar{y} = E[y]$:

$$\Pr(y|f) = \prod_{i=1}^{M} \frac{\bar{y}_i^{y_i} e^{-\bar{y}_i}}{y_i!} . \tag{1.1}$$

The corresponding log-likelihood, after dropping constants, is

$$L(y|f) = \sum_{i=1}^{M} y_i \log(\bar{y}_i) - \bar{y}_i . \tag{1.2}$$

The mean $\bar{y}$ is related to the image f through the affine transform:

$$\bar{y} = Pf + r + s , \tag{1.3}$$

in which $P \in \mathbb{R}^{M \times N}$ is the forward projection matrix, whose elements, $p_{i,j}$, contain the probabilities of detecting an emission from pixel site j at detector pair i. In this general data acquisition model, the measured data, $\bar{y}$, are also assumed to be corrupted by additive random coincidences, r, and scattered coincidences, s. For simplicity, we shall assume that both r and s are zeros in this paper.

In the Bayesian framework, an *a priori* probability distribution is proposed for f so that the image is treated as a realization of a random field. When we only assume short range correlations among image pixel values, the random field corresponding to the prior distribution is called a Markov random field (MRF). The probability distribution of a random field has the form

$$\Pr(f|\lambda) = \frac{1}{Z} e^{-U_\lambda(f)} , \tag{1.4}$$

where Z is the normalizing constant, U_λ is the energy function that depends on a hyperparameter λ (or a set of hyperparameters, in some cases) and takes the form

$$U_\lambda(f) = \lambda \sum_i U_i(f). \tag{1.5}$$

Here each $U_i(f)$ is a local energy function depending on the value of $f(i)$ and the values of $f(k)$ for $k \in \partial(i)$, where $\partial(i)$ is a neighborhood system of pixel i.

A commonly used family of priors is the generalized Gaussian MRF (GGMRF) [2] which enjoys the nice property of convexity. The local energy function for GGMRF is

$$U_i^{GG}(f) = \sum_{k\in\partial(i)} w_{i,k}|f(i) - f(k)|^p, \qquad 1 \le p \le 2, \tag{1.6}$$

where $w_{i,k} > 0$ are the weights that control the space resolution properties [17]. Some non-convex energy functions are also studied. An important example of such energies is the thin plate (TP) prior which has the local energy function given by

$$U_i^{TP}(f) = f_{hh}(i)^2 + 2f_{hv}(i)^2 + f_{vv}(i)^2. \tag{1.7}$$

The TP energy function is a discrete approximation of bending energy of a thin-plate. Here and hereafter we denote by $f_h(i)$, $f_v(i)$, $f_{hh}(i)$, $f_{hv}(i)$ and $f_{vv}(i)$ the discretized version of the first and second order partial derivatives of a image f at pixel i along the vertical or horizontal directions. Both of these models are based on the assumption that images are equally smooth in all directions and therefore they tend to smooth out local regions uniformly, causing lost of edge information.

Other priors, such as the weak membrane (WM) and weak plate (WP) priors, extend the smoothing models by allowing for spatial discontinuities [18–21]. In these models, line processes [22] are introduced. For example, the local energy for WM can be written as

$$U_i^{WM}(f) = (1 - l_v(i))f_h(i)^2 + (1 - l_h(i))f_v(i)^2 + \alpha(l_h(i) + l_v(i)). \tag{1.8}$$

In this function, the binary variables l_h and l_v form horizontal and vertical line processes, respectively. The last term in the energy function penalizes the creation of the discontinuities, charging an amount of α at each such site. A disvantage of such a prior is that it dramatically increases the computational complexity due to the presence of binary variables.

When a prior distribution is incorporated with data y, one obtains the *posterior* distribution:

$$\Pr(f|y,\lambda) = \frac{\Pr(y|f)\Pr(f|\lambda)}{\Pr(y)}. \tag{1.9}$$

The MAP approach [22] in this framework is to estimate the parameter f by maximizing the posterior probability, or equivalently to find

$$f^* = \arg\min_{f\ge 0}[-L(y|f) + U_\lambda(f)]. \tag{1.10}$$

This minimization problem is commonly solved using iterative methods. Many algorithms have been developed in the recent years. Most of them are successors of the original EM algorithm [14]. An alternative approach is the ICD method [11] in which the pixels are updated sequentially. Given the current estimate $f^{(k)}$, the update for the ith pixel is given by

$$f_i^{(k+1)} = \arg\min_{z \geq 0} \left[-L(y|f) + U_\lambda(f)\right]_{f=f_i^{(k)}[z]}, \tag{1.11}$$

where $f_i^{(k+1)} = f^{(k+1)}(i)$ and

$$f_i^{(k)}[z] = \{f_1^{(k+1)}, \cdots, f_{i-1}^{(k+1)}, z, f_{i+1}^{(k)}, \cdots, f_N^{(k)}\}. \tag{1.12}$$

6.2 MATERIALS AND METHODS

6.2.1 A Mechanical Image Model

Since the choice of the prior energy function $U_\lambda(f)$ is crucial for reconstructing tomographic images, we try to understand the principle for the design of image priors by building a mechanical image model.

We consider the transverse motion of a surface consisting of a collection of *pixtrons*. A single pixtron is an analogue of a particle in physics with unknown mass (or charge) which may vary with respect to the time t. The image intensity $f(x,t)$ at location $x \in \Omega \subset \mathbb{R}^2$ and time $t \in \mathbb{R}$ is modeled as the vertical displacement of a pixtron, or the average of displacements of several pixtrons. For a fixed time t, we will often omit the variable t and denote $f(x)$ the image instead of $f(x,t)$.

A tomographic reconstruction from a given tomographic data y is interpreted, in this mechanical model, as a computer experiment in which the system of pixtrons is placed inside a field of potential energy given by $L(y|f)$. The image prior $U_\lambda(f)$ is then the kinetic energy of the system of pixtrons

$$U_\lambda(f,t) = \frac{1}{2}\int_\Omega \lambda(x,t)\left[\frac{\partial f}{\partial t}(x,t)\right]^2 dx\,, \tag{2.1}$$

where the nonnegative function $\lambda(x,t)$ is the mass of the pixtron at location $x \in \Omega$ and time t.

If the tomographic data y is determinstic and complete, each pixtron is trapped by the potential energy to its equilibrium position such that $Pf = y$. Under such a circumstance, the image prior $U_\lambda(f,t)$ is redundant, and we can consider that pixtrons are massless, $\lambda(x,t) \equiv 0$, or equivalently, the image $f(x,t)$ is independent of time, that is, $\dfrac{\partial f}{\partial t} \equiv 0$. From either point of view, the total kinetic energy at any given time t is zero.

If the tomographic data y is random or incomplete, certain pixtrons in the system gain freedom of motion along the vertical direction. In this case, the *least action* principle for dynamical systems asserts that the system of pixtrons always tends toward its lowest energy configuration, and therefore, the optimal estimate $\hat{f}$ of the reconstruction is

$$\hat{f}(x,t) = \arg\min_{f \geq 0} \int_0^T \left[-L(y|f) + U_\lambda(f,t)\right] dt\,, \tag{2.2}$$

for a given positive number $T \in \mathbb{R}$. For given boundary conditions of $f(x,t)$ at $t = 0$ and $t = T$, solving the minimization problem (2.2) with the total variation method leads to a wave equation for $f(x,t)$. The solution of this equation describes precisely the deformation process of the image from $\hat{f}(x,0)$ to $\hat{f}(x,T)$ during the time interval $[0,T]$.

In tomographic reconstruction, our goal is to find the final image $\hat{f}(x,T) = f^*(x)$ given by the Bayesian principle (1.10) with an initial guess $\hat{f}(x,0)$. Due to the fact that the boundary condition $\hat{f}(x,T) = f^*(x)$ is indeterminate, the variational problem (2.2) itself is ill-posed. We thus weaken the least action principle in following two aspects.

First, instead of using the total variation method to solve the dynamical problem, we impose an *a priori* conservation law for the Lagrangian formulation of the variational problem. The choice of such a prior can be made from our knowledge of image processing. For instance, from an axiomatic approach of the multiscale analysis for image processing [10], we can assume that the velocity field $\frac{\partial f}{\partial t}(x,t)$ is of the form:

$$\frac{\partial f}{\partial t} = F(\nabla^2 f, \nabla f, t)\,, \tag{2.3}$$

for a continuous function F. In the next subsection, we will consider the kinetic energy induced from a level-set evolution of the image $f(x,t)$.

Second, we assume that the time-dependent energy function $U_\lambda(f,t)$ with the velocity field given by (2.3) possesses the following growth condition:

$$U_\lambda(f,t_1) \geq U_\lambda(f,t_2)\,, \qquad \text{whenever} \quad t_1 \leq t_2, \tag{2.4}$$

for each fixed function $f(x)$ independent on t. We then apply the direct minimizing method to approximate the solution of (2.2) with the boundary conditions $f(x,0) = f^{(0)}(x)$ which is an initial guess and $f(x,T) = f^*(x)$ which satisfies (1.10). We divide the interval $[0,T]$ into K subintervals by introducing the points:

$$0 = t^{(0)} < t^{(1)} < \cdots < t^{(K-1)} < t^{(K)} = T\,, \tag{2.5}$$

and replace the function $f(\cdot,t)$ by the "polygonal line" with vertices

$$(t^{(0)}, f^{(0)}),\ (t^{(1)}, f^{(1)}),\ \ldots\ (t^{(K-1)}, f^{(K-1)}),\ (t^{(K)}, f^{(K)}), \tag{2.6}$$

where $f^{(k)}(x) = f(x,t^{(k)})$.

We claim that if $\{\bar{f}^{(k)} \colon k = 0,\ldots,K\}$ is the minimizer of the function

$$J(f^{(0)},\ldots,f^{(K)}) = \sum_{k=0}^{K} [-L(y|f^{(k)}) + U_\lambda(f^{(k)},t^{(k)})]\,, \tag{2.7}$$

then it is necessary to satisfy the following monotonicity condition:

$$[-L(y|\bar{f}^{(k-1)}) + U_\lambda(\bar{f}^{(k-1)},t^{(k-1)})] \geq [-L(y|\bar{f}^{(k)}) + U_\lambda(\bar{f}^{(k)},t^{(k)})]\,, \tag{2.8}$$

for all $1 \leq k \leq K$. Indeed, suppose that there is an index k $(0 < k < K)$ such that

$$-L(y|\bar{f}^{(k)}) + U_\lambda(\bar{f}^{(k)},t^{(k)}) > -L(y|\bar{f}^{(k-1)}) + U_\lambda(\bar{f}^{(k-1)},t^{(k-1)})\,, \tag{2.9}$$

then by taking $\tilde{f}^{(k)} = \bar{f}^{(k-1)}$ and using the growth condition (2.4), we have

$$-L(y|\bar{f}^{(k-1)}) + U_\lambda(\bar{f}^{(k-1)},t^{(k-1)}) \geq -L(y|\tilde{f}^{(k)}) + U_\lambda(\tilde{f}^{(k)},t^{(k)})\,, \tag{2.10}$$

where we have used the fact that $L(y|\bar{f}^{(k-1)}) = L(y|\tilde{f}^{(k)})$ since $L(y|f)$ is independent of t. This shows that

$$J(\bar{f}^{(0)},\ldots,\bar{f}^{(K)}) > J(\bar{f}^{(0)},\ldots,\bar{f}^{(k-1)},\tilde{f}^{(k)},\bar{f}^{(k+1)},,\ldots,\bar{f}^{(K)}), \tag{2.11}$$

which contradicts the assumption that $\{\bar{f}^{(k)} \colon k = 0,\ldots,K\}$ is the minimizer of $J(f^{(0)},\ldots,f^{(K)})$.

In practice, because the desired boundary condition $f^*(x)$ is unknown, one often searches for a sequence $\{f^{(k)}: k = 0, \ldots, K\}$ only satisfying the monotonicity condition (2.8) without requiring $f^{(K)}(x) = f^*(x)$. In particular, the sequence obtained from the ICD method (1.11) can be viewed as such a sequence.

6.2.2 Kinetic Energy Induced from Level-Set Evolution

In this subsection, we propose a novel image prior for Bayesian tomographic reconstruction. The prior is based on the mechanical image model discussed in the previous subsection and the level-set evolution driven by the mean curvature motion.

For each fixed $c \in \mathbb{R}$, the c-level set Γ_c of $f(x,t)$ at time $t = 0$ is defined by

$$\Gamma_c = \{x \in \Omega: f(x, t = 0) = c\}\,. \tag{2.12}$$

We consider the evolution of Γ_c due to the transverse motion of pixtrons in Ω. To do this, we let $x(t)$ be a differentiable trajectory of a point x on Γ_c for which the following equation is satisfied:

$$\frac{dx}{dt}(t) = \beta \frac{\nabla f}{|\nabla f|}\,, \tag{2.13}$$

with a speed function β. Substituting the expression in the equation

$$\frac{\partial f}{\partial t}(x(t), t) + \nabla f(x(t), t) \cdot \frac{dx}{dt}(t) = 0\,, \tag{2.14}$$

we obtain

$$\frac{\partial f}{\partial t} = -\beta |\nabla f|\,. \tag{2.15}$$

An important case is $\beta \equiv 1$, which means the curve $x(t)$ is moving along the normal direction at the unit speed. We also assume that all the pixtrons have the same mass $2\lambda_G(t)$ at a given time t. In this case, the kinetic energy (2.1) is reduced to

$$U_{\lambda_G}(f,t) = \frac{1}{2} \int_\Omega 2\lambda_G(t) \left[\frac{\partial f}{\partial t}(x,t)\right]^2 dx = \lambda_G(t) \int_\Omega |\nabla f(x,t)|^2 dx\,. \tag{2.16}$$

In practice, the Gaussian MRF with the local energy function given by (1.6) with $p = 2$ can be considered as a implementation of (2.16). It is well known that GMRF is not efficient in preserving edges (see [2]).

In order to preserve edges in the image f via MAP reconstruction, we consider the mean curvature evolution of level sets [7]. The mean curvature motion drives each level set Γ_c of f at a speed proportional to its normal mean curvature field κ_{Γ_c}. By a straightforward computation [23, §5.4.5], we have

$$\kappa_{\Gamma_c} = \frac{\nabla_\xi^2 f}{|\nabla f|}\,, \tag{2.17}$$

where

$$\nabla_\xi^2 f = |\nabla f|\, \nabla \cdot \left(\frac{\nabla f}{|\nabla f|}\right) \tag{2.18}$$

is the second directional derivative along the direction ξ orthogonal to the gradient of f. It then follows from (2.15) with $\beta = \kappa_{\Gamma_c}$ that

$$\frac{\partial f}{\partial t} = -\nabla_\xi^2 f\,. \tag{2.19}$$

Assuming the spatially homogenous mass distribution $2\lambda_E(t)$ of pixtrons, the kinetic energy induced from the level-set evolution driven by the mean curvature motion is given by

$$U_{\lambda_E}(f,t) = \frac{1}{2}\int_\Omega 2\lambda_E(t)\left[\frac{\partial f}{\partial t}(x,t)\right]^2 dx = \lambda_E(t)\int_\Omega [\nabla_\xi^2 f(x,t)]^2 dx\,. \tag{2.20}$$

For MAP reconstruction, $U_{\lambda_E}(f,t)$ allows us to smooth the images discriminatorily so that it only encourages the smoothing in the directions along edges while penalizing any attempt of blurring across edges. In particular, it favors piecewise regions where the image distribution f is of the form $u_{a,b}(x) = u(a\cdot x + b)$ for some function u, constant vector a and constant scalar b, due to the fact that $\nabla_\xi^2 u_{a,b} \equiv 0$.

For edge-preserving image denoising, one needs the joint effort of both $U_{\lambda_G}(f)$ and $U_{\lambda_E}(f)$, we thus propose the following image prior based on our image model:

$$U_{\lambda,\mu}(f) = U_{\lambda_G}(f) + U_{\lambda_E}(f) = \lambda\int_\Omega \left(\mu|\nabla f(x)|^2 + (1-\mu)[\nabla_\xi^2 f(x)]^2\right)dx\,, \tag{2.21}$$

where $\lambda = \lambda_G + \lambda_E$ and $\mu = \lambda_G/(\lambda_G+\lambda_E)$.

6.2.3 Numerical Implementations

To compute the energy function $U_{\lambda,\mu}(f)$ numerically, we first discretize the second-order directional derivative using the 8-point neighborhood system at each pixel $x\in\Omega$.

For an arbitrary fixed point $x = (x_1, x_2)$ and a direction vector (a,b) with $a^2+b^2 = 1$, the second-order directional derivative of f along the direction (a,b) is given by the second-order derivative of the function $g(t) = f(x_1 + at, x_2 + bt)$:

$$\frac{d^2 g}{dt^2}(0) = a^2\frac{\partial^2 f}{\partial x_1^2}(x) + 2ab\frac{\partial^2 f}{\partial x_1\partial x_2}(x) + b^2\frac{\partial^2 f}{\partial x_2^2}(x)\,. \tag{2.22}$$

We approximate this derivative using a 9-point difference formula:

$$\frac{d^2 g}{dt^2}(0) \approx \sum_{k=-1}^{1}\sum_{l=-1}^{1}\alpha_{k,l} f(x_1 + kh, x_2 + lh), \tag{2.23}$$

where h is the step size in both directions. To determine the values of $\alpha_{k,l}$, we substitute the fourth-order Taylor series of $f(x_1+kh, x_2+lh)$, $k,l = -1,0,1$, into (2.23) and compare the coefficients of the partial derivatives in this expression to those in (2.22). By solving the resulting system of linear equations we obtain

$$\begin{aligned}
\alpha_{0,0} &= -\frac{2c}{h^2}\,;\\
\alpha_{0,1} = \alpha_{0,-1} &= \frac{c-a^2}{h^2}\,;\\
\alpha_{1,0} = \alpha_{-1,0} &= \frac{b^2-c}{h^2}\,;\\
\alpha_{1,1} = \alpha_{-1,-1} &= \frac{1+ab-c}{2h^2}\,;\\
\alpha_{1,-1} = \alpha_{-1,1} &= \frac{1-ab-c}{2h^2}\,,
\end{aligned}$$

where c is a free constant (we have 9 unknowns and 8 equations). The remainder of this approximation is $R(x')h^2$ for some point $x' = (x_1', x_2')$, where $R(x')$ is given by

$$R(x') = \frac{1}{2}(1-c)\frac{\partial^4 f}{\partial^2 x_1 \partial^2 x_2}(x') + \frac{1}{12}a^2\frac{\partial^4 f}{\partial x_1^4}(x')$$
$$+ \frac{1}{3}ab\left[\frac{\partial^4 f}{\partial x_1^3 \partial x_2}(x') + \frac{\partial^4 f}{\partial x_1 \partial x_2^3}(x')\right] + \frac{1}{12}b^2\frac{\partial^4 f}{\partial x_2^4}(x').$$

We then take $c = 1$ to eliminate the first term of $R(x')$.

Suppose the image f has the size $N = n \times n$. For each grid index (i,j), $1 \le i,j \le n$ we let $k = in + j$. Then it follows from the above computations that the difference formula for the second-order directional derivative of pixel k along the direction (a,b) can be written as

$$D^2_{(a,b)} f(k) = a^2 f_{hh}(k) + 2ab f_{hv}(k) + b^2 f_{vv}(k)\,, \tag{2.24}$$

where, with $f_{i,j} = f(i,j)$,

$$f_{hh}(k) = \frac{1}{h^2}(f_{i+1,j} - 2f_{i,j} + f_{i-1,j})\,;$$
$$f_{vv}(k) = \frac{1}{h^2}(f_{i,j+1} - 2f_{i,j} + f_{i,j-1})\,;$$
$$f_{hv}(k) = \frac{1}{4h^2}(f_{i+1,j+1} - f_{i+1,j-1} - f_{i-1,j+1} + f_{i-1,j-1})\,.$$

We next take (a,b) to be the direction vector ξ perpendicular to the gradient of f at $x = (x_1, x_2) \in \Omega$:

$$(a,b) = \xi = \frac{\left(\frac{\partial f}{\partial x_2}, -\frac{\partial f}{\partial x_1}\right)}{\sqrt{\left(\frac{\partial f}{\partial x_1}\right)^2 + \left(\frac{\partial f}{\partial x_2}\right)^2}}\,. \tag{2.25}$$

To approximate the first-order partial derivatives with all of 8 neighbors at each pixel k, we use the following difference formulas:

$$f_h(k) = \frac{1}{4h}\left[f_{i+1,j} - f_{i-1,j} + \frac{1}{2}(f_{i+1,j+1} - f_{i-1,j-1} - f_{i-1,j+1} + f_{i+1,j-1})\right],$$
$$f_v(k) = \frac{1}{4h}\left[f_{i,j+1} - f_{i,j-1} + \frac{1}{2}(f_{i+1,j+1} - f_{i-1,j-1} + f_{i-1,j+1} - f_{i+1,j-1})\right].$$

In our numerical computations, we let ε be a small positive number close to machine precision of double floating point and use the following modified approximation of $\nabla^2_\xi f$:

$$f_{\xi\xi}(k) = \frac{f_v(k)^2 f_{hh}(k) - 2f_h(k)f_v(k)f_{hv}(k) + f_h(k)^2 f_{vv}(k)}{f_h(k)^2 + f_v(k)^2 + \varepsilon}\,. \tag{2.26}$$

Therefore $f_{\xi\xi}(k)$ does not always agree with the value of $D^2_\xi f(k)$ obtained from (2.24), if $f_h(k)^2 + f_v(k)^2 = 0$.

With the first term in (2.21) implemented as GMRF, which is given by (1.6) with $p = 2$, the discretization of (2.21) can be read as

$$U_{\lambda,\mu}(f) = \lambda \sum_{i=1}^{N}\left[\mu \sum_{k\in\partial(i)} w_{i,k}\left[f(i) - f(k)\right]^2 + (1-\mu)\sum_{k\in\partial(i)} f_{\xi\xi}(k)^2\right], \tag{2.27}$$

where $w_{i,k} = (4+2\sqrt{2})^{-1}$ for the nearest neighbors k of i and $w_{i,k} = (4+4\sqrt{2})^{-1}$ for diagonal neighbors k of i [2]. We note that the computation of the second inner sum of the energy function requires the values of 24 neighbors for each pixel.

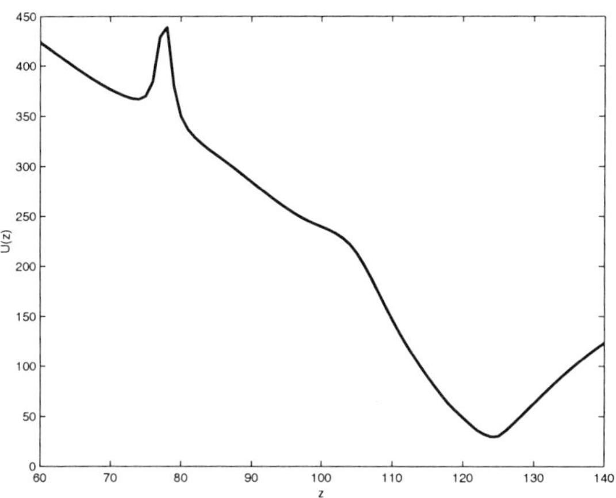

Figure 6.1. An example of one-dimensional prior functions

In the implementation of our reconstruction algorithm, we have adapted the ICD approach (1.11) in which each pixel is updated in sequence by minimizing the corresponding one-dimensional log-posterior function. Since the prior is a nonconvex function, there could be many local minima and maxima in a given searching interval. An example of one-dimensional prior functions is plotted in Figure 6.1, in which there are two minima and one maximum. Therefore in our current implementation, we apply Brent's method to search for a local minimum of the one-dimensional target function, $-L(y|f) + U_\lambda(f)$. Brent's method only requires computation of the values of the target function. To further simplify the computations, we follow Bouman and Sauer [2] to approximate the log-likelihood function (1.2) by the quadratic function for the emission case:

$$L(y|f) \approx -\frac{1}{2}(y - Pf)^T D(y - Pf) + c(y) \tag{2.28}$$

where P is the forward projection matrix, $D = \text{diag}\{y_i^{-1}\}$, and $c(y)$ is a function of data independent of the parameter set f and therefore can be ignored in later computations.

6.3 RESULTS AND DISCUSSION

6.3.1 Simulation Results

Our algorithm has been tested with the simulated emission data, which were posted on the web site `http://dynamo.ecn.purdue.edu/~bouman/software/tomography/`. This set of data was generated from a magnetic resonance imaging (MRI) reconstruction image. Figure 6.2(a) shows the original phantom of size 256 × 256 pixels. The cross-section was assumed to be of size 40.5cm square with an average emission rate of 0.2mm^{-1} and a maximum emission rate of 0.7mm^{-1}. The projection data was calculated at 128 evenly spaced angles each with 256 parallel projections. The photon noise were simulated by Poisson random variables with the appropriate means.

We have performed three sets of tests. First of all, a good prior should be capable of preserving significant anatomic features of the original image, which is referred as the morphological principle. Figure 6.2 shows the test results with noise-free projection data and $\lambda = 1$. We observe that the MAP reconstruction with Gaussian prior (2.16) destroys most of the significant edges after only 5 iterations, while the reconstruction with the proposed edge prior (2.20) keeps almost all sharp edges. For a large number of iterations up to 100, the Gaussian image almost becomes a single gray scale as shown in Figure 6.2(e), while the edge prior continues to preserve most of significant features of the original image, Figure 6.2(f).

The second test demonstrates the limitation of the static selection of hyperparameters λ and μ. Figures 6.3(c)-(f) show the MAP reconstructions with 6 iterations, in which we have fixed $\lambda = 0.001$, and $\mu = 1.0$, 0.8, 0.5 and 0.2, respectively. The resulting images are either too fuzzy as can be seen in Figures 6.3(c) and (d) because of a large amount of Gaussian smoothing; or appear to be too edgy, as is shown in Figures 6.3(e) and (f) because the edge prior "enhances" every possible edge including the noise. One of the causes of such phenomena is due to the fact that Brent's method used in our current implementation only searches for a "convenient" local minimum during the ICD updating.

Table 6.1. Dynamic setting of hyperparameters for $U_{\lambda,\mu}(f)$

No. of Iterations	1	2	3	4	5	6
λ	0.2	0.02	0.002	0.0002	0.0002	0.0002
μ	1.0	1.0	1.0	1.0	0.0	0.0

Finally, Figure 6.3(a) shows the filtered backprojection reconstruction without using any smoothing filter. The reconstruction contains strong noise by which significant details of anatomy are concealed. Figure 6.3(b) shows the MAP reconstruction after 6 iterations with the proposed prior and a dynamic setting of hyperparameters. The selection of hyperparameters λ and μ for this experiment is listed in Table 6.1. The strategy for the selection is to use the convexity of the Gaussian prior to obtain a fast descent of the reconstructed image to a neighborhood of a "global" minimum during the first few iterations, and then to take advantage of the edge prior to enhance edges of the image during the last 2 iterations. In comparison with published reconstructions from the same data set [24, 25, 5], the image Figure 6.3(b) produced by the proposed algorithm contains richer and sharper edges as well as fewer artifacts. Further quantitative evaluation is still under the way.

6.3.2 Discussion

Our purpose in introducing an explicit mechanical image model is to facilitate adaption of many efficient methods for image processing either using nonlinear PDEs or total variation in the context of Bayesian tomographic reconstruction. Two examples presented in this paper have shown that both old and new image priors could be derived based on our mechanical image model. Other existing image prior models may be derived in a similar fashion (for example, the thin-plate spline model proposed in [3], see (1.7)). New image priors with different characters can also be derived based on this image model. For instance, in order to further "enhance" edges during an image reconstruction, one can consider the "relative mass ratio" μ in (2.21) to be a function of $|\nabla f|$, which separates the behaviors of pixtrons closed to an edge from those in a smooth region of the image. Therefore, rather than using (2.21), we can investigate an image prior analogue to the anisotropic diffusion proposed for image processing by Alvarez, Lions and Morel [9]:

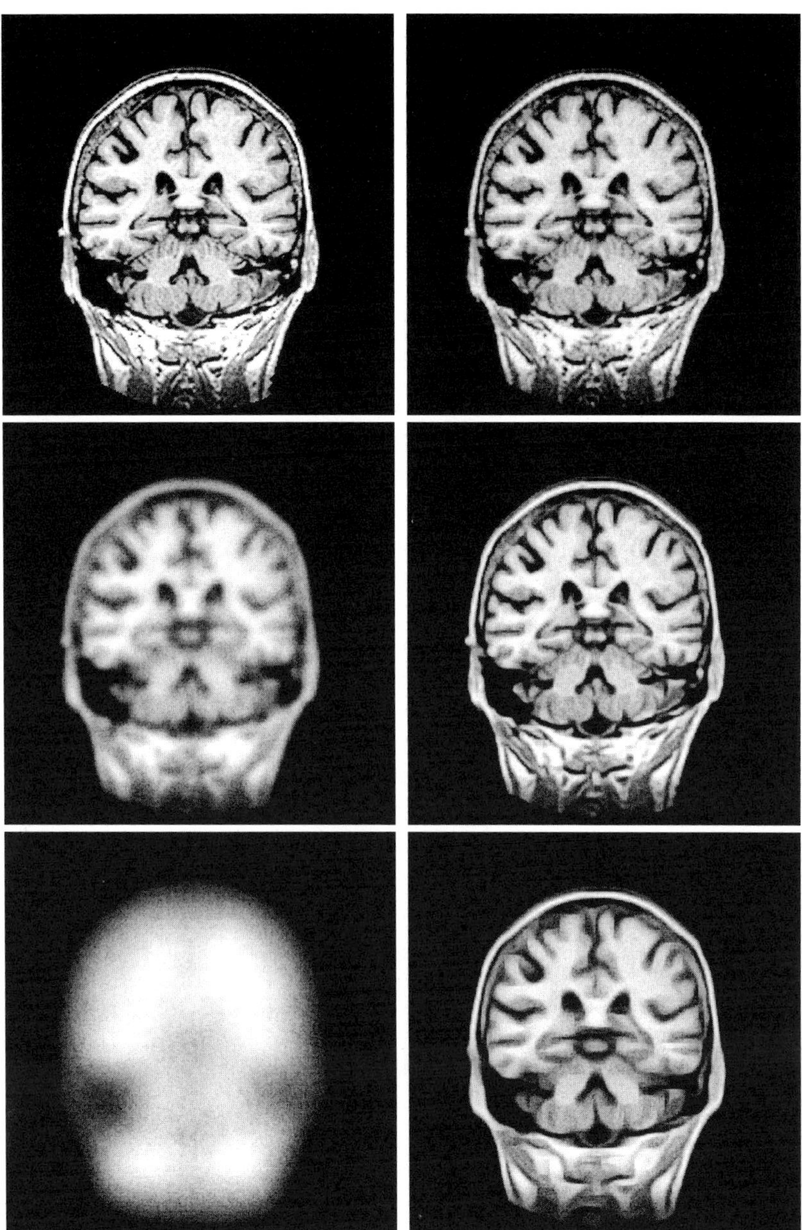

Figure 6.2. Tests with noise-free data (subfigures are labelled from top-left to bottom-right). (a) Original image. (b) Reconstruction with filtered-backprojection. (c)-(f) MAP reconstructions with (c) Gaussian prior, 5 iterations; (d) edge prior, 5 iterations; (e) Gaussian prior, 100 iterations; and (f) edge prior, 100 iterations

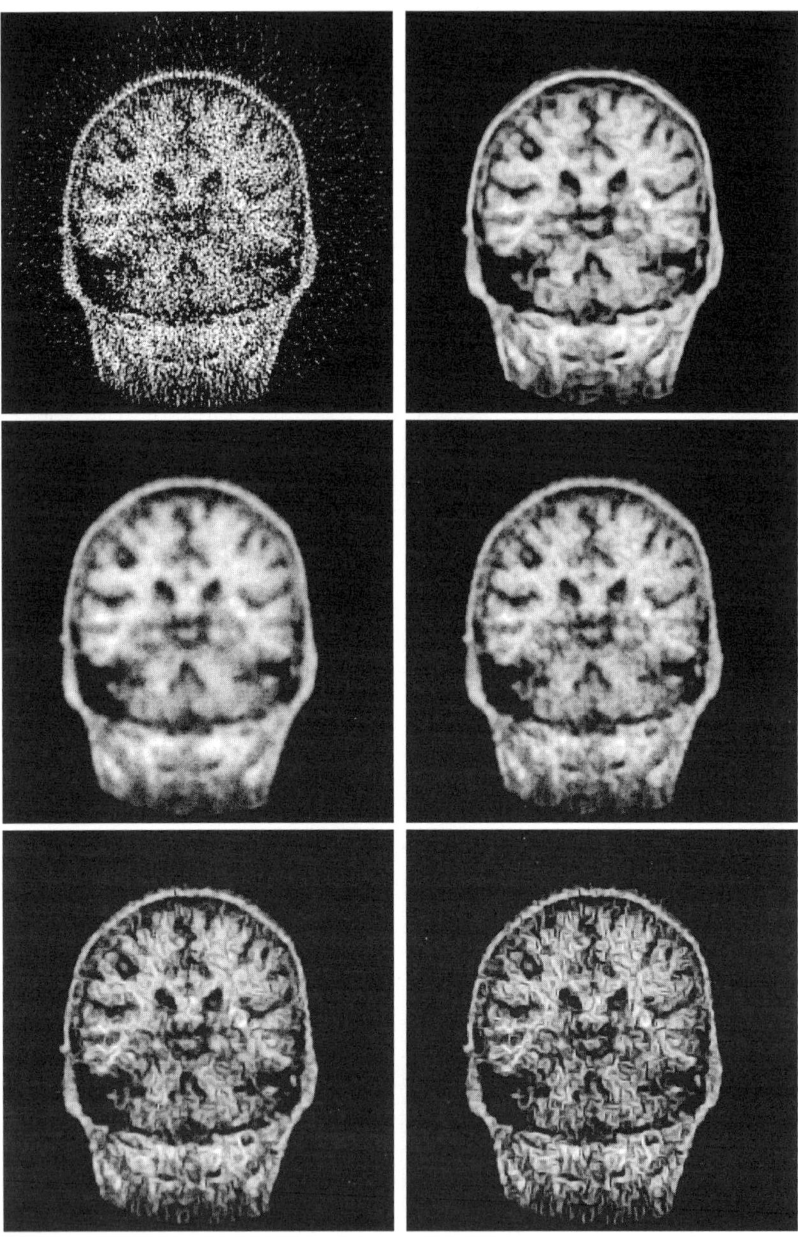

Figure 6.3. Tests with noisy data (subfigures are labelled from top-left to bottom-right). (a) Reconstruction with filtered-backprojection. (b) MAP reconstruction with dynamic hyper-paramter setting listed in Table 6.1. (c)-(f) MAP reconstructions with $\lambda = 0.001$, and (c) $\mu = 1.0$; (d) $\mu = 0.8$; (e) $\mu = 0.5$; and (f) $\mu = 0.2$

$$U_\lambda(f) = \lambda \int_\Omega \left(g(|G * \nabla f|)|\nabla f|^2 + [1 - g(|G * \nabla f|)] \left[\nabla_\xi^2 f\right]^2 \right) dx\,, \tag{3.1}$$

where the function $g(s) \geq 0$ is a nonincreasing function satisfying $g(0) = 1$, and G is a convolution kernel (for example, a Gaussian function). The relative mass distribution $\mu = g(|G * \nabla f|)$ in (3.1) controls the kinetic energy of each pixtron: if $|\nabla f|$ has a small mean in a neighborhood of a pixel x, this pixel x is considered an interior point of a smooth region of the image and the pixtron is therefore more actively moving towards the average position of its neighbors; if $|\nabla f|$ has a small mean in the neighborhood, x is considered as an edge pixel and then the kinetic energy of the pixtron will be so low that it is more likely to be trapped by the potential field $L(y|f)$, since $g(s)$ is small for large s.

More importantly, we believe that this mechanical image model may motivate a more systematic approach for Bayesian tomographic reconstruction, not only for developing new families of image priors to suit variety of applications, but also for hyperparameter estimations, since the "physical" meanings of these priors are very clear in our image model.

6.3.3 Conclusion

In conclusion, we have proposed an explicit mechanical image model for Bayesian tomographic reconstruction. A new image prior based on the mean curvature motion has been derived from this image model and tested with simulated tomographic data. The performance of the new image prior meets the requirements of our design. Improving image priors and more quantified tests are the focus of our further work.

ACKNOWLEDGEMENTS

This work was supported in part by the UM Research Board (#S-3-40641), University of Missouri, USA.

REFERENCES

[1] R. Leahy and C. Byrne, "Recent developments in iterative image reconstruction for PET and SPECT," *IEEE Trans. Med. Imag.*, vol. 19, no. 4, pp. 257–260, 2000.

[2] C. Bouman and K. Sauer, "A generalized gaussian image model for edge-preserving map estimation," *IEEE Trans. Med. Imag.*, vol. 2, no. 3, pp. 296–310, 1993.

[3] S. J. Lee, A. Rangarajan, and G. Gindi, "Bayesian image reconstruction in SPECT using higher order mechanical models as priors," *IEEE Trans. Med. Imag.*, vol. 4, no. 4, pp. 669–680, 1995.

[4] E. Jonsson, S. Huang, and T. Chan, "Total variation regularization in positron emission tomography," Reports 98-48, U.C.L.A. Computational and Applied Mathematics, November 1998.

[5] T. Frese, C. A. Bouman, and K. Sauer, "Adaptive wavelet graph model for Bayesian tomographic reconstruction," *preprint*, 2001.

[6] D. F. Yu and J. A. Fessler, "Edge-preserving tomographic reconstruction with nonlocal regularization," *preprint*, 2001.

[7] S. Osher and J. A. Sethian, "Fronts propagating with curvature-dependent speed: algorithms based on Hamilton-Jacobi formulation," *Journal of Computation Physics*, vol. 79, pp. 21 – 49, 1988.

paper. This extension, as well as extension to vector-valued (i.e. multi-) wavelets, can be considered as special cases of a more general consideration of tight spline-wavelet frames with arbitrary nested knot sequences that allow multiple (i.e. stacked) knots. In particular, when m knots are stacked at $x = a$ for m^{th} order spline functions, we have splines and spline-wavelet tight frames on a half interval $[a, \infty)$, and if, in addition, another m knots are stacked at $x = b > a$, the theory applies to a bounded interval $[a, b]$. This study can be considered as the spline approach to construction of non-stationary wavelets on bounded intervals. It is a summary of our recent joint work with W. He.

Most of the results presented in this paper are valid for bi-frames, and particularly, sibling frames (i.e. bi-frames associated with the same refinable function), when tightness is sacrificed to achieve certain additional desirable properties. We will only consider the properties of symmetry, shift-invariance, and inter-orthogonality.

7.1 INTRODUCTION

The first key ingredient in the construction of the Daubechies scaling functions and wavelets is construction of the two-scale Laurent polynomial symbols

$$P_D(z) = \left(\frac{1+z}{2}\right)^m S(z)$$

with $\deg S \leq m-1$ and $S(1) = 1$, to meet the orthogonality design criterion

$$|P_D(z)|^2 + |P_D(-z)|^2 = 1, \quad |z| = 1. \tag{1.1}$$

Hence, by considering the corresponding symbol

$$Q_D(x) = -z\, \overline{P_D(-z)}, \tag{1.2}$$

we have the Daubechies wavelet ψ_D, with Fourier transform given by

$$\widehat{\psi}_D(\omega) := Q_D\left(e^{-i\omega/2}\right) \widehat{\phi}_D\left(\frac{\omega}{2}\right), \tag{1.3}$$

where ϕ_D is the Daubechies scaling function defined by

$$\widehat{\phi}_D(\omega) := \prod_{k=1}^{\infty} P_D\left(e^{-i\omega/2^k}\right),$$

which obviously satisfies

$$\widehat{\phi}_D(\omega) = P_D\left(e^{-i\omega/2}\right) \widehat{\phi}_D\left(\frac{\omega}{2}\right). \tag{1.4}$$

Here and throughout this paper, the Fourier transform is defined by

$$\widehat{f}(\omega) = \int_{-\infty}^{\infty} f(x)\, e^{-i\omega x}\, dx.$$

Other important ingredients in the Daubechies paper [22] include the proof of convergence of the above infinite product that defines $\widehat{\phi}_D$ and identification of the smoothness class in terms of $m \geq 2$ to which ϕ_D belongs.

A major portion of this current paper is concerned with cardinal B-splines N_m, defined by m-fold convolution of the characteristic function of the unit interval $[0,1]$, $m \geq 2$, with two-scale polynomial symbol

$$P(z) = \left(\frac{1+z}{2}\right)^m. \tag{1.5}$$

Hence, there is no need to consider convergence of infinite products or smoothness properties. On the other hand, the orthogonality design criterion (1.1) for P_D is now replaced by the inequality

$$\begin{aligned} |P(z)|^2 + |P(-z)|^2 &= \cos^{2m}\left(\frac{\omega}{4}\right) + \sin^{2m}\left(\frac{\omega}{4}\right) \\ &\leq \left(\cos^2\frac{\omega}{4} + \sin^2\frac{\omega}{4}\right)^m = 1, \quad z = e^{-i\omega/2}. \end{aligned} \tag{1.6}$$

As a consequence, for $m \geq 2$, there does not exist a Laurent polynomial Q such that the matrix

$$M_{P,Q}(z) := \begin{bmatrix} P(z) & Q(z) \\ P(-z) & Q(-z) \end{bmatrix} \tag{1.7}$$

is unitary for $|z| = 1$. Observe that the choice of Q_D in (1.2) corresponding to P_D in (1.1) in the construction of Daubechies wavelets is to achieve such "unitary matrix extension" criterion, namely:

$$M^*_{P_D Q_D}(z) M_{P_D Q_D}(z) = 1, \quad |z| = 1, \tag{1.8}$$

where the $(1,1)$ entry of the matrix product on the left-hand side of (1.8) is precisely the orthogonality design criterion (1.1). Here and throughout, the asterisk notation in (1.8) denotes complex conjugation of matrix transposition.

By using $M^*_{P_D Q_D}(z)$ as the right inverse of $M_{P_D Q_D}(z)$, an equivalent formulation of (1.8) is given by

$$\begin{cases} |P_D(z)|^2 + |Q_D(z)|^2 = 1; \\ P_D(z)\overline{P_D(-z)} + Q_D(z)\overline{Q_D(-z)} = 0, \quad |z| = 1. \end{cases} \tag{1.9}$$

So, with $P(z)$ in (1.5) in place of $P_D(z)$, to compensate for being short of satisfying the orthogonality design criterion (1.1), as shown in (1.6), it is still feasible to design two or more Laurent polynomials $Q^1, \cdots, Q^L$, $L \geq 2$, such that

$$\begin{cases} |P(z)|^2 + \sum_{\ell=1}^L |Q^\ell(z)|^2 = 1; \\ P(z)\overline{P(-z)} + \sum_{\ell=1}^L Q^\ell(z)\overline{Q^\ell(-z)} = 0, \quad |z| = 1. \end{cases} \tag{1.10}$$

This is called the unitary extension principle (UEP) by Ron and Shen in [50]. Indeed, for Q^ℓ, $\ell = 1, \ldots, L$, that satisfy (1.10), the spline-wavelets ψ^ℓ, $\ell = 1, \ldots, L$, defined by

$$\widehat{\psi}^\ell(\omega) = Q^\ell(e^{-i\omega/2})\, \widehat{N}_m\left(\frac{\omega}{2}\right) \tag{1.11}$$

do generate a tight frame

$$\Psi := \{\psi^\ell_{j,k} : j,k \in \mathbf{Z}, \ell = 1, \ldots, L\} \tag{1.12}$$

of $L^2 := L^2(\mathbf{R})$, with frame bound constant 1; that is,

$$\sum_{\ell=1}^{L} \sum_{j,k\in\mathbf{Z}} |\langle f, \psi_{j,k}^{\ell}\rangle|^2 = ||f||^2, \; f \in L^2. \tag{1.13}$$

Here and throughout, the standard notation

$$g_{j,k}(x) := 2^{j/2}\, g(2^j x - k) \tag{1.14}$$

is used. Of course, the notion of tight frames as defined in (1.13) is a natural generalization of orthonormal wavelets, where the only additional requirement is that the wavelets must have unit norm. In other words, frame redundancy is achieved when the norms of the wavelets are allowed to be less than 1.

In Ron and Shen [52], $L = m$ Laurent polynomials $Q^1, \cdots, Q^m$ were constructed to satisfy the UEP (1.10), for any $m \geq 2$. This number was later reduced to $L = 2$ for all $m \geq 2$ in Chui and He [13]. Hence, instead of one generator ψ_D for the Daubechies orthonormal wavelets, we need two generators ψ^1 and ψ^2 of compactly supported tight frames of cardinal splines of order $m \geq 2$. Of course, additional redundancy can be achieved by applying the Second Oversampling Theorem of Chui and Shi [19]. Recall that for integer dilation $d \geq 2$, oversampling by $p \geq 1$ preserves tight frames, provided that p is relatively prime to d.

Unfortunately, independent of the number L of frame generators being used, and for all integer dilations $d \geq 2$, at least one of the cardinal spline tight frame generators ψ^ℓ has exactly one vanishing moment for all $m \geq 2$. This can be seen easily, for $d = 2$ say, from the UEP (1.10) itself, since

$$\sum_{\ell=1}^{L} |Q^\ell(z)|^2 = 1 - \left|\frac{1+z}{2}\right|^{2m} = |1-z|^2 |R(z)|^2, \quad |z| = 1, \tag{1.15}$$

where R is some Laurent polynomial with $R(1) \neq 0$. This is somewhat disappointing, since vanishing moments of higher order contribute to the great success of applications of wavelets to signal processing, particularly in areas that benefit from local extraction of multi-scale details.

In a recent joint work [14] with W. He, we introduced the notion of vanishing moment recovery (VMR) functions for the construction of compactly supported tight frames, in terms of the mth order B-spline N_m, that possess vanishing moments of order m. Again, it was shown that two frame generators ψ^1 and ψ^2 as in (1.11) suffice. In [14], it was shown that the VMR functions are necessarily quotients of two Laurent polynomials, but a Laurent polynomial $S(z)$, that satisfies the "positivity condition":

$$\frac{1}{S(z^2)} - \frac{|P(z)|^2}{S(z)} - \frac{|P(-z)|^2}{S(-z)} \geq 0, \quad |z| = 1, \tag{1.16}$$

already suffices. In other words, by choosing a Laurent polynomial $S(z)$ that satisfies (1.16), the "modified UEP"

$$\begin{cases} S(z^2)|P(z)|^2 + \sum_{\ell=1}^{L} |Q^\ell(z)|^2 = S(z); \\ S(z^2)P(z)\overline{P(-z)} + \sum_{\ell=1}^{L} Q^\ell(z)\overline{Q^\ell(-z)} = 0, \quad |z| = 1, \end{cases} \tag{1.17}$$

has Laurent polynomial solutions $Q^1, \cdots, Q^L$, even for $L = 2$, and the compactly supported spline-wavelet tight frame generators ψ^1 and ψ^2, as defined in (1.11), do have vanishing moments of order m, provided that $S(z)$ satisfies the additional condition

$$S(z) = \frac{1}{E_m(z)} + O(|1-z|^2 m), \quad \text{near } z = 1, \tag{1.18}$$

where

$$E_m(z) := \sum_{k=-m+1}^{m-1} N_{2m}(m+k) z^k \tag{1.19}$$

denotes the Euler-Frobenius polynomial associated with N_m. Details and related results will be discussed in this paper.

Consideration of the modified UEP (1.17) in our work [14] was inspired by an earlier work of Ron and Shen [50], in which it is shown the "fundamental function of multiresolution"

$$\begin{aligned} \Theta(\omega) &:= \frac{1}{|\hat{\phi}(\omega)|^2} \sum_{j=1}^{\infty} \sum_{\ell=1}^{L} \overline{\hat{\psi}^\ell(2^j\omega)\widehat{\widetilde{\psi}^\ell}(2^j\omega)} \\ &= \sum_{j=1}^{\infty} \sum_{\ell=1}^{L} Q^\ell(z^{2^j})\overline{\widetilde{Q}^\ell(z^{2^j})} \prod_{k=1}^{j-1} |P(z^{2^k})|^2, \end{aligned} \tag{1.20}$$

with $\phi = N_m$ in our consideration, satisfies (1.17) with $S(e^{-i\omega}) = \Theta(\omega)$. In a parallel independent work [25], Daubechies, Han, Ron, and Shen also addressed the same issue of achieving additional vanishing moments. We will also address certain results in [25] in this paper.

Furthermore, we will also address the extension from dilation factor 2 to arbitrary integer dilations. The results on this topic to be discussed are extracted from our joint work [17] with W. He and Q. Sun. Of course, this consideration is a special case of the general spline setting with arbitrary nested knot sequences, namely: by inserting $M-1$ equally-spaced knots between consecutive knots, where M is the dilation factor. By considering equally-spaced stacked knots of the same multiplicity $r \geq 2$, multi-wavelet tight frames of spline functions of "multiplicity r" also result from a study of spline tight frames of mth order splines on arbitrary nested knot sequences. When m knots are stacked at the two end-points of an interval, it also leads to tight frames of mth order splines on a bounded interval. This topic will constitute another major topic of this survey article, for which we will report on our joint work [16] with W. He.

The following survey is divided into four major sections. In Section 2, the general topic of wavelet frames of splines will be studied. Here, the notions of VMR functions and sibling frames are introduced and elaborated. Spline wavelet frames with multiple knots which are equally spaced on the real line are discussed in Section 3. Such wavelet frames are also called multi-wavelets in the literature. To extend the study to splines with nested sequences of knots, the Fourier approach no longer applies. The notion of approximate duals is therefore introduced in Section 4 to facilitate the transition from Fourier-domain to time-domain considerations. The results in this section are applied in Section 5 for constructing tight frames of spline functions with non-uniform knots.

7.2 CHARACTERIZATION OF WAVELET SPLINE FRAMES

The space of all cardinal splines of order $m \in \mathbf{N}$ is defined by

$$V_0 = \text{clos span}\{N_m(. - k);\ k \in \mathbf{Z}\} \tag{2.1}$$

where the closure is taken in $L^2(\mathbf{R})$ and N_m is the cardinal B-spline of order m (degree $m-1$) with knots $0, 1, \ldots, m$. Its Fourier transform is given by

$$\widehat{N}_m(\omega) = \left(\frac{1 - e^{-i\omega}}{i\omega} \right)^m . \tag{2.2}$$

The integer shifts of N_m are stable in the sense that

$$D_m \|\{c_k\}_{k\in\mathbf{Z}}\|_{\ell_2} \leq \left\| \sum_{k\in\mathbf{Z}} c_k N_m(. - k) \right\|_{L^2} \leq \|\{c_k\}_{k\in\mathbf{Z}}\|_{\ell_2} \tag{2.3}$$

holds for all ℓ_2 sequences $\{c_k\}$, where $D_m > 0$ is a constant.

For integer dilation factor $M \geq 2$, the relation $\widehat{N}_m(M\omega) = P_{m,M}(e^{-i\omega})\widehat{N}_m(\omega)$ holds with the Laurent polynomial symbol

$$P_{m,M}(z) = \left(\frac{1 - z^M}{M(1-z)} \right)^m = \left(\frac{1 + z + \cdots + z^{M-1}}{M} \right)^m . \tag{2.4}$$

Note that (1.5) refers to the special case $M = 2$. The zero properties of $P_{m,M}$ at the Mth roots of unity are obvious from the formula

$$P_{m,M}(z) = \frac{1}{M} \prod_{k=1}^{M-1} (1 - w_M^k z)^m, \qquad w_M := e^{i2\pi/M}. \tag{2.5}$$

It is a well-known fact that the scaled spaces

$$S_h = \{f(./h);\ f \in V_0\} \tag{2.6}$$

provide L^2-approximation order m; i.e. the error estimate

$$\|f - \mathcal{P}_h f\|_{L^2} \leq C h^m |f|_m \tag{2.7}$$

holds for all functions f in the Sobolev space $H^m(\mathbf{R})$, with a constant C that does not depend on f. Here, we denote by $|f|_m$ the Sobolev semi-norm and by $\mathcal{P}_h$ the orthogonal projection (in $L^2(\mathbf{R})$) onto S_h.

The *spline multiresolution analysis* is defined by the sequence of nested spaces

$$\cdots \subset V_{-1} \subset V_0 \subset V_1 \subset \cdots \tag{2.8}$$

where we make use of the notation $V_j := S_{h_j}$, $h_j := M^{-j}$. The results in [23, 57] show that the zero properties of $P_{m,M}$ in (2.5) can be derived from the approximation order result (2.7).

In this section, we give a complete characterization of certain families of wavelet frames that are derived from the spline multiresolution analysis $\{V_j\}$. Results and examples for $M = 2$ are contained in [14, 25], and results for general M are given in [17].

In order to agree on the notations to be used throughout the paper, we let $L \in \mathbf{N}$ and

$$\{\psi^1, \ldots, \psi^L\} \subset V_1, \qquad \{\tilde{\psi}^1, \ldots, \tilde{\psi}^L\} \subset V_1 \tag{2.9}$$

be real-valued compactly supported splines with knots in $\frac{1}{M}\mathbf{Z}$. Since V_1 is a shift-invariant space, with respect to shifts in $\frac{1}{M}\mathbf{Z}$, there exist unique Laurent polynomials Q^ℓ, $\widetilde{Q}^\ell$, $1 \leq \ell \leq L$, such that

$$\widehat{\psi^\ell}(M\omega) = Q^\ell(e^{-i\omega})\hat{N}_m(\omega), \qquad \widehat{\widetilde{\psi}^\ell}(M\omega) = \widetilde{Q}^\ell(e^{-i\omega})\hat{N}_m(\omega). \tag{2.10}$$

Here, the coefficients of the Laurent polynomials Q^ℓ and $\widetilde{Q}^\ell$ are real. We say that the functions ψ^ℓ have *vanishing moments of order* $\mu \in \mathbf{N}$, if

$$\int_{-\infty}^{\infty} x^\nu \psi^\ell(x)\,dx = 0, \qquad 0 \le \nu \le \mu - 1. \tag{2.11}$$

Since $\widehat{N}_m(0) = 1$, this property is equivalent to the condition that the Laurent polynomials Q^ℓ have zeros of order at least μ at $z = 1$. In other words, there exist Laurent polynomials q^ℓ with real coefficients, such that

$$Q^\ell(z) = (1-z)^\mu q^\ell(z), \qquad \ell = 1, \ldots, L. \tag{2.12}$$

The same arguments are valid for $\widetilde{\psi}^\ell$, of course.

The wavelet frames of $L^2(\mathbf{R})$ that we are going to analyze in this section, are the two families of shifts and dilates,

$$\Psi := \{\psi^\ell_{j,k};\ 1 \le \ell \le L,\ j,k \in \mathbf{Z}\}, \tag{2.13}$$

$$\widetilde{\Psi} := \{\tilde{\psi}^\ell_{j,k};\ 1 \le \ell \le N,\ j,k \in \mathbf{Z}\}. \tag{2.14}$$

Here, we make use of the notation

$$f_{j,k}(x) = M^{j/2} f(M^j x - k), \qquad j,k \in \mathbf{Z}, \tag{2.15}$$

where the dilation factor $M \in \mathbf{N}$, $M \ge 2$, is defined in the context.

Definition 7.2.1(a) The family Ψ is a **Bessel family**, if there exists a constant $B > 0$, such that

$$\sum_{\ell=1}^{L} \sum_{j,k\in\mathbf{Z}} \left|\langle f, \psi^\ell_{j,k}\rangle\right|^2 \le B\|f\|^2_{L^2} \tag{2.16}$$

for all $f \in L^2(\mathbf{R})$.

(b) The family Ψ is a **wavelet frame**, if there exist constants $B \ge A > 0$, such that

$$A\|f\|^2_{L^2} \le \sum_{\ell=1}^{L} \sum_{j,k\in\mathbf{Z}} \left|\langle f, \psi^\ell_{j,k}\rangle\right|^2 \le B\|f\|^2_{L^2} \tag{2.17}$$

for all $f \in L^2(\mathbf{R})$. If both frame constants can be chosen to be equal, that is $A = B$, Ψ is called a **tight frame**, and the tight frame is said to be **normalized** if $A = B = 1$.

(c) The families Ψ and $\widetilde{\Psi}$ are called **sibling frames**, if they are Bessel families and if the duality relation

$$\langle f, g\rangle = \sum_{\ell=1}^{L} \sum_{j,k\in\mathbf{Z}} \langle f, \psi^\ell_{j,k}\rangle\ \langle \tilde{\psi}^\ell_{j,k}, g\rangle \tag{2.18}$$

is satisfied for all $f, g \in L^2(\mathbf{R})$.

The results in [19] imply that both families Ψ and $\widetilde{\Psi}$ are Bessel families, if we only assume that every ψ^ℓ and $\widetilde{\psi}^\ell$ has at least one vanishing moment. We also note that both

$$f \mapsto Qf(x) = \sum_{k \in \mathbf{Z}} \lambda_k(f) N_m(x-k), \qquad \lambda_k(f) := \langle f, N_m^S(\,.\,-k)\rangle,$$

is bounded on $L^2(\mathbf{R})$, reproduces all polynomials of degree up to $m-1$, and the functionals λ_k have local support. We will emphasize in Section 4 that the notions of VMR functions and approximate duals are closely related, for each of the settings of arbitrary integer dilations of N_m, splines with multiple knots of equal multiplicity, and splines with non-uniform knots.

Remark 7.2.3 In [50], the definition of the *fundamental function* of the MRA-frame Ψ is given by

$$\Theta(\omega) := \frac{1}{|\hat{\phi}(\omega)|^2} \sum_{j=1}^{\infty} \sum_{\ell=1}^{L} |\hat{\psi}^\ell(2^j\omega)|^2 = \sum_{j=0}^{\infty} \sum_{\ell=1}^{L} |Q^\ell(e^{-i2^j\omega})|^2 \prod_{k=0}^{j-1} |P(e^{-i2^k\omega})|^2. \tag{2.29}$$

Identities (2.19) and (2.20) show that the VMR function S and the fundamental function Θ are related by

$$S(e^{-i\omega}) = \Theta(\omega), \qquad \omega \in \mathbf{R}.$$

However, while Θ seems to depend on all scaling levels $j \geq 1$, through its definition (2.29), 7.2.3 shows that it can be found by a "mono-scale" construction based on the autocorrelation symbol Φ of the (cardinal) B-spline N_m.

How can a VMR Laurent polynomial S be constructed, so that S is nonnegative on $\mathbb{T}$, satisfies (2.25), and admits the definition of Laurent polynomials $Q^1, \ldots, Q^L$ that satisfy both identities (2.19) and (2.20)? An explicit solution is presented in [25] for general order m, and in [14] for splines of low order. Before we describe the solution in [25], we put the identities (2.19) and (2.20) into the equivalent form

$$\mathcal{M}(z) = \mathcal{Q}(1/z)\mathcal{Q}(z)^T, \tag{2.30}$$

where

$$\mathcal{M}(z) = \begin{bmatrix} S(z) - S(z^2)|P(z)|^2 & -S(z^2)P(1/z)P(-z) \\ -S(z^2)P(z)P(-1/z) & S(-z) - S(z^2)|P(-z)|^2 \end{bmatrix} \tag{2.31}$$

and

$$\mathcal{Q}(z) = \begin{bmatrix} Q^1(z) & \cdots & Q^L(z) \\ Q^1(-z) & \cdots & Q^L(-z) \end{bmatrix}. \tag{2.32}$$

From this form, we find that a necessary condition for the existence of $Q^1, \ldots, Q^L$ in (2.19)–(2.20) is the property that the matrix $\mathcal{M}(z)$ must be positive semi-definite for all $z \in \mathbb{T}$. New factorization methods for Laurent polynomial matrices in [14] and [35] show that the positive semi-definiteness of $\mathcal{M}(z)$ is also sufficient for the existence of two Laurent polynomials Q^1, Q^2 that define a square matrix $\mathcal{Q}$ and a factorization (2.30). Another simple observation aids in controlling the definiteness of $\mathcal{M}(z)$: if the determinant of the matrix is nonnegative, i.e.

$$\det \mathcal{M}(z) = S(z)S(-z) - S(z^2)(S(-z)|P(z)|^2 + S(z)|P(-z)|^2) \geq 0 \tag{2.33}$$

for all $z \in \mathbb{T}$, then the first diagonal entry is also nonnegative. In summary, the inequality (2.33) is equivalent to the positive semi-definiteness of $\mathcal{M}(z)$, which again is equivalent to the existence of two Laurent polynomials Q^1 and Q^2 such that (2.19) and (2.20) are satisfied (with $L = 2$). An equivalent form of (2.33) is given, due to the positivity of S, by (1.16).

Our next result highlights the construction of a suitable VMR function S for the cardinal B-spline N_m, such that two frame generators with a prescribed order of vanishing moments can be constructed.

Theorem 7.2.4 *Let $m \in \mathbf{N}$ and N_m be the cardinal B-spline of order m with two-scale symbol P in (2.4) and autocorrelation symbol Φ in (2.23). For every $1 \leq \mu \leq m$, there exists a unique Laurent polynomial*

$$S(z) = s_0 + \sum_{k=1}^{\mu-1} s_k(z^k + z^{-k}), \tag{2.34}$$

with real coefficients s_k, which is nonnegative on $\mathbb{T}$ and satisfies $S(1) = 1$, the approximation property (2.25), and the determinant condition (2.33).

The explicit solution to (2.25) of degree $\mu - 1$ is given in [14] by means of the expansion

$$S(z) = \sum_{k=0}^{\mu-1} u_k \left(\frac{2 - z - z^{-1}}{4} \right)^k. \tag{2.35}$$

Here, u_k are the coefficients of the series

$$\frac{1}{\Phi(e^{-i\omega})} = \sum_{k=0}^{\infty} u_k \sin(\omega/2)^{2k}. \tag{2.36}$$

For $0 \leq k \leq m - 1$, u_k can be computed recursively by

$$u_0 = 1, \qquad u_k = \frac{1}{4^k - 1} \sum_{\ell=0}^{k-1} (-1)^{k-1-\ell} 4^\ell u_\ell \binom{m+\ell}{k-\ell}.$$

A similar approach in [25] defines the coefficients $\widetilde{u}_k$ in

$$\frac{1}{|\widehat{N}_m(\omega)|^2} = \left(\frac{\arcsin(\sin(\omega/2))}{\sin(\omega/2)} \right)^{2m} = \sum_{k=0}^{\infty} \widetilde{u}_k \sin(\omega/2)^{2k}, \qquad \omega \to 0.$$

The Strang-Fix conditions imply that $u_k = \widetilde{u}_k$ for all $0 \leq k \leq m - 1$. Indeed, the Poisson Summation Formula and (2.2) give

$$\Phi(e^{-i\omega}) = \sum_{k \in \mathbf{Z}} |\widehat{N}_m(\omega + 2\pi k)|^2 = |\widehat{N}_m(\omega)|^2 + \mathcal{O}(|\omega|^{2m}), \qquad \omega \to 0.$$

Therefore, the coefficients u_k in (2.36), with $0 \leq k \leq m-1$, can be computed by $2m$-fold convolution of the coefficient sequence of the series on the right-hand side of

$$\frac{\arcsin(\sin(\omega/2))}{\sin(\omega/2)} = 1 + \sum_{k=1}^{\infty} \frac{(2k-1)!!}{(2k)!!(2k+1)} \sin(\omega/2)^{2k}.$$

Here, the notation $n!!$ denotes the product $n(n-2) \cdots 1$ for odd n and $n(n-2) \cdots 2$ for even n. This method for the construction of S shows that all coefficients u_k, $0 \leq k \leq \mu - 1$, in

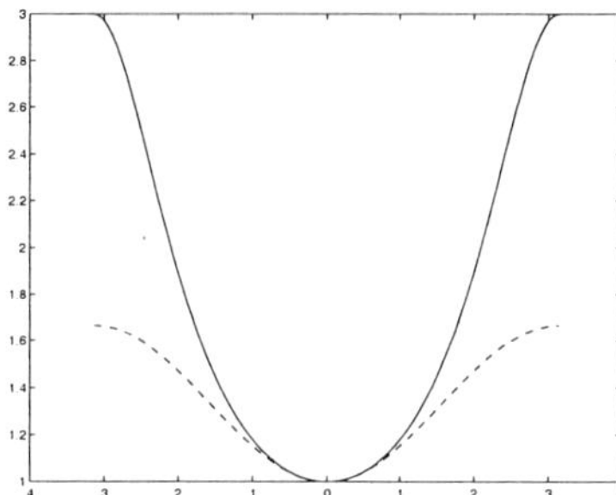

Figure 7.1. Approximation of $1/\Phi$ (solid line) by the VMR function S (dashed line) for $m = \mu = 2$. Approximation at $\omega = 0$ is of fourth order

(2.36) are strictly positive and, consequently, $S(z) > 0$ for all $z \in \mathbb{T}$. Moreover, (2.25) is a direct consequence of this construction, while (2.33) is proven in Proposition 3.5 of [25] by a somewhat sophisticated argument. We show, in Section 5, that our method for the construction of spline frames with non-uniform knots provides a new and comprehensive method for the definition of approximate duals that satisfy all properties in 7.2.4. Our new method provides approximate duals also for MRA constructions with integer dilation factor $M > 2$, splines with multiple knots of equal multiplicity, and non-uniform B-splines.

Table 7.1 gives the coefficients u_k in (2.36) of S for low order B-splines. Figure 1 shows the approximation of $1/\Phi(e^{-i\omega})$ by the trigonometric polynomial $S(e^{-i\omega})$ for $m = \mu = 2$, which is of fourth order at $\omega = 0$.

Table 7.1. Coefficients u_k of the VMR function S for the B-spline N_m

m	s_0	s_1	s_2	s_3	s_4
1	1				
2	1	2/3			
3	1	1	13/15		
4	1	4/3	62/45	1244/945	
5	1	5/3	2	134/63	2021/945

Remark 7.2.4 The result in Proposition 3.5 of [25] differs from the necessary and sufficient condition (2.33). In [25], the authors show that

$$A(z) := S(z) - S(z^2)(|P(z)|^2 + |P(-z)|^2) \geq 0 \tag{2.37}$$

for all $z \in \mathbb{T}$. The connection to $\det \mathcal{M}(z)$ can be established as follows. If $S(-z) \geq S(z) \geq 0$, we have

$$\det \mathcal{M}(z) \geq S(-z)A(z) \geq 0,$$

and if $S(z) \geq S(-z) \geq 0$, we have

$$\det \mathcal{M}(z) \geq S(z)A(-z) \geq 0.$$

The result (2.37) is strictly stronger than the necessary and sufficient condition (2.33), unless $S(z) = S(-z)$ for all $z \in \mathbb{T}$, which only holds for $\mu = 1$.

Finally, we can conclude that there always exists a pair of generators $\psi^1, \psi^2 \in V_1$ with μ vanishing moments, $1 \leq \mu \leq m$, such that the family Ψ is a tight frame of $L^2(\mathbf{R})$.

Theorem 7.2.5 *Let $1 \leq \mu \leq m$, and S be a VMR Laurent polynomial, which is non-negative on $\mathbb{T}$ and satisfies (2.25) and (2.33). Then there exist two Laurent polynomials q^1, q^2, such that $Q^1(z) = (1-z)^\mu q^1(z)$ and $Q^2(z) = (1-z)^\mu q^2(z)$ define functions $\psi^1, \psi^2 \in V_1$ with μ vanishing moments, that constitute a normalized tight frame Ψ of $L^2(\mathbf{R})$.*

Two elementary steps are required prior to constructing the factorization of the matrix $\mathcal{M}$ in (2.30). First, the factors $(1-z)^\mu$ are cancelled, and then a conversion to polyphase form is applied. These operations do not affect the semi-definiteness of the matrix. The following example illustrates this procedure.

Example 7.2.1 Let $m = \mu = 2$ and $P(z) = ((1+z)/2)^2$. The VMR Laurent polynomial S from Table 7.1 and Figure 1 is $S(z) = (8 - z - z^{-1})/6$. The matrix $\mathcal{M}$ is given by

$$\mathcal{M}(z) = \begin{bmatrix} (1-z^{-1})^2 & 0 \\ 0 & (1+z^{-1})^2 \end{bmatrix} \mathcal{M}_0(z) \begin{bmatrix} (1-z)^2 & 0 \\ 0 & (1+z)^2 \end{bmatrix},$$

where

$$\mathcal{M}_0(z) = \frac{1}{96} \begin{bmatrix} 24 + 8(z+z^{-1}) + z^2 + z^{-2} & -(8 - z^2 - z^{-2}) \\ -(8 - z^2 - z^{-2}) & 24 - 8(z+z^{-1}) + z^2 + z^{-2} \end{bmatrix}.$$

In accordance with 7.2.5, we find that

$$\det \mathcal{M}(e^{-i\omega}) = \frac{2}{3}\sin^4 \omega \geq 0.$$

Hence, the condition (2.33) is satisfied. Making use of the conversion to polyphase form gives

$$\frac{1}{4}\begin{bmatrix} 1 & 1 \\ z & -z \end{bmatrix} \mathcal{M}_0(z) \begin{bmatrix} 1 & z^{-1} \\ 1 & -z^{-1} \end{bmatrix} = \frac{1}{96}\begin{bmatrix} 8 + z^2 + z^{-2} & 4 + 4z^{-2} \\ 4 + 4z^2 & 16 \end{bmatrix}.$$

We denote the matrix on the right-hand side by $\mathcal{M}_1(z)$. The factorization from [14] gives

$$\mathcal{M}_1(z) = \begin{bmatrix} \frac{1}{4} & \frac{1}{4\sqrt{6}}(1+z^{-2}) \\ 0 & \frac{1}{\sqrt{6}} \end{bmatrix} \begin{bmatrix} \frac{1}{4} & 0 \\ \frac{1}{4\sqrt{6}}(1+z^2) & \frac{1}{\sqrt{6}} \end{bmatrix}.$$

$\mathbb{T}$. The positivity of A in (2.37), with θ substituted for S, is shown, and the three frame generators are defined by the Laurent polynomials

$$Q^1(z) = z\theta_1(z^2)P(-1/z), \qquad Q^2, Q^3(z) = \frac{1}{2}(\alpha(z^2) \pm \alpha(1/z^2))P(z).$$

Other triplets $\{\psi^1, \psi^2, \psi^3\}$ of symmetric/anti-symmetric generators of a normalized tight frame are constructed in [25]. These generators have fewer nonzero coefficients than the ones found by the aforementioned constructions. We present, in Section 5, a new method for the construction of tight frames of splines which we developed in our recent joint work [16] with W. He. This method yields a triplet of symmetric generators for $m = \mu = 4$ whose filter lengths are 7, 9, and 11, respectively. The graphs are shown in Section 5, Figure 10.

Finally, we wish to comment on the number L of generators that are needed for the construction of a tight frame Ψ. If $\mathcal{M}(z)$ has full rank at some $z \in \mathbb{T}$, then at least 2 Laurent polynomials Q^1, Q^2 are needed for the factorization (2.30). This means that at least two functions ψ^1, ψ^2 are needed in order to generate a tight frame Ψ of $L^2(\mathbf{R})$. It was shown in [14, Theorem 9] and [25, Theorem 3.8] that the only case where **one** compactly supported spline function $\psi \in V_1$ generates a tight frame of $L^2(\mathbf{R})$ is the case $m = 1$; examples of such frames are the orthonormal Haar basis ψ_H and dilates $\psi_H(./n)$ with odd n. These examples are known from the First Oversampling Theorem in [19]. For all other values of $m \geq 2$, however, no compactly supported spline function $\psi \in V_1$ exists, whose dilates and translates generate a tight frame of $L^2(\mathbf{R})$.

Remark 7.2.6 Most results discussed in this section remain valid for the general setting with the B-spline N_m replaced by a compactly supported refinable function (w.r.t. dilation by 2), which is piecewise Lip$^\alpha$ for some $\alpha > 0$, has nonvanishing integral over $\mathbf{R}$, and whose integer shifts $\{\phi(. - k);\ k \in \mathbf{Z}\}$ are a Riesz basis of the space V_0. In particular, there always exists a nonnegative VMR Laurent polynomial S that satisfies the positivity condition (2.33) and defines a quasi-interpolation operator as in (2.28), which reproduces all polynomials in the span of the integer shifts of ϕ. This result was shown in [14, Theorem 5] by a sophisticated analysis of the positivity condition (1.16), see also (2.33). The formulation (1.16) exhibits the relation of this condition and the "transfer operator"

$$T_{|P|^2} f(\omega) := |P(e^{-i\omega/2}|^2 f(\omega/2) + |P(-e^{-i\omega/2}|^2 f(\omega/2 + \pi),$$

which is an operator on $L^2((0, 2\pi))$ and maps a certain space of trigonometric polynomials (that depends on the degree of the two-scale Laurent polynomial) into itself. The spectrum of this operator was analyzed in [14] in order to show the existence of a VMR Laurent polynomial S which satisfies (1.16). Therefore, pairs $\psi^1, \psi^2 \in V_1$ of generators of a tight frame of $L^2(\mathbf{R})$ can always be constructed. If the integer shifts of ϕ are not stable, then the characterization of tight wavelet frames in 7.2.2 remains valid, if we allow S to be a quotient of two Laurent polynomials with real coefficients and real values for all $z \in \mathbb{T}$, and with no pole at $z = 1$. This case is further analyzed in [14].

7.2.2 Non-tight sibling frames with dilation factor 2

We begin this section with a characterization of compactly supported sibling frames of cardinal splines. Recall Definition 1(c) for the notation of sibling frames and duality. In analogy with Theorems 1 and 2, we obtain the following.

Theorem 7.2.6 *Let Q^ℓ, $\widetilde{Q}^\ell$, $1 \le \ell \le L$, be Laurent polynomials with real coefficients vanishing at $z = 1$. The functions ψ^ℓ, $\widetilde{\psi}^\ell$, $1 \le \ell \le L$, defined in (2.10) generate sibling frames of $L^2(\mathbf{R})$, with respect to dilation by 2 and integer shifts, if and only if there exists a Laurent polynomial S with real coefficients, which satisfies $S(1) = 1$ and*

$$S(z^2)|P(z)|^2 + \sum_{\ell=1}^{L} Q^\ell(z)\widetilde{Q}^\ell(1/z) = S(z); \tag{2.38}$$

$$S(z^2)P(z)P(-1/z) + \sum_{\ell=1}^{L} Q^\ell(z)\widetilde{Q}^\ell(-1/z) = 0. \tag{2.39}$$

Moreover, if all of the functions ψ^ℓ, $\widetilde{\psi}^\ell$, $1 \le \ell \le L$, have μ vanishing moments, for some $1 \le \mu \le m$, then S satisfies the approximation property (2.25). Conversely, if S satisfies (2.25), where $1 \le \mu \le m$, then there exist four compactly supported functions $\psi^\ell, \widetilde{\psi}^\ell \in V_1$, $\ell = 1, 2$, which have μ vanishing moments and generate sibling frames of $L^2(\mathbf{R})$.

For a proof of this result, we refer to Theorems 1 and 2 in [14], where more general MRA's are considered. The construction of sibling frames is performed in a similar manner that was already described for tight frames. The identities (2.38) and (2.39) are reformulated as the matrix equation

$$\mathcal{M}(z) = \mathcal{Q}(1/z)\widetilde{\mathcal{Q}}(z)^T, \tag{2.40}$$

where $\mathcal{M}(z)$ is the matrix in (2.31) and $\mathcal{Q}(z)$, $\widetilde{\mathcal{Q}}(z)$ are defined as in (2.32). Clearly, there is no constraint of positive definiteness of $\mathcal{M}(z)$, in order that the factorization (2.40) exists. The rank of $\mathcal{M}(z)$, $z \in \mathbb{T}$, is a lower bound for the number of columns of the matrices $\mathcal{Q}(z)$ and $\widetilde{\mathcal{Q}}(z)$. As mentioned in the previous section, except for $m = \mu = 1$, there exists no Laurent polynomial S which yields $\operatorname{rank}\mathcal{M}(z) = 1$ for all $z \in \mathbb{T}$. Therefore, the minimal number of frame generators in 7.2.6 is $L = 1$ for $m = \mu = 1$ and $L = 2$ for $m \ge 2$ and any $1 \le \mu \le m$.

The factorization (2.40) allows for much greater flexibility in finding the matrices $\mathcal{Q}(z)$ and $\widetilde{\mathcal{Q}}(z)$. A simple factorization, if S satisfies (2.25) and $\mu \le m$, can be obtained by mimicking the first two steps that appear in Example 1. This gives

$$\mathcal{M}(z) = \underbrace{\begin{bmatrix} (1-z^{-1})^\mu & z^{-1}(1-z^{-1})^\mu \\ (1+z^{-1})^\mu & -z^{-1}(1+z^{-1})^\mu \end{bmatrix}}_{=:\,\mathcal{Q}(1/z)} \underbrace{\mathcal{M}_1(z) \begin{bmatrix} (1-z)^\mu & (1+z)^\mu \\ z(1-z)^\mu & -z(1+z)^\mu \end{bmatrix}}_{=:\,\widetilde{\mathcal{Q}}(z)^T}, \tag{2.41}$$

where the matrix $\mathcal{M}_1(z)$ has Laurent polynomial entries of even powers of z. This type of factorization yields the Laurent polynomials

$$Q^1(z) = (1-z)^\mu, \qquad Q^2(z) = z(1-z)^\mu,$$

and, by simple calculations, we obtain, for $\ell = 1, 2$, that

$$\widetilde{Q}^\ell(z) = z^{\ell-1}\frac{(1-z)^\mu}{2}\left[\frac{S(z) - S(z^2)|P(z)|^2}{(2-z-1/z)^\mu} \pm (-1)^{\mu+1}2^{-2m}(z-1/z)^{m-\mu}S(z^2)\right].$$

Note that only the "trivial" factors (that are due to vanishing moments and transformation to polyphase form) are involved. Hence, no complicated polynomial factorization is required in order to define the Laurent polynomials Q^1, Q^2 and $\widetilde{Q}^1, \widetilde{Q}^2$ of the sibling frames. If S is the VMR Laurent polynomial in 7.2.4, then the following properties can be easily verified:

- ψ^ℓ has support $\frac{\ell-1}{2} + [0, (m+\mu)/2]$ and is even (odd) with respect to its center, if μ is even (odd). Since the B-splines in V_1 satisfy the variation diminishing property (see [4]), ψ_1 and ψ_2 are minimally supported splines in V_1 with μ vanishing moments. Moreover, ψ^2 is a shift by 1/2 of ψ^1.
- The support of $\widetilde{\psi}^\ell$ is contained in the interval $\frac{\ell-1}{2} + [-(2\mu+m)/2+1, (3\mu+m)/2-1]$. Both functions are even (odd) with respect to the center of this interval, if μ is even (odd) and $m-\mu$ is even. No factorization (2.40) has been found yet, where $m-\mu$ is odd and $\psi^{1,2}$ are the minimally supported splines in V_1 with μ vanishing moments.

Remark 7.2.7 It is worthwhile to mention that both generators ψ^1, ψ^2 have the same parity. This property is unavoidable, if we choose ψ^1, ψ^2 to be minimally supported splines in V_1 with the same order of vanishing moments. By distributing the order of vanishing moments unevenly (order μ for ψ^1, $\widetilde{\psi}^1$, and $\mu \pm 1$ for ψ^2, $\widetilde{\psi}^2$, for example) it is possible to create frames with two generators of different parity. Instead of having the shift-invariance property mentioned above, such frames may possibly give rise to better shift-invariance of the frame decomposition. This was analyzed experimentally by Kingsbury [40, 41] and Selesnick [54] for other types of frames. Preliminary investigations concerning spline frames are contained in [3].

Remark 7.2.8 One may ask if the upper bound m for the order of vanishing moments in 7.2.6 can be relaxed, at least for one family of the sibling frames. The answer to this question is negative, as we show next, even if the order of approximation of the VMR function S in (2.25) exceeds $2m$. More precisely, if the functions $\psi^\ell, \widetilde{\psi}^\ell \in V_1$ generate sibling frames of $L^2(\mathbf{R})$, there always exist ℓ, ℓ', $1 \le \ell, \ell' \le L$, such that ψ^ℓ and $\widetilde{\psi}^{\ell'}$ have at most m vanishing moments. This can be seen by analyzing the identity (2.39) in a neighborhood of $z = 1$ and -1. Indeed, equation (2.39) gives

$$\sum_i Q^i(1/z)\widetilde{Q}_i(-z) = -S(z^2)P(1/z)P(-z) = -\left(\frac{1-z}{2}\right)^m [1 + O(|z-1|)].$$

This shows that not all Q^ℓ can have zeros of order greater than m at $z = 1$. Likewise, the same technique applied at $z = -1$ shows that not all $\widetilde{Q}^\ell$ can have zeros of order greater than m at $z = 1$. This restriction can be avoided, if we turn to more general constructions of so-called bi-frames, see [25], which are based on two different sets of multiresolution analyses $\{V_j\}$ and $\{\widetilde{V}_j\}$.

Example 7.2.4 The "trivial" factorization of the matrix $\mathcal{M}(z)$ in (2.40) can be replaced by a more balanced factorization, where the lengths of the supports of ψ^ℓ and $\widetilde{\psi}^\ell$ are comparable. For $m = 3$, some calculations based on some type of Euclidean algorithm for Laurent polynomials yields the following two-scale symbols for the generators of sibling frames with 3 vanishing moments:

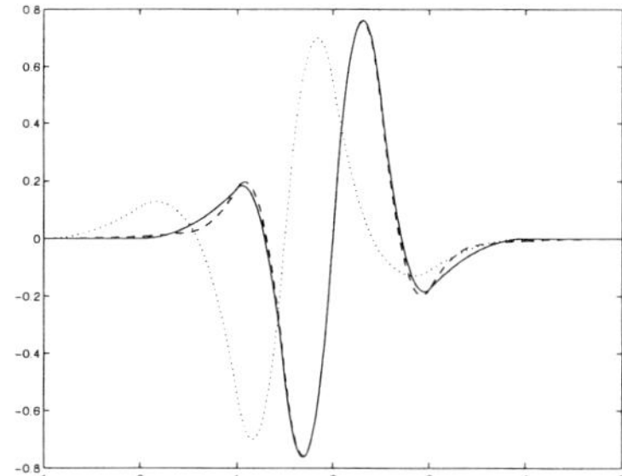

Figure 7.4. Generators $\psi^1 = \tilde{\psi}^1$ (dotted line) and ψ^2 (solid line), $\tilde{\psi}^2$ (dashed line) of quadratic spline sibling frame with three vanishing moments.

$$
\begin{aligned}
Q^1(z) &= \tfrac{1}{3\sqrt{1545}} \left(\tfrac{1-z}{2}\right)^3 \left(361 + 156(z+z^{-1}) + 26(z^2+z^{-2})\right),\\
\tilde{Q}^1(z) &= Q^1(z),\\
Q^2(z) &= \tfrac{1}{6}\sqrt{\tfrac{2719}{618}} \left(\tfrac{1-z}{2}\right)^3 (1+6z+z^2),\\
\tilde{Q}^2(z) &= \left(1 + \tfrac{247}{5438}(z^2+z^{-2})\right) Q^2(z).
\end{aligned}
$$

We denote such a sibling frame as a "twin" in [14], since not only are all four generators taken from the same spline space V_1, but the integer shifts of ψ^ℓ and $\tilde{\psi}^\ell$, $\ell = 1, 2$, span the same subspace of V_1. The graphs of these functions are shown in Figure 4. The supports are supp$\psi_1 = [-1, 4]$, supp$\psi_2 = [0, 4]$, and supp$\tilde{\psi}_2 = [-1, 5]$. Note that although $\tilde{\psi}_2$ is a linear combination of integer shifts of ψ_2, their graphs look almost identical. Furthermore, the approximate shift-invariance $\psi_2 \approx \psi_1(. - 1/2)$ can be observed.

Example 7.2.5 For cubic spline sibling frames with 4 vanishing moments, we obtain, by using a similar method,

$$
\begin{aligned}
Q^1(z) &= \tfrac{1}{4}\left(\tfrac{1-z}{2}\right)^4 (22 + 8(z+z^{-1}) + z^2 + z^{-2}),\\
Q^2(z) &= \tfrac{1}{2}\left(\tfrac{1-z}{2}\right)^4 (1 + 8z + z^2),\\
\tilde{Q}^1(z) &= \tfrac{1}{18900}\left(\tfrac{1-z}{2}\right)^4 \Big(132666 + 94712(z+z^{-1}) +\\
&\qquad 44494(z^2+z^{-2}) + 12440(z^3+z^{-3}) + 1555(z^4+z^{-4})\Big),\\
\tilde{Q}^2(z) &= \tfrac{1}{9450}\left(\tfrac{1-z}{2}\right)^4 \Big(61024z + 33045(1+z^2) +\\
&\qquad 9952(z^{-1}+z^3) + 1244(z^{-2}+z^4)\Big).
\end{aligned}
$$

The functions $\psi^\ell, \tilde{\psi}^\ell$, $\ell = 1, 2$, are shown in Figure 5.

Instead of symmetry or anti-symmetry, some other advantageous property can be realized by imposing certain side conditions on the Laurent polynomials Q^ℓ and $\tilde{Q}^\ell$. If the correlation of the coefficients of the frame decomposition on a fixed scaling level must be reduced, the following type of orthogonality relations may be useful.

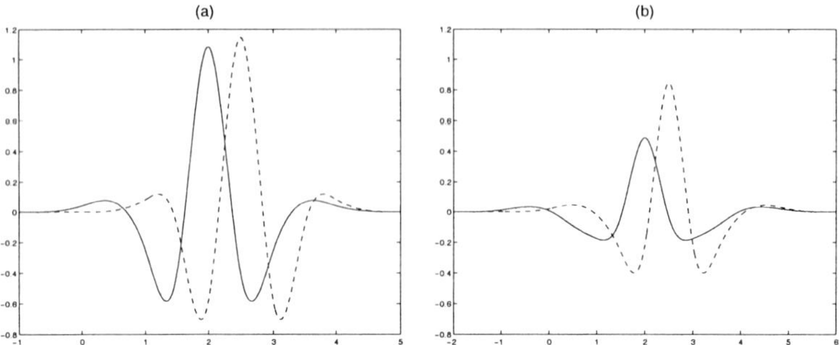

Figure 7.5. Generators ψ^1, ψ^2 (left) and $\widetilde{\psi}^1, \widetilde{\psi}^2$ (right) of cubic spline sibling frame with four vanishing moments.

Definition 7.2.7 The family $\Psi = \{\psi_1, \dots, \psi_L\} \subset L^2(\mathbf{R})$ is *inter-orthogonal*, if the spaces

$$W^\ell = \operatorname{clos\,span}\{\psi^\ell(. - k);\ k \in \mathbf{Z}\}, \quad \ell = 1, \dots, L,$$

are mutually orthogonal.

By making use of the auto-correlation symbol Φ in (2.23) of the B-spline N_m, we find that inter-orthogonality for $\psi^\ell \in V_1$ is characterized by the identity

$$Q^\ell(z)Q^j(1/z)\Phi(z) + Q^\ell(-z)Q^j(-z)\Phi(-z) = 0, \quad z \in \mathbb{T}, \ j \neq \ell. \tag{2.42}$$

A standard argument from linear algebra is employed in [14], in order to show that, for scaling by 2, at most 2 functions ψ^ℓ can be inter-orthogonal. On the other hand, the existence of inter-orthogonal functions $\psi^{1,2}$, which constitute the generators of the primal side of sibling frames, requires some nontrivial analysis. In Theorem 3 of [14], we prove the following.

Theorem 7.2.8 *There exist sibling frames with generators $\psi_1, \psi_2 \in V_1$ and $\tilde{\psi}_1, \tilde{\psi}_2 \in V_1$, such that all of the four functions have compact support and the maximum number m of vanishing moments, and that (ψ_1, ψ_2) is inter-orthogonal.*

Example 7.2.6 For the cardinal B-spline N_2 and $\mu = 2$ we use the same VMR Laurent polynomial $S(z) = 1 + (2 - z - z^{-1})/6$ as in Example 2. The two-scale symbols of the inter-orthogonal frame generators ψ^1 and ψ^2 are formulated as $Q^\ell(z) = ((1-z)/2)^2 q^\ell(z)$, $\ell = 1, 2$, where

$$q^1(z) = \left(\frac{2 + z + z^{-1}}{4}\right)^2 q_0(z), \qquad q^2(z) = \frac{z(4 - z - z^{-1})}{6} q_0(-1/z),$$

and $q_0(z) = az^2 + bz + c$ with

$$\begin{aligned}
a &= 1/4 + \tfrac{1}{12}\sqrt{57} - \tfrac{1}{12}\sqrt{42 + 6\sqrt{57}} \approx 0.1005, \\
b &= 1/2 - \tfrac{1}{6}\sqrt{57} \approx -0.7583, \\
c &= 1/4 + \tfrac{1}{12}\sqrt{57} + \tfrac{1}{12}\sqrt{42 + 6\sqrt{57}} \approx 1.6578.
\end{aligned}$$

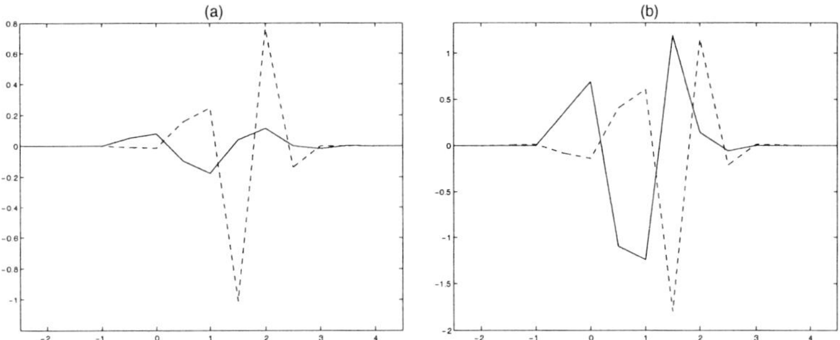

Figure 7.6. Generators ψ^1, ψ^2 (left) and $\widetilde{\psi}^1, \widetilde{\psi}^2$ (right) of linear spline sibling frames with two vanishing moments. The functions ψ^1, ψ^2 are inter-orthogonal

The two-scale symbols for the dual pair are obtained in the form of $\tilde{Q}^\ell(z) = ((1-z)/2)^2\tilde{q}^\ell(z)$, $\ell = 1, 2$, where

$$\tilde{q}^1(z) : = -zS(z^2)q^2(1/z) - zA(z)q^2(-1/z),$$
$$\tilde{q}^2(z) : \ = zA(z)q^1(-1/z) + zS(z^2)q^1(1/z),$$

and $A(z) = (24 + 8(z + z^{-1}) + z^2 + z^{-2})/24$. The graphs of these functions are shown in Figure 6.

7.2.3 Frames with integer dilation factor

As noted at the beginning of Section 2, every B-spline N_m is refinable with respect to arbitrary integer dilation factor $M \geq 2$. We will report in this section that most results from Sections 2.1 and 2.2 have a proper extension to this setting. Some of these results appeared in [17, 34]. However, the main theorem, where we extend the result of 7.2.5, is new and appears here for the first time.

We fix $m, M \in \mathbf{N}$, $M \geq 2$, and recall from (2.4) that

$$P(z) := P_{m,M}(z) = \left(\frac{1 - z^M}{M(1-z)}\right)^m = \left(\frac{1 + z + \cdots + z^{M-1}}{M}\right)^m$$

is the Laurent polynomial two-scale symbol of N_m with respect to dilation by M. The functions $\psi^\ell \in V_1$ (which is the space of splines of order m with simple knots in $(1/M)\mathbf{Z}$) are defined in (2.10) by Laurent polynomials Q^ℓ, such that

$$\widehat{\psi^\ell}(M\omega) = Q^\ell(e^{-i\omega})\hat{N}_m(\omega), \qquad 1 \leq \ell \leq L.$$

We let $w_M = e^{i2\pi/M}$. The following generalization of Theorems 1 and 2 is given in [17].

Theorem 7.2.9 *The compactly supported functions $\psi^\ell \in V_1$, $1 \leq \ell \leq L$, generate a normalized tight frame of $L^2(\mathbf{R})$, if and only if there exists a Laurent polynomial $S(z)$ with real coefficients, such that $S(1) = 1$, S is nonnegative on $\mathbb{T}$, and the identity*

$$S(z^M)P(z)P(w_M^k z^{-1}) + \sum_{\ell=1}^{L} Q^\ell(z)Q^\ell(w_M^k z^{-1}) = \delta_{k0}S(z), \tag{2.43}$$

holds for all $k = 0, \ldots, M-1$.

A variant of the proof for the sufficiency in the above theorem is given in [17], which establishes an important identity for the inner products $\langle f, \phi_{j,k}\rangle$ and $\langle f, \phi^S_{j,k}\rangle$, where we let

$$\phi_{j,k} := M^{j/2} N_m(M^j. - k), \qquad \phi^S_{j,k} := M^{j/2} N^S_m(M^j. - k),$$

and N^S_m is the approximate dual in (2.26). This identity, namely

$$\sum_{k\in\mathbf{Z}} \langle f, \phi_{j+1,k}\rangle\langle \phi^S_{j+1,k}, f\rangle = \sum_{k\in\mathbf{Z}} \langle f, \phi_{j,k}\rangle\langle \phi^S_{j,k}, f\rangle + \sum_{\ell=1}^{L}\sum_{k\in\mathbf{Z}} |\langle f, \psi^\ell_{j,k}\rangle|^2, \tag{2.44}$$

holds for all $j \in \mathbf{Z}$. It is related to the characterization of tight frames in [50] and was first established, for the special case $S \equiv 1$, in [13]. It is also the guiding identity for constructions of tight frame in subsequent sections.

If vanishing moments of the functions ψ^ℓ are analyzed, we again make use of the representation

$$Q^\ell(z) = (1-z)^\mu q^\ell(z), \quad 1 \le \ell \le L,$$

where q^ℓ are Laurent polynomials and μ is the order of vanishing moments of ψ^ℓ. An almost identical argument that leads to the statement of 7.2.3 can be employed. The auto-correlation symbol Φ of N_m was defined in (2.23). Note that the dilation parameter M has no effect on the definition of Φ. In more generality than (2.24), however, the identity

$$\sum_{k=0}^{M-1} |P(w^k_M z)|^2 \Phi(w^k_M z) = \Phi(z^M), \qquad z \in \mathbb{T}, \tag{2.45}$$

is valid. The zero properties of $P = P_{m,M}$ are described in (2.5). These properties, together with (2.45), yield the following extension of 7.2.3.

Theorem 7.2.10 *Let $m, M \in \mathbf{N}$, $1 \le \mu \le m$, and $M \ge 2$. The functions $\psi^1, \ldots, \psi^L$ in (2.10) have vanishing moments of order μ and generate a normalized tight frame Ψ of $L^2(\mathbf{R})$, with dilation factor M, if and only if there exists a Laurent polynomial S that satisfies (2.25) together with all conditions in 7.2.9. Moreover, no tight frame with generators $\psi^\ell \in V_1$ exists, where all functions ψ^ℓ have more than m vanishing moments.*

We already noticed that the auto-correlation symbol Φ does not depend on the dilation factor M, of course. Likewise, the approximation property (2.25) that relates S and Φ does not depend on M. It is natural to ask if the same VMR Laurent polynomial S, that was chosen for the construction of tight spline frames with 2 generators for $M = 2$, can be utilized for the construction of dilation M frames. In other words, is there a "universal" VMR function S, for a given B-spline N_m, such that the equations (2.43) admit Laurent polynomial solutions Q^ℓ, $1 \le \ell \le L$, and can we choose $L = M$? In [17] it was observed that S from Table 7.1 could be utilized for piecewise linear splines ($m = 2$) and dilation factors $M = 2, 3, 4$. Moreover, the equations (2.43) can be equivalently written as a matrix equation, where the matrix

$$\mathcal{M}(z) := \begin{bmatrix} S(z) & & \beta 0 \\ & \ddots & \\ \beta 0 & & S(w_M^{M-1}z)) \end{bmatrix} - \qquad (2.46)$$
$$S(z^M) \begin{bmatrix} P(1/z) \\ \vdots \\ P(w_M^{1-M}/z) \end{bmatrix} \begin{bmatrix} P(z) \cdots P(w_M^{M-1}z) \end{bmatrix}.$$

is involved. The critical part of the result was still unsettled, if this matrix is positive semi-definite. A similar simplification of this problem as for $M = 2$ is obtained in [17, Theorem 4.1]: the matrix is positive semi-definite, if and only if its determinant

$$\prod_{k=0}^{M-1} S(w_M^k z) - S(z^M) \sum_{k=0}^{M-1} |P(w_M^k z)|^2 \prod_{j=0, j\neq k}^{M-1} S(w_M^j z)$$

is nonnegative for all $z \in \mathbb{T}$. If this inequality can be established, then the newly developed factorization technique in [35] reveals the matrix of Laurent polynomials

$$\mathcal{Q}(z) = \begin{bmatrix} Q^1(z) & Q^1(w_M z) & \cdots & Q^1(w_M^{M-1}z) \\ \vdots & \vdots & & \vdots \\ Q^M(z) & Q^M(w_M z) & \cdots & Q^M(w_M^{M-1}z) \end{bmatrix}$$

whose first column contains the Laurent polynomials for the definition of the frame generators $\psi^1, \ldots, \psi^M$. As an application of our more general approach to tight frames of splines on non-uniform knot sequences, we are now in a position to give a complete answer to these questions. The following result describes a special case of 7.4.4 of Section 4, which is proved in [16].

Theorem 7.2.11 *Let $M \in \mathbf{N}$ with $M \geq 2$. If we choose S to be the VMR Laurent polynomial in (2.35), then the matrix $\mathcal{M}(z)$ in (2.46) is positive semi-definite for all $z \in \mathbb{T}$. Moreover, there exist M compactly supported functions $\psi^\ell \in V_1$, $1 \leq \ell \leq M$, that have μ vanishing moments and generate a tight frame of splines with dilation factor M.*

The following example illustrates the previous results.

Example 7.2.7 For piecewise linear splines ($m = 2$), we consider dilation factors $M = 3$ and $M = 4$ separately. The two-scale symbols for these two cases are

$$P_{2,3}(z) = \frac{1}{9}(1 + z + z^2)^2, \qquad P_{2,4}(z) = \frac{1}{16}(1 + z + z^2 + z^3)^2.$$

The VMR Laurent polynomial for two vanishing moments is, as before, $S(z) = 1 + \frac{1}{6}(2 - z - z^{-1})$. The Laurent polynomials Q^ℓ, $1 \leq \ell \leq M$, rounded to four decimals are given by

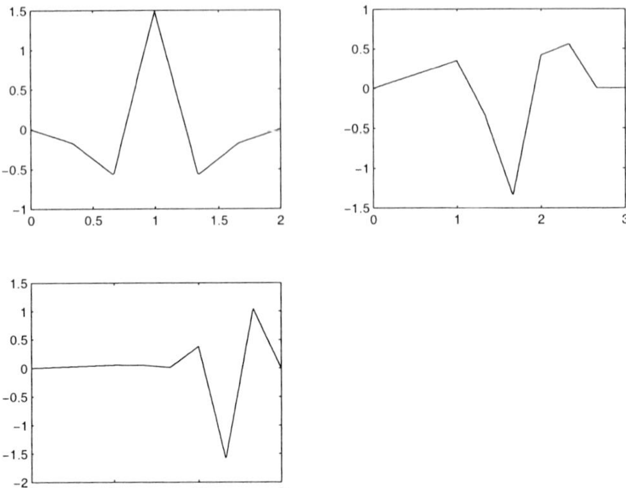

Figure 7.7. Generators ψ^1, ψ^2, ψ^3 of 3-dilation tight frame of piecewise linear splines with two vanishing moments

$$
\begin{aligned}
Q^1(z) &= -(1-z)^2(.0574 + .3059z + .0574z^2),\\
Q^2(z) &= (1-z)^2(.0389 + .1555z + .3887z^2 + .5125z^3 + .1863z^4),\\
Q^3(z) &= (1-z)^2(.0059 + .0236z + .0589z^2 + .1113z^3 + .1675z^4 + .3492z^5)
\end{aligned}
$$

for the case $M = 3$, and

$$
\begin{aligned}
Q^1(z) &= -(1-z)^2(.0265 + .1061z + .3339z^2 + .0907z^3 + .0073z^4)\\
Q^2(z) &= (1-z)^2(.0145 + .0581z + .1508z^2 + .3038z^3 - .0179z^4)\\
Q^3(z) &= (1-z)^2(.0188 + .0751z + .1933z^2 + .3556z^3 + .3671z^4)\\
Q^4(z) &= -(1-z)^2(.0209z^{-4} + .0835z^{-3} + .2087z^{-2} + .4173z^{-1} + .5942\\
&\qquad + .6240z + .3120z^2 + .1248z^3 + .0312z^4)
\end{aligned}
$$

for $M = 4$. The graphs for $M = 3$ are shown in Figure 7.7, and those for $M = 4$ are shown in Figure 7.8.

Finally, we present a general result about the existence of tight M-dilation frames with only $M - 1$ generators and end this section with a conjecture.

Theorem 7.2.12 *Let $\{V_j\}_{j\in\mathbf{Z}}$ be an MRA generated by an M-dilation compactly supported scaling function ϕ with Laurent polynomial two-scale symbol $P(z)$. Then there exist compactly supported functions $\psi_1, \ldots, \psi_{M-1} \in V_1$ that are generators of a normalized tight frame, if and only if there exists a Laurent polynomial B such that $B(1) = 1$, $B(z^M)/B(z)$ is a Laurent polynomial, and $B(z^M)P(z)/B(z)$ is an M-CQF, meaning that, in terms of a generic Laurent polynomial H,*

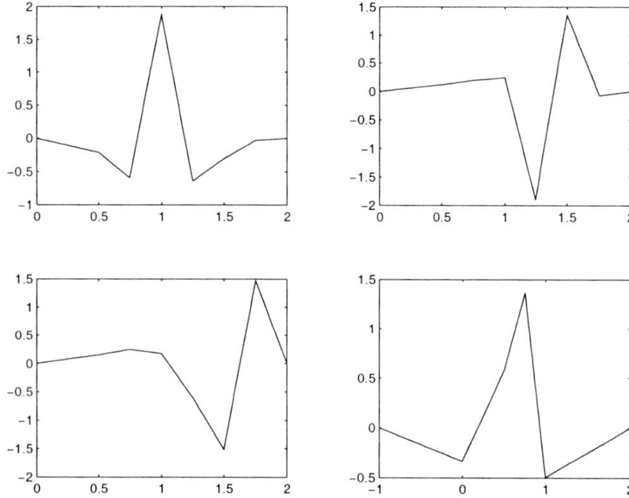

Figure 7.8. Generators $\psi^1, \ldots, \psi^4$ of 4-dilation tight frame of piecewise linear splines with two vanishing moments

$$\sum_{k=0}^{M-1} \left| H(w_M^k z) \right|^2 = 1, \qquad z \in \mathbb{T}.$$

Until now it has not been shown if this result rules out the existence of tight frames with dilation M and $M-1$ generators in the spline space V_1 of splines of order $m \geq 2$. Only the case $M = 2$ is settled, with a negative result, as mentioned at the end of Section 2.1.

Conjecture 1. *For any $m \geq 2$ and $M \geq 2$, there exists no family of $M-1$ compactly supported splines $\psi^\ell \in V_1$ of order m, $1 \leq \ell \leq M-1$, that generate an M-dilation tight frame of $L^2(\mathbf{R})$.*

7.3 WAVELET FRAMES OF SPLINES WITH MULTIPLE KNOTS

The theoretical development of wavelet bases for spline spaces with multiple knots of equal multiplicity evolved under the terminology of "multiwavelets". Here, the underlying MRA is generated by finitely many compactly supported functions $\phi^1, \ldots, \phi^r \in L^2(\mathbf{R})$, which define the function spaces

$$V_j = \text{clos span}\{\phi^\ell_{j,k};\ k \in \mathbf{Z},\ 1 \leq \ell \leq r\},$$

where the subscript notation means the same dilation and shift as before; hence,

$$\phi^\ell_{j,k}(x) = M^{j/2} \phi^\ell (M^j x - k).$$

As before, the dilation factor M is supposed to be an integer greater than or equal to 2. The number $r \in \mathbf{N}$ is called the *multiplicity* of the MRA. Stability of the shifts of the functions ϕ^ℓ (in the usual sense of Riesz bases) is equivalent to the positive definiteness of the "Gramian"

$$\Phi(z) = \Phi(e^{-i\omega}) := \begin{bmatrix} [\hat{\phi}^1, \hat{\phi}^1](\omega) & \cdots & [\hat{\phi}^1, \hat{\phi}^r](\omega) \\ \vdots & & \vdots \\ [\hat{\phi}^r, \hat{\phi}^1](\omega) & \cdots & [\hat{\phi}^r, \hat{\phi}^r](\omega) \end{bmatrix}, \qquad \omega \in \mathbf{R}, \tag{3.1}$$

where we make use of the so-called bracket product

$$[\hat{f}, \hat{g}](\omega) = \sum_{k \in \mathbf{Z}} \hat{f}(\omega + 2\pi k)\overline{\hat{g}(\omega + 2\pi k)}.$$

For more details we refer to [6, 7].

Refinability of the function family $(\phi^1, \ldots, \phi^r)$ is expressed in terms of the matrix refinement equation

$$\begin{bmatrix} \hat{\phi}^1(2\omega) \\ \vdots \\ \hat{\phi}^r(2\omega) \end{bmatrix} = \begin{bmatrix} P_{11}(z) & \cdots & P_{1r}(z) \\ \vdots & & \vdots \\ P_{r1}(z) & \cdots & P_{rr}(z) \end{bmatrix} \begin{bmatrix} \hat{\phi}^1(\omega) \\ \vdots \\ \hat{\phi}^r(\omega) \end{bmatrix}, \qquad z = e^{-i\omega}, \tag{3.2}$$

where $P := [P_{mn}]$ is a Laurent polynomial matrix of dimension $r \times r$. The study of approximation properties of the spaces V_j and smoothness properties of the functions ϕ^ℓ, which are based on certain sum rules of the matrix P in (3.2), are performed in the literature, see [36, 47, 48]. The extension of the formula (2.24) is given by

$$\sum_{k=0}^{M-1} P(w_M^k z)\Phi(w_M^k z)P^*(w_M^k z) = \Phi(z^M), \qquad z \in \mathbb{T}, \tag{3.3}$$

where $w_M = e^{-i2\pi/M}$.

Wavelets or frame generators are compactly supported functions in V_1, that can be defined by

$$\begin{bmatrix} \hat{\psi}^1(2\omega) \\ \vdots \\ \hat{\psi}^L(2\omega) \end{bmatrix} = \begin{bmatrix} Q_{1,1}(z) & \cdots & Q_{1,r}(z) \\ \vdots & & \vdots \\ Q_{L,1}(z) & \cdots & Q_{L,r}(z) \end{bmatrix} \begin{bmatrix} \hat{\phi}^1(\omega) \\ \vdots \\ \hat{\phi}^r(\omega) \end{bmatrix}, \qquad z = e^{-i\omega}. \tag{3.4}$$

The matrix $Q := [Q_{mn}]$ is a Laurent polynomial matrix of dimension $L \times r$, where L denotes the number of generators of the wavelet basis or frame. A general method for the construction of wavelets from an MRA of multiplicity r can be found in [31].

A new method for the construction of non-tight frames with arbitrary vanishing moments was recently presented in [34]. The key step for finding an appropriate VMR function S was tackled by making use of a linear transformation of the set of generators ϕ^ℓ of the MRA; this transformation resembles the approach of finding a "superfunction" in V_0 for the purpose of studying the approximation order of V_j, see [8]. In this section, we not only present a method for the construction of *tight* frames with maximum vanishing moments, but our result also leads to frame generators with much shorter supports than the existing constructions. The principal steps of our method are similar to the case of ordinary MRA's described in Section 2. The construction of the VMR function, however, depends heavily on our results developed jointly with W. He for the construction of tight frames of splines with non-uniform knots. In the paper [16], we are able to avoid several difficult steps of the Fourier-domain approach, that involve factorization techniques for the Laurent polynomial matrix P in connection with sum rules or the characterization of symmetry, see [34]. It turns out that reference to the shift-invariant structure of the spaces V_j is of lesser importance for these constructions, and a time-domain approach leads to a simpler analysis of VMR functions. The full description of the time-domain approach is given in Sections 4 and 5. Here, we draw from those results and present them in a Fourier-domain framework.

We begin our discussion by agreeing on the notations for splines with knots of fixed multiplicity. Let $r \in \mathbf{N}$ and $m \geq r$. We consider m^{th} order B-splines with respect to r-fold integer knots

$$\dots t_{-1} = -1 < t_0 = \cdots = t_{r-1} = 0 < t_r = \cdots = t_{2r-1} = 1 < t_{2r} \dots$$

and denote by $N_{m,r;k}$, the B-spline of order m with r-fold knots, whose first and last knots are t_k and t_{k+m}. For example, for $r = 2$ and $m = 4$ we obtain the B-splines

$$N_{4,2;0} = N(\cdot; 0, 0, 1, 1, 2), \quad N_{4,2;1} = N(\cdot; 0, 1, 1, 2, 2), \tag{3.5}$$

and

$$N_{4,2;2k} = N_{4,2;0}(\cdot - k), \qquad N_{4,2;2k+1} = N_{4,2;1}(\cdot - k),$$

that provide a Riesz basis of the space V_0 of cubic splines with double integer knots. Basic definitions and properties of B-splines are described in [4, 28]. Here we only note that $N_{m,r;k}$ has support $[t_k, t_{k+m}]$, is strictly positive in the interior of this interval, and has continuous derivatives up to order $m-r-1$ and a piecewise continuous and bounded $m-r^{\text{th}}$ derivative. For arbitrary integer dilation $M \geq 2$, the B-splines

$$\phi^\ell = N_{m,r;\ell-1}, \qquad 1 \leq \ell \leq r,$$

generate an MRA of multiplicity r. Moreover, they are stable in the sense that their integer shifts constitute a Riesz basis of V_0.

It is not at all trivial to write down the expression for the matrix P in the refinement equation (3.2) for B-splines with multiple knots. The two-scale representation in the time-domain can be computed by means of a recursive algorithm, the so-called "Oslo algorithm", or by using the explicit formulas in [18]. A recursive scheme in the Fourier-domain was developed by Plonka [46], see also [43], which makes use of the recursion for the derivative of $N_{m,r;\ell}$ given below, see (3.6). For example, if we let $m = 4$, $r = 2$, and $M = 2$, compute the coefficients of the two-scale equation in the time-domain and convert the result to the frequency domain, we obtain

$$P_{4,2,2} = \frac{1}{16} \begin{bmatrix} z^2 + 6z + 2 & 2z + 5 \\ 5z^2 + 2z & 2z^2 + 6z + 1 \end{bmatrix}.$$

Identity (3.3) can be verified for this case, with

$$\Phi_{4,2} = \frac{1}{560} \begin{bmatrix} 128 + 9(z + 1/z) & 80 + z + 53/z \\ 80 + 53z + 1/z & 128 + 9(z + 1/z) \end{bmatrix}.$$

It is worthwhile to note, that all coefficients of the two-scale relation are positive. A much stronger property of the B-splines, which assures total positivity of related matrices in the time-domain, is the key ingredient in our proof of the positivity result (1.16) in [16].

Remark 7.3.1 Another set of generators of V_0, for $m = 4$ and $r = 2$, are the cubic Hermite splines $\tilde{\phi}^1 = N_0 + N_1$ and $\tilde{\phi}^2 = (N_1 - N_0)/3$. Their refinement equation and subsequent wavelet constructions are given in [31]. All results concerning the B-spline basis can be easily converted to the Hermite spline basis by means of a simple linear transformation.

It is well known that derivatives of B-splines are linear combinations of B-splines of lower order which are defined on the same knot sequence. More precisely, the identity

$$N'_{m,r;k}(x) = (m-1)\left(\frac{N_{m-1,r;k}(x)}{t_{k+m-1} - t_k} - \frac{N_{m-1,r;k+1}(x)}{t_{k+m} - t_{k+1}}\right) \tag{3.6}$$

holds for all $x \in \mathbf{R}$ if $m - 1 > r$, and for all $x \in \mathbf{R} \setminus \mathbf{Z}$, if $m - 1 = r$. Simple observations show that there is a polynomial matrix $D_{m,r}$, of dimension $r \times r$ and degree at most 1, such that

$$[(i\omega)\hat{N}_{m,r;k}(\omega)]_{0 \le k \le r-1} = D_{m,r}(z)[\hat{N}_{m-1,r;k}(\omega)]_{0 \le k \le r-1}, \qquad z = e^{-i\omega}. \tag{3.7}$$

In fact, equation (3.6) shows, that $D_{m,r}(\omega)$ has constants in its main and upper diagonal, a linear monomial in its lower left corner, and zeros elsewhere.

It is quite obvious that a compactly supported spline function s in V_0 has μ vanishing moments, if and only if it is the μth derivative of a compactly supported spline of order $m + \mu$ on the same knot sequence. This results in the following characterization in terms of Laurent polynomial coefficients of the Fourier transform of the spline s. We use the notation $\phi^\ell = N_{m,r,\ell-1}$, $1 \le \ell \le r$, for simplicity.

Proposition 7.3.1 *Let $m, r, \mu \in \mathbf{N}$, $m \ge r$, and s be a compactly supported spline function in V_0, with*

$$\hat{s}(\omega) = [Q^1(\omega), \ldots, Q^r(\omega)]\,[\hat{\phi}^1(\omega), \ldots, \hat{\phi}^r(\omega)]^T$$

and Laurent polynomials Q^ℓ, $1 \le \ell \le r$. Then s has μ vanishing moments, if and only if there exist Laurent polynomials $\tilde{Q}^\ell$, $1 \le \ell \le r$, such that

$$[Q^1(\omega), \ldots, Q^r(\omega)]^T = D_{m+1,r}(z)^T \cdots D_{m+\mu,r}(z)^T [\tilde{Q}^1(\omega), \ldots, \tilde{Q}^r(\omega)]^T. \tag{3.8}$$

We can now give the main result about tight spline frames with knots of multiplicity r. The following result was stated in a similar form in [34, Theorem 3.1]. The main ideas of the proof are already contained in [13, 24].

Theorem 7.3.1 *Let $m \ge r \in \mathbf{N}$ and $\{V_j\}_{j \in \mathbf{Z}}$ be the MRA generated by B-splines of order m and r-fold integer knots. Let $\psi^\ell \in V_1$, $1 \le \ell \le L$, be compactly supported splines and $Q = [Q_{mn}]$ the Laurent polynomial matrix of dimension $L \times r$ in (3.4). Also, assume that all the Laurent polynomials have real coefficients, and that each ψ^ℓ has at least one vanishing moment. If there exists a Laurent polynomial matrix S of dimension $r \times r$, which is hermitian and positive definite for all $z \in \mathbf{T}$ and satisfies the identities*

$$[\hat{\phi}^1(0), \ldots, \hat{\phi}^r(0)]S(1)[\hat{\phi}^1(0), \ldots, \hat{\phi}^r(0)]^T = 1; \tag{3.9}$$

$$P(z)^* S(z^M)P(w_M^k z) + Q(z)^* Q(w_M^k z) = \delta_{k0} S(z), \tag{3.10}$$

for all $z \in \mathbb{T}$ *and all* $k = 0, \ldots, M-1$, *then the functions* ψ^ℓ, $1 \le \ell \le L$, *generate a normalized tight frame of* $L^2(\mathbf{R})$.

Proof: The proof makes use of an adapted version of identity (2.44). Since our theorem gives an improved version of Theorem 3.1 in [34], we give a brief sketch of the proof, but leave out the technical details. The assumptions on the functions ψ^ℓ are sufficient, namely that the family $\{\psi_{j,k}^\ell\}_{\ell,j,k}$ is a Bessel family. For $f \in L^2(\mathbf{R})$, we define the periodic function vector

$$a_j(f)(\omega) := \{\, [M^{j/2}\hat{f}(M^j \cdot), \hat{\phi}^\ell](\omega);\ 1 \le \ell \le r\}, \tag{3.11}$$

which is a row vector of dimension r. Moreover, the function vector $[\phi^{S,\ell}]_{1\le \ell \le r}$ is introduced, with

$$[\hat{\phi}^{S,\ell}(\omega)] = S(e^{-i\omega})[\hat{\phi}^\ell(\omega)].$$

By standard arguments, involving the Parseval Identity and typical periodization techniques as in [23, Chapter 3], and with $z := e^{-i\omega/M}$, it follows that

$$\begin{aligned}
&\sum_{\ell=1}^{r}\sum_{k\in\mathbf{Z}} \langle f, \phi_{j,k}^\ell\rangle \langle \phi_{j,k}^{S,\ell}, f\rangle \\
&= \frac{1}{2\pi}\int_0^{2\pi} a_j(f)(\omega) S(z^M) a_j(f)(\omega)^*\, d\omega \\
&= \frac{1}{2M\pi}\int_0^{2\pi} b(\omega) \begin{bmatrix} P(z)^* \\ \vdots \\ P(w_M^{M-1}z)^* \end{bmatrix} S(z^M)[P(z), \ldots, P(w_M^{M-1}z)] b(\omega)^*\, d\omega,
\end{aligned} \tag{3.12}$$

where we make use of the short-hand notation

$$b(\omega) := [a_{j+1}(\omega/M), \ldots, a_{j+1}((\omega + (M-1)2\pi)/M)]$$

which is a row vector of dimension rM. Similarly, if Q is a Laurent polynomial matrix of dimension $L \times r$ and ψ^ℓ, $1 \le \ell \le L$, satisfy (3.4), then

$$\begin{aligned}
&\sum_{\ell=1}^{L}\sum_{k\in\mathbf{Z}} |\langle f, \psi_{j,k}^\ell\rangle|^2 \\
&= \frac{1}{2M\pi}\int_0^{2\pi} b(\omega) \begin{bmatrix} Q(z)^* \\ \vdots \\ Q(w_M^{M-1}z)^* \end{bmatrix} [Q(z), \ldots, Q(w_M^{M-1}z)] b(\omega)^*\, d\omega.
\end{aligned} \tag{3.13}$$

The identity (3.10) implies that

$$\begin{aligned}
\sum_{\ell=1}^{r}\sum_{k\in\mathbf{Z}}\langle f,\phi_{j,k}^{\ell}\rangle\langle\phi_{j,k}^{S,\ell},f\rangle &+ \textstyle\sum_{\ell=1}^{L}\sum_{k\in\mathbf{Z}}|\langle f,\psi_{j,k}^{\ell}\rangle|^2 \\
&= \frac{1}{2M\pi}\int_0^{2\pi} b(\omega)\begin{bmatrix} S(z) & & \beta 0 \\ & \ddots & \\ \beta 0 & & S(w_M^{M-1}z)\end{bmatrix} b(\omega)^*\,d\omega \\
&= \frac{1}{2M\pi}\int_0^{2M\pi} a_{j+1}(f)(\omega/M)S(z)a_{j+1}(f)(\omega/M)^*\,d\omega \\
&= \frac{1}{2\pi}\int_0^{2\pi} a_{j+1}(f)(\omega)S(z^M)a_{j+1}(f)(\omega)^*\,d\omega \\
&= \sum_{\ell=1}^{r}\sum_{k\in\mathbf{Z}}\langle f,\phi_{j+1,k}^{\ell}\rangle\langle\phi_{j+1,k}^{S,\ell},f\rangle.
\end{aligned} \tag{3.14}$$

This is the desired generalization of the telescoping identity (2.44) to the case of multiply generated MRA. In order to show that the functions ψ^ℓ, $1\le\ell\le L$, generate a normalized tight frame, we only need to analyze the limits

$$\lim_{j\to\pm\infty}\sum_{\ell=1}^{r}\sum_{k\in\mathbf{Z}}\langle f,\phi_{j,k}^{\ell}\rangle\langle\phi_{j,k}^{S,\ell},f\rangle. \tag{3.15}$$

That the limit for j tending to $-\infty$ is zero can be shown by precisely the standard arguments as for ordinary MRA's. The limit for j tending to ∞ can be analyzed in the Fourier domain. Its value, for all band-limited functions f, is

$$[\hat\phi^1(0),\ldots,\hat\phi^r(0)]S(1)[\hat\phi^1(0),\ldots,\hat\phi^r(0)]^T\|f\|^2.$$

By making use of identity (3.9), we conclude that

$$\sum_{j\in\mathbf{Z}}\sum_{\ell=1}^{L}\sum_{k\in\mathbf{Z}}|\langle f,\psi_{j,k}^{\ell}\rangle|^2=\|f\|^2$$

holds for all band-limited $f\in L^2(\mathbf{R})$. Since the functions $\{\psi_{j,k}^\ell\}_{\ell,j,k}$ are a Bessel family, standard continuity arguments yield the same identity for all $f\in L^2(\mathbf{R})$.

Remark 7.3.2 The formulation of Theorem 3.1 in [34] differs in two ways from our theorem. First, the definition of the frame generators ψ^ℓ, $1\le\ell\le L$, has an unnecessary restriction on the number of generators: equation (3.5) in [34] uses a matrix formulation of (3.4) where L must be a multiple of r, and consequently the equation (3.3) in that paper has a more special form than our equation (3.10). Secondly, they require a third condition

$$P(1)^*S(1)[\hat\phi^1(0),\ldots,\hat\phi^r(0)]^T=S(1)[\hat\phi^1(0),\ldots,\hat\phi^r(0)]^T.$$

It turns out, however, that this identity is a consequence of (3.10), with $k=0$ and $z=1$, and the assumption that ψ^ℓ has at least one vanishing moment. Indeed, these two conditions together with (3.2), (3.4) yield

$$P(1)^*S(1)\begin{bmatrix}\hat{\phi}^1(0)\\ \vdots\\ \hat{\phi}^r(0)\end{bmatrix} = (P(1)^*S(1)P(1)+Q(1)^*Q(1))\begin{bmatrix}\hat{\phi}^1(0)\\ \vdots\\ \hat{\phi}^r(0)\end{bmatrix} = S(1)\begin{bmatrix}\hat{\phi}^1(0)\\ \vdots\\ \hat{\phi}^r(0)\end{bmatrix}.$$

Other more important differences concerning the construction of the VMR Laurent polynomial matrix S and the frame generators are described in the remainder of this section.

Remark 7.3.3 Note that we only claim the sufficiency in the previous theorem. To our knowledge, no proof for the necessity of the existence of the VMR Laurent polynomial matrix S has yet been published.

The explicit construction of tight frames of splines with multiple knots follows the procedure that was laid out in Section 2. The identity (3.10) is written as a matrix equation

$$\mathcal{M}(z) = \begin{bmatrix} Q(z)^* \\ \vdots \\ Q(w_M^{M-1}z)^* \end{bmatrix} [Q(z),\ldots,Q(w_M^{M-1}z)] \tag{3.16}$$

where

$$\mathcal{M}(z) = \begin{bmatrix} S(z) & & \beta 0 \\ & \ddots & \\ \beta 0 & & S(w_M^{M-1}z)) \end{bmatrix} - \begin{bmatrix} P(z)^* \\ \vdots \\ P(w_M^{M-1})^* \end{bmatrix} S(z^M)\left[P(z) \cdots P(w_M^{M-1}z)\right]. \tag{3.17}$$

Hence, the VMR Laurent polynomial matrix $S(z)$ must be defined, such that (3.9) holds and that the matrix $\mathcal{M}(z)$ in (3.17) is positive semi-definite. Moreover, in order for the functions ψ^ℓ to have at least $\mu \geq 1$ vanishing moments, the matrix $\mathcal{M}(z)$ admits the factorization

$$\mathcal{M}(z) = \begin{bmatrix} D(z)^* & & \beta 0 \\ & \ddots & \\ \beta 0 & & D(w_M^{M-1}z)^* \end{bmatrix} \mathcal{M}_0(z) \begin{bmatrix} D(z) & & \beta 0 \\ & \ddots & \\ \beta 0 & & D(w_M^{M-1}z) \end{bmatrix} \tag{3.18}$$

where $\mathcal{M}_0$ is a Laurent polynomial matrix of dimension $rM \times rM$ and

$$D(z) := D_{m+1,r}(z)\cdots D_{m+\mu,r}(z). \tag{3.19}$$

The methods in [35] for the factorization of the positive semi-definite matrix $\mathcal{M}_0(z)$ then allow us to find the Laurent polynomial matrix $Q = D\tilde{Q}$ for the definition of ψ^ℓ in (3.4).

The key observation drawn from the time-domain approach in Section 5 is that the VMR matrix S should satisfy an approximation property similar to (2.25), which we brought into play in Section 2.1. This identity reads as

$$S(z) - \Phi(z)^{-1} = D(z)^* X(z) D(z), \tag{3.20}$$

where X is a hermitian matrix of Laurent series, which are continuous on $\mathbb{T}$ and whose coefficients decay exponentially. As in Section 2.1, we see that condition (3.20) does not depend on the scaling factor. Hence, the same VMR matrix can be employed for different integer dilation factors M. Let us consider the identity (3.9) in 7.3.1. Since

$$D(1)[\hat{\phi}^1(0), \ldots, \hat{\phi}^r(0)]^T = 0,$$

we observe that (3.20) implies

$$\begin{bmatrix} \hat{\phi}^1(0) \\ \vdots \\ \hat{\phi}^r(0) \end{bmatrix}^* S(1) \begin{bmatrix} \hat{\phi}^1(0) \\ \vdots \\ \hat{\phi}^r(0) \end{bmatrix} = \begin{bmatrix} \hat{\phi}^1(0) \\ \vdots \\ \hat{\phi}^r(0) \end{bmatrix}^* \Phi(1)^{-1} \begin{bmatrix} \hat{\phi}^1(0) \\ \vdots \\ \hat{\phi}^r(0) \end{bmatrix} = \begin{bmatrix} \hat{\phi}^1(0) \\ \vdots \\ \hat{\phi}^r(0) \end{bmatrix}^* \begin{bmatrix} \widehat{\tilde{\phi}^1}(0) \\ \vdots \\ \widehat{\tilde{\phi}^r}(0) \end{bmatrix},$$

where $\tilde{\phi}^\ell$ are the generators of the dual basis of V_0. The last expression is equal to 1, as a result of the fact that the spline space contains all constant polynomials.

Although some of the theoretical development is still lacking in the Fourier domain, we can give the following result, which is a consequence of our general approach in Section 5.

Theorem 7.3.2 *Let $m \geq r \in \mathbf{N}$ and $1 \leq \mu \leq m$. There exists a unique Laurent polynomial matrix S of the form*

$$\begin{aligned} S(z) = U_0 + & D_{m+1,r}(z)^* U_1 D_{m+1,r}(z) + \cdots + \\ & D_{m+\mu-1,r}(z)^* \cdots D_{m+1,r}(z)^* U_{\mu-1} D_{m+1,r}(z) \cdots D_{m+\mu-1,r}(z), \end{aligned} \tag{3.21}$$

where U_k, $0 \leq k \leq \mu - 1$, are diagonal matrices of dimension $r \times r$ with positive constant diagonal entries, such that (3.20) is satisfied, where X is a hermitian matrix of Laurent series, which are continuous on $\mathbb{T}$ and whose coefficients decay exponentially. Moreover, the matrix $\mathcal{M}(z)$ in (3.17) is positive semi-definite on $\mathbb{T}$ and admits a factorization as in (3.18).

As mentioned before, our proof is solely based on the time-domain approach in [16] that we sketch in Section 4. The matrix S is identified with a bi-infinite block Toeplitz matrix with blocks of dimension $r \times r$. The resulting matrix is real, symmetric and has bandwidth μ. The proof of the semi-definiteness of the matrix $\mathcal{M}$ is based on the variation diminishing properties of B-splines. We are not aware of an analogous formulation of this property in the Fourier domain. It would still be of some interest to find a different proof of 7.3.2 based directly on the Fourier approach.

Remark 7.3.4 The definition of the diagonal matrices U_k in 7.3.2 is made explicit in the next section. More details about approximate duals of B-splines are also explained in Section 4. The partial sum of the first $\nu \leq \mu$ elements on the right-hand side of (3.21) defines the corresponding VMR matrix, with ν in place of μ. The diagonal entry $u_{nn}^{(k)}$ of U_k, $1 \leq n \leq r$, is based on a combinatorial expression that depends on the knots of the B-spline $N_{m+k,r;n-1}$ of order $m + k$. This spline has an $(r - n + 1)$-fold knot at the left endpoint, interior knots of multiplicity r, and a knot of multiplicity $\nu_n := (m+k+n) \bmod r$ at the right endpoint of its support. Equal values for $u_{nn}^{(k)}$ and $u_{pp}^{(k)}$ are obtained if the multiplicities at the left and right endpoints are interchanged. It is yet too complicated to give a short-hand expression for all diagonal entries of U_k in terms of k, r, and m. Therefore, we work with the general formula (4.39) which is also valid for non-uniform knots.

The results about matrix factorization of positive semi-definite Laurent polynomial matrices in [35] can be applied in order to give the following continuation of 7.3.2.

Theorem 7.3.3 *Let $m \geq r \in \mathbf{N}$, $1 \leq \mu \leq m$ and $M \in \mathbf{N}$ with $M \geq 2$ be given. For the VMR Laurent polynomial matrix S in (3.21), there exists a Laurent polynomial matrix Q of dimension $r \times r$, such that (3.10) is satisfied and Q has a factorization of the form*

$$Q(z) = D_{m+1,r}(z) \cdots D_{m+\mu-1,r}(z)\tilde{Q}(z),$$

where $\tilde{Q}$ is a Laurent polynomial matrix. Moreover, the functions ψ^ℓ, $1 \leq \ell \leq r$, in (3.4) are generators of a normalized tight frame of $L^2(\mathbf{R})$ with dilation factor M.

We include the following example as an illustration of the method.

Example 7.3.1 We wish to construct a normalized tight frame of cubic splines with double knots and four vanishing moments. Hence, we choose $m = \mu = 4$ and $r = 2$. Instead of the matrix factorization in [35], which would lead to 4 non-symmetric generators, we have developed another factorization for this special case, where we obtain 5 generators $\psi^1, \ldots, \psi^5$ with 3 generators being symmetric and 2 being anti-symmetric. The matrices U_k in (3.21) are given by

$$U_0 = \begin{bmatrix} 2 & 0 \\ 0 & 2 \end{bmatrix}, \quad U_1 = \begin{bmatrix} \frac{1}{3} & 0 \\ 0 & \frac{1}{9} \end{bmatrix}, \quad U_2 = \begin{bmatrix} \frac{11}{900} & 0 \\ 0 & \frac{11}{900} \end{bmatrix}, \quad U_3 = \begin{bmatrix} \frac{43}{32400} & 0 \\ 0 & \frac{1}{2700} \end{bmatrix},$$

see also (4.39). The matrices $D_{m+k,2}$ in (3.7) are

$$D_{5,2}(z) = D_{7,2}(z) = \begin{bmatrix} 2 & -2 \\ -2z & 2 \end{bmatrix}, \quad D_{6,2}(z) = \begin{bmatrix} \frac{5}{2} & -\frac{5}{3} \\ -\frac{5z}{2} & \frac{5}{3} \end{bmatrix},$$

$$D_{8,2}(z) = \begin{bmatrix} \frac{7}{3} & -\frac{7}{4} \\ -\frac{7z}{3} & \frac{7}{4} \end{bmatrix}.$$

Hence, we obtain the VMR matrix

$$S(z) = \frac{1}{2916} \begin{bmatrix} 18088 + 1689(z + 1/z) & -9432 - 387z - 5767/z - 48/z^2 \\ -9432 - 387/z - 5767z - 48z^2 & 18088 + 1689(z + 1/z) \end{bmatrix}. \tag{3.22}$$

Identity (3.18) is satisfied with

$$\mathcal{M}_0(z) = \frac{1}{1024 \cdot 81 \cdot 25 \cdot 49} \begin{bmatrix} A(z) & B(z)^* \\ B(z) & A(-z) \end{bmatrix}$$

where the 2×2 matrices A and B are defined by their entries

$$\begin{aligned}
A_{11}(z) &= 293080 + 106528(z + 1/z) + 8583(z^2 + 1/z^2) + 192(z^3 + 1/z^3), \\
A_{21}(z) &= 210136 + 219856z + 44743z^2 + 1920z^3 + 48z^4 + 28656/z + 1155/z^2, \\
A_{12}(z) &= A_{21}(1/z), \qquad A_{22}(z) = A_{11}(z), \\
B_{11}(z) &= -37416 - 12784(z - 1/z) + 903(z^2 + 1/z^2) - 192(z^3 - 1/z^3), \\
B_{21}(z) &= -38440 + 6343z^2 + 48z^4 - 381/z^2, \\
B_{12}(z) &= B_{21}(1/z), \qquad B_{22}(z) = B_{11}(1/z).
\end{aligned}$$

Some reduction of the degree of the Laurent polynomials can be achieved by elementary transformations. An analytic solution in the time-domain was developed by assuming certain symmetry relations. The polynomial matrix $\tilde{Q}$ (before multiplication by $D_{5,1} \cdots D_{8,1}$) has dimension 5×2 and is given by

$$\tilde{Q}(z) = \frac{1}{32 \cdot 9 \cdot 5 \cdot 7} \left[\begin{matrix} 70.135914z^2 + 280.543657z \\ 175.866405z^2 + 141.746653z + 46.809907 \\ -129.165281z^2 + 6.955361z + 42.475507 \\ 10.188783z^4 + 203.775425z^3 + 254.515242z^2 + 40.755053z \\ -7.470651z^4 - 149.413582z^3 + 79.795457z^2 + 29.882684 \end{matrix} \right.$$

$$\left. \begin{matrix} 280.543657z + 70.135914 \\ 46.809907z^2 + 141.746653z + 175.866405 \\ -42.475507z^2 - 6.955361z + 129.165281 \\ 40.755053z^3 + 254.515242z^2 + 203.775425z + 10.188783 \\ -29.882684z^3 - 79.795457z^2 + 149.413582z + 7.470651 \end{matrix} \right].$$

The frame generators ψ^ℓ, $1 \le \ell \le 5$ are obtained from

$$\begin{bmatrix} \hat{\psi}^1(2\omega) \\ \vdots \\ \hat{\psi}^5(2\omega) \end{bmatrix} = D_{5,1}(z) \cdots D_{8,1}(z)\tilde{Q}(z) \begin{bmatrix} \hat{N}_{4,2;0}(\omega) \\ \hat{N}_{4,2;1}(\omega) \end{bmatrix}, \qquad z = e^{-i\omega}.$$

The graphs of the 5 generators are shown in Figure 9. Further properties are listed in Table 7.2.

7.4 THE COMMON LINK: APPROXIMATE DUALS

As mentioned in Remark 2, the definition of the VMR Laurent polynomial (matrix) S in the previous sections is closely related to an approximate dual basis $[N_m^S(\cdot - k);\ k \in \mathbf{Z}]$

Table 7.2. Properties of the 5 generators of a normalized tight frame of cubic splines with double knots

	support	symmetry	left boundary is single knot	right boundary is single knot
ψ^1	$[0,2]$	even	yes	yes
ψ^2	$[0,2]$	even	no	no
ψ^3	$[0,2]$	odd	no	no
ψ^4	$[0,3]$	even	yes	yes
ψ^5	$[0,3]$	odd	yes	yes

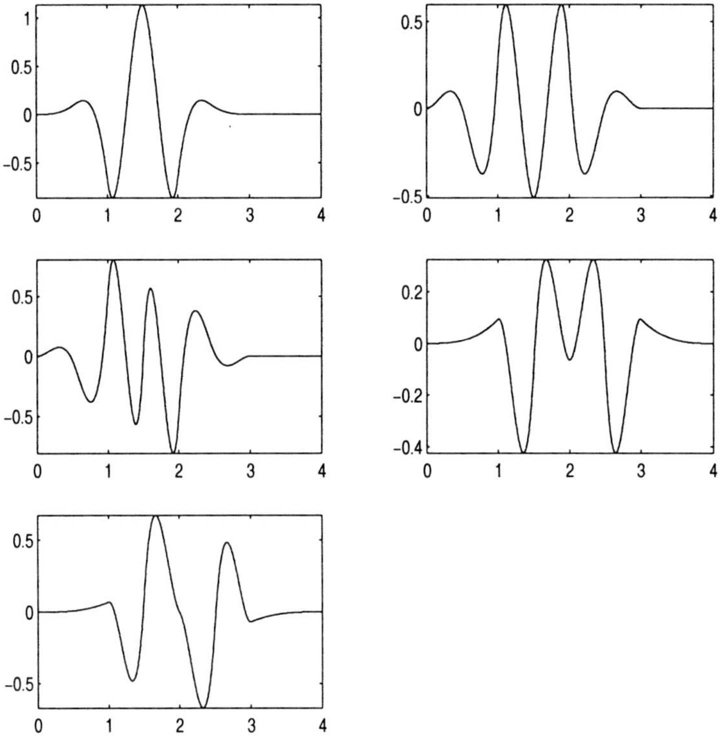

Figure 7.9. Symmetric and anti-symmetric generators of tight frame of cubic splines with double knots and 4 vanishing moments

of the B-spline basis $[N_m(\cdot - k);\ k \in \mathbf{Z}]$. In this section, we introduce the notion of approximate duals of B-splines with arbitrary non-uniform knot sequences and describe their explicit construction in the time-domain. The analogue of the positivity result (1.16) is shown, and several examples are included.

7.4.1 Background on univariate B-splines

To facilitate our presentation and in order to introduce certain necessary notations, we highlight several relevant results on B-splines. When the interval I of interest to be considered is all of $\mathbf{R}$, then the knot vector (sequence)

$$\mathbf{t} = \{t_k,\ k \in \mathbf{Z}\}$$

is assumed to satisfy

$$t_k \leq t_{k+1} \quad \text{and} \quad t_k < t_{k+m} \qquad \text{for all} \quad k, \tag{4.1}$$

and

$$\lim_{j \to \pm\infty} t_k = \pm\infty. \tag{4.2}$$

Moreover, we also assume that the knots grow at most polynomially, that is

$$|t_k| \leq C|k|^s$$

for some constants $C, s > 0$. These are the only restrictions on the knot vectors when we consider splines of order m on $I := \mathbf{R}$.

When $I = [a, \infty)$, then the only modification is that m knots are stacked at the boundary point a, namely $t_{-m+1} = \cdots = t_0 = a$, and that the knot vector becomes

$$\mathbf{t} = \{a, \ldots, a, t_1, t_2, \ldots\}, \tag{4.3}$$

with the boundary knot of multiplicity m. Finally, if $I = [a, b]$ is a bounded interval, then another m knots are stacked at the right-hand boundary point b, namely $t_{N+1} = t_{N+2} = \cdots = t_{N+m} = b$, and the knot vector becomes

$$\mathbf{t} = \{a, \ldots, a, t_1, \ldots, t_N, b, \ldots, b\}, \tag{4.4}$$

with both boundary knots of multiplicity m. Of course, the case of $I := (-\infty, b]$ is treated accordingly.

Each B-spline of order m with knot sequence $\mathbf{t}$ is indexed by its first (active) knot; hence,

$$\begin{aligned} N_{\mathbf{t};m,k}(x) &= N_m(x; t_k, \ldots, t_{k+m})(x) \\ &= (t_{k+m} - t_k)\left[t_k, t_{k+1}, \ldots, t_{k+m} \mid (x - .)_+^{m-1}\right], \end{aligned}$$

where the notation $[t_k, t_{k+1}, \ldots, t_{k+m} \mid f]$ denotes the mth order divided difference of f relative to the nodes $t_k, \ldots, t_{k+m}$, and $(x)_+^r := \max(x^r, 0)$ is the truncated power function. We note that knot vectors are ordered sets whose elements may have multiplicities larger than one. The conditions in (4.1)-(4.2) assure that all knots have multiplicity not exceeding m. In the following, we also use K_m to denote the set of all indices of B-splines $N_{\mathbf{t};m,k}$ defined for the knot vector $\mathbf{t}$.

It is well-known that $N_{\mathbf{t};m,k}$ has support $[t_k, t_{k+m}]$, is strictly positive inside this interval, and is a polynomial of degree $m-1$ in each subinterval (t_i, t_{i+1}), $k \leq i \leq k+m-1$. Moreover, it has $m - \mu_i - 1$ continuous derivatives at t_i, if μ_i is the multiplicity of t_i in the subsequence $t_k, \ldots, t_{k+m}$. Choosing the knot locations and multiplicities adaptively allows for the great flexibility and approximation power of spline functions, that has made B-splines an indispensable tool in many areas of Applied Mathematics. The B-splines $N_{\mathbf{t};m,k}$ provide a stable basis of the spline space $S_{\mathbf{t};m} \cap L^\infty(I)$ that consists of all bounded piecewise polynomials of degree $m-1$ with "breakpoints" $t_k \in \mathbf{t}$ and smoothness $m - \mu_k - 1$ at t_k. Stability with respect to other L^p-norms, $1 \leq p < \infty$, is achieved by considering the L^p-normalized B-splines

$$N^p_{\mathbf{t};m,k} := d^{-1/p}_{\mathbf{t};m,k} N_{\mathbf{t};m,k},$$

where

$$d_{\mathbf{t};m,k} := \frac{t_{k+m} - t_k}{m} \tag{4.5}$$

denotes the average knot distance of the B-spline $N_{\mathbf{t};m,k}$. Results on the stability estimates are given in [28].

The short-hand notations

$$\Phi_{\mathbf{t};m} := [N_{\mathbf{t};m,k}]_{k \in \mathrm{K}_m} \tag{4.6}$$

and

$$\Phi^B_{\mathbf{t};m} := [d^{-1/2}_{\mathbf{t};m,k} N_{\mathbf{t};m,k}]_{k \in \mathrm{K}_m}$$

for the L^∞ and L^2-normalized B-spline bases, respectively, will be useful in the sequel. The family $\Phi^B_{\mathbf{t};m}$ is a Riesz basis of $V_{\mathbf{t}} := S_{\mathbf{t};m} \cap L^2(I)$, and the stability constants are bounded away from zero and infinity uniformly for all knot vectors $\mathbf{t}$ that satisfy (4.1)-(4.2), see [4, 28].

The B-splines also provide a partition of unity and, more generally, Marsden's identity

$$\frac{(y-x)^s}{s!} = \sum_k g^{(m-1-s)}_{\mathbf{t};m,k}(y) N_{\mathbf{t};m,k}(x), \qquad 0 \leq s \leq m-1, \tag{4.7}$$

where

$$g_{\mathbf{t};m,k}(y) = \frac{1}{(m-1)!}(y - t_{k+1}) \dots (y - t_{k+m-1})$$

is a polynomial of degree $m-1$ that depends only on the interior knots of $N_{\mathbf{t};m,k}$. This identity assures that polynomials have a locally finite expansion in terms of the B-spline basis $\Phi_{\mathbf{t};m}$. Approximation properties of B-splines result from this property.

In addition, B-splines satisfy several recurrence relations. The identity for the derivative

$$N'_{\mathbf{t};r+1,k}(x) = d^{-1}_{\mathbf{t};r,k} N_{\mathbf{t};r,k}(x) - d^{-1}_{\mathbf{t};r,k+1} N_{\mathbf{t};r,k+1}(x) \tag{4.8}$$

is valid, whenever $t_k < t_{k+r}$ and $t_{k+1} < t_{k+r+1}$ are satisfied, for every x that differs from an r-fold knot of any of the two splines on the right-hand side. This identity will be of eminent importance for our subsequent investigation. Equation (3.6) gives this identity for a special case of B-splines with multiple knots at the integers. By means of the bi-infinite matrix

$$D_{\mathbf{t};r} := \begin{bmatrix} \ddots & \ddots & & \\ & -d^{-1}_{\mathbf{t};r,k} & d^{-1}_{\mathbf{t};r,k} & \\ & & -d^{-1}_{\mathbf{t};r,k+1} & d^{-1}_{\mathbf{t};r,k+1} \\ & & \ddots & \ddots \end{bmatrix}, \tag{4.9}$$

with zeros outside the two diagonals, we obtain

$$\frac{d}{dx}\Phi_{\mathbf{t};r+1}(x) = \Phi_{t;r}(x)D_{\mathbf{t};r}. \tag{4.10}$$

This identity is valid a.e., whenever the knot sequence satisfies (4.1) (with r substituted for m). If we consider the interval $[a, \infty)$ and the knot vector in (4.3), then the matrix

$$D_{\mathbf{t};r} := \begin{bmatrix} d_{\mathbf{t};r,-m+1}^{-1} & & \\ -d_{\mathbf{t};r,-m+2}^{-1} & d_{\mathbf{t};r,-m+2}^{-1} & \\ & \ddots & \ddots \end{bmatrix} \tag{4.11}$$

yields (4.10) for any $r \geq m$. For a bounded interval $[a, b]$ and the knot vector in (4.4) we replace the matrix $D_{\mathbf{t};r}$ with

$$D_{\mathbf{t};r} := \begin{bmatrix} d_{\mathbf{t};r,-m+1}^{-1} & & & \\ -d_{\mathbf{t};r,-m+2}^{-1} & d_{\mathbf{t};r,-m+2}^{-1} & & \\ & \ddots & \ddots & \\ & & \ddots & d_{\mathbf{t};r,N+m-r-1}^{-1} \\ & & & -d_{\mathbf{t};r,N+m-r}^{-1} \end{bmatrix}, \tag{4.12}$$

which is a matrix with $N + 2m - r$ rows and $N + 2m - r - 1$ columns. We also introduce the abbreviation

$$E_{\mathbf{t};m,\nu} := D_{\mathbf{t};m} \cdots D_{\mathbf{t};m+\nu-1}, \tag{4.13}$$

in order to write

$$\frac{d^\nu}{dx^\nu}\Phi_{\mathbf{t};m+\nu}(x) = \Phi_{\mathbf{t};m}(x)E_{\mathbf{t};m,\nu}. \tag{4.14}$$

The recursion for the L^2-normalized splines reads as

$$\frac{d^\nu}{dx^\nu}\Phi^B_{\mathbf{t};m+\nu}(x) = \Phi^B_{\mathbf{t};m}(x) \underbrace{\operatorname{diag}\left[d^{1/2}_{\mathbf{t};m,k}\right]_{k\in\mathbf{K}_m} E_{\mathbf{t};m,\nu} \operatorname{diag}\left[d^{-1/2}_{\mathbf{t};m+\nu,k}\right]_{k\in\mathbf{K}_{m+\nu}}}_{=E^B_{\mathbf{t};m,\nu}}. \tag{4.15}$$

It is clear that $E_{\mathbf{t};m,\nu}$ and $E^B_{\mathbf{t};m,\nu}$ are banded matrices with precisely $\mu + 1$ nonzero diagonals.

The identities (4.14) and (4.15) are particularly useful in order to express the order of vanishing moments of splines in $S_{\mathbf{t};m} \cap L^2(I)$. A spline $s \in S_{\mathbf{t},m} \cap L^2(I)$ has μ vanishing moments (and compact support or exponential decay, if it is defined on an unbounded interval), if and only if it is the μth derivative of a spline S of order $m + \mu$ with respect to the same knot vector $\mathbf{t}$. The spline S is hereby defined uniquely, if we require S to have compact support or exponential decay as well, in the case of an unbounded interval, or to have zero values of derivatives $S^{(\nu)}(a)$ and $S^{(\nu)}(b)$, $0 \leq \nu \leq \mu - 1$, in the case of a bounded interval. Hence, the knots a and/or b of the B-splines of order $m + \mu$ that represent S have multiplicity at most m (and not $m + \mu$). We obtain the following result.

Lemma 7.4.1 *Let*

$$s = \Phi^B_{\mathbf{t};m}\mathbf{c} = \sum_{k\in\mathbf{K}_m} c_k N^B_{\mathbf{t};m,k}$$

be given, where (a) the entries of $\mathbf{c}$ *decay exponentially or (b)* $c_k = 0$ *for all* $k < i_1$ *and/or* $k > i_2$. *Then* s *has* μ *vanishing moments, if and only if there exists a column vector* $\mathbf{d}$ *such that*

$$\mathbf{c} = E^B_{\mathbf{t};m,\mu}\mathbf{d}, \tag{4.16}$$

and the entries of $\mathbf{d}$ *decay exponentially in case (a) and satisfy* $d_k = 0$ *for all* $k < i_1$ *and/or* $k > i_2 - \mu$ *in case (b).*
For the case $I = [a, b]$*, the same result holds when the superscript* B *is dropped.*

Finally, we describe the refinability of the B-spline basis. Consider two knot vectors

$$\mathbf{t}_j \subset \mathbf{t}_{j+1}, \qquad j \geq 0, \tag{4.17}$$

that satisfy conditions (4.1)–(4.2). Note that the subset notation is used for ordered sets: new knots of multiplicity $\leq m$ can be inserted into $\mathbf{t}_j$, or the multiplicity $\mu_k < m$ of an existing knot $t_k^{(j)}$ in $\mathbf{t}_j$ can be increased. (We use superscript (j), in order to denote knots in $\mathbf{t}_j$. We also drop $\mathbf{t}$ in the index of B-splines and write $N_{j;m,k}$, etc.)

Remark 7.4.1 One type of knot refinement is defined by the insertion of a new knot of the same multiplicity into each knot interval of $\mathbf{t}_j$; this is called "two-threaded" refinement in [26]. If new knots are placed halfway between old knots, we obtain quasi-uniform refinements, and if, in addition, $\mathbf{t}_0 = \mathbf{Z}$, we are in the situation described in Section 2. In general, we do not assume any of these special types of refinements. The only additional restriction on the refinement is that the number of knots inserted between $t_k^{(j)}$ and $t_{k+1}^{(j)}$ is bounded by a constant n_j that may vary with j.

The B-spline bases $\Phi_{j;m}$ and $\Phi_{j+1;m}$ satisfy the refinement equation

$$\Phi_{j;m} = \Phi_{j+1;m} P_{j;m} \tag{4.18}$$

with a real matrix $P_{j;m}$ whose entries are nonnegative and whose row sums equal 1. Moreover, the matrix is sparse in the following sense. We define strictly increasing sequences $\ell(k)$ and $\eta(k)$ such that

$$\{t_k^{(j)}, \ldots, t_{k+m}^{(j)}\} \subset \{t_{\ell(k)}^{(j+1)}, \ldots, t_{\eta(k)+m}^{(j+1)}\}. \tag{4.19}$$

Here, $\ell(k)$ is the maximal index and $\eta(k)$ is the minimal index with the property (4.19), and the subset notation is employed for ordered sets. Then the entries $p_{i,k}$ in the k-th column of $P_{j,m}$ are zero, if $i < \ell(k)$ or $i > \eta(k)$. In other words, only the B-splines of $\Phi_{j+1;m}$, whose support is contained in the support of $N_{j;m,k}$, have a nonzero coefficient in the refinement relation of $N_{j;m,k}$. The restriction on the knot insertion in the previous remark guarantees, that at most $mn_j + 1$ entries can be nonzero in every column of $P_{j;m}$. In the special case, where $\mathbf{t}_{j+1} \setminus \mathbf{t}_j = \{\tau\}$ is a singleton and $\tau \in [t_k^{(j)}, t_{k+1}^{(j)})$, the matrix $P_{j,m}$ has the form

$$P_{j,m} = \begin{bmatrix} \ddots & & & & & & \\ & 1 & & & & & \\ & & 1 & & & & \\ & & b_{k-m+2} & a_{k-m+2} & & & \\ & & & \ddots & \ddots & & \\ & & & & b_k & a_k & \\ & & & & & 1 & \\ & & & & & & 1 \\ & & & & & & & \ddots \end{bmatrix}, \tag{4.20}$$

where

$$a_i = \frac{\tau - t_i^{(j)}}{t_{i+m-1}^{(j)} - t_i^{(j)}} \in [0,1], \quad b_i = 1 - a_i, \quad i = k-m+2, \dots, k. \tag{4.21}$$

Here, a_i is the element with row and column index i. For insertion of more than one knot, the matrix $P_{j,m}$ can be written as a finite product of matrices that are block diagonal with blocks of the form (4.20). Another important algorithm for the computation of $P_{j;m}$ by a recursion on m is the "Oslo-algorithm", see [29]. Note that the L^2-normalized basis satisfies the refinement equation

$$\Phi_{j;m}^B = \Phi_{j+1;m}^B P_{j;m}^B, \quad \text{where} \quad P_{j;m}^B = \text{diag}\left[d_{j+1;m,k}^{1/2}\right]_k P_{j;m} \text{diag}\left[d_{j;m,k}^{-1/2}\right]_k. \tag{4.22}$$

Refinable differentiable function vectors are known to satisfy a "commutation" relation [43], which reads as

$$D_{j+1,m} P_{j,m+1} = P_{j,m} D_{j,m} \tag{4.23}$$

for the B-spline basis. For higher order derivatives we obtain

$$E_{j+1;m,\nu} P_{j,m+\nu} = P_{j,m} E_{j;m,\nu}, \qquad E_{j+1;m,\nu}^B P_{j,m+\nu}^B = P_{j,m}^B E_{j;m,\nu}^B. \tag{4.24}$$

7.4.2 A particular polynomial

Similar to the Marsden coefficients in (4.7), we define multivariate polynomials

$$F_\nu(x_1, \dots, x_r) = \frac{2^{-\nu}}{\nu!} \sum_{\substack{1 \le i_1, \dots, i_{2\nu} \le r, \\ i_1, \dots, i_{2\nu} \text{ distinct}}} \prod_{j=1}^{\nu} (x_{i_{2j-1}} - x_{i_{2j}})^2, \tag{4.25}$$

for $\nu \in \mathbf{N}$, and $F_0(x_1, \dots, x_r) \equiv 1$. Without causing any confusion, we make use of the same symbol F_ν for different numbers of arguments. It is obvious from (4.25), that F_ν equals zero, if $r < 2\nu$, and for all $r \ge 2\nu$ the recursion

$$F_\nu(x_1, \dots, x_r) = F_\nu(x_1, \dots, x_{r-1}) + \sum_{i=1}^{r-1} (x_r - x_i)^2 F_{\nu-1}(x_1, \dots, \widehat{x}_i, \dots, x_{r-1}) \tag{4.26}$$

is valid, where the notation $\widehat{x}_i$ is used to denote that x_i is left out from the list of arguments. In particular, for any $r \ge 2\nu$, F_ν is a homogeneous polynomial of total degree 2ν, which is symmetric in its variables and is invariant under a shift of the arguments $(x_1, \dots, x_r) \mapsto (x_1 - c, \dots, x_r - c)$. Its coordinate degree in each of its variables is 2. These properties are enough in order to assure, that F_ν can be written in terms of the centered moments of its arguments

$$\sigma_\ell := \frac{1}{r}\sum_{k=1}^{r}(x_k - \overline{x})^\ell, \qquad \ell \in \mathbf{N},$$

where $\overline{x} = (x_1 + \cdots + x_r)/r$. This result and representations up to $\nu = 10$ were worked out by our Summer Intern Tim Huegerich, an undergraduate student from Rice University in Houston. For $1 \le \nu \le 3$, we have

$$\begin{aligned}
F_1(x_1, \ldots, x_r) &= r^2\sigma_2, \\
F_2(x_1, \ldots, x_r) &= \frac{r^2(r^2 - 3r + 3)}{2}\,\sigma_2^2 - \frac{r^2(r-1)}{2}\,\sigma_4, \\
F_3(x_1, \ldots, x_r) &= \frac{(r-2)(r-1)r^2}{3}\,\sigma_6 - \frac{(r-2)(r^2 - 5r + 10)r^2}{2}\,\sigma_4\sigma_2 - \\
&\quad \frac{(3r^2 - 15r + 20)r^2}{3}\,\sigma_3^2 + \frac{(r-2)(r^2 - 7r + 15)r^3}{6}\,\sigma_2^3.
\end{aligned}$$

7.4.3 Explicit form of an approximate dual

We now approach the task of constructing approximate duals of the B-spline basis. The Gramian matrix of the L^2-normalized B-splines is

$$\Gamma = \int_I \Phi^B_{\mathbf{t};m}(x)^T \Phi^B_{\mathbf{t};m}(x)\,dx = \left[(d_{\mathbf{t};m,k} d_{\mathbf{t};m,\ell})^{-1/2}\langle N_{\mathbf{t};m,k}, N_{\mathbf{t};m,\ell}\rangle\right]_{k,\ell\in \mathrm{K}_m}. \qquad (4.27)$$

This defines an spd banded matrix, whose upper and lower operator bounds on ℓ^2 are the Riesz bounds of $\Phi^B_{\mathbf{t};m}$. The matrix is totally positive, see [28]. Its inverse Γ^{-1} is a full matrix (if $m \ge 2$), whose entries decay exponentially; namely, Demko's result [27] assures that

$$\Gamma^{-1}_{k,\ell} \le c\left(\frac{\kappa^2 - 1}{\kappa^2 + 1}\right)^{|k-\ell|/4r},$$

where κ is the condition number of Γ (in ℓ^2) and r its bandwidth.

The dual Riesz basis of the spline space $S_{\mathbf{t},m} \cap L^2(I)$ is given by $\widetilde{\Phi} = \Phi^B_{\mathbf{t};m}\Gamma^{-1}$. The dual basis functions have global support in I for $m \ge 2$. The kernel

$$K(x, y) = \Phi^B_{\mathbf{t};m}(x)(\widetilde{\Phi}(y))^T \qquad (4.28)$$

defines the kernel of the orthoprojection

$$\Pi_{\mathbf{t};m} f := \int_I f(y) K(x, y)\,dy, \qquad (4.29)$$

which maps $L^2(I)$ into $S_{\mathbf{t};m} \cap L^2(I)$. The result of the recent proof of de Boor's conjecture by A. Shadrin [55] states that there exists a constant C_m that neither depends on the knot vector nor depends on the interval I, such that this operator has operator bound C_m, if it is considered as an operator on any $L^p(I)$, for $1 \le p \le \infty$. Equivalently, the kernel K in (4.28) satisfies

$$\sup_{x\in I}\int_I |K(x, y)|\,dy \le C_m. \qquad (4.30)$$

Moreover, if the knots increase at most polynomially, that is, with constants $C, r > 0$ we have

$$|t_k| \le C(1 + |k|)^r \quad \text{for all } k, \qquad (4.31)$$

then the kernel $K(x, y)$ decays exponentially and the operator in Eq. (4.29) reproduces all polynomials of degree up to $m - 1$.

The terminology of approximate duals of B-splines (or other Riesz bases) is introduced in our paper [16] with W. He. For any matrix S, with row and column indices in K_m, we define the function vector

$$\Phi^S_{\mathbf{t};m} = [N^S_{\mathbf{t};m,k}]_{k\in\mathrm{K}_m} := \Phi^B_{\mathbf{t};m}\, S \tag{4.32}$$

and the kernel

$$K^S(x,y) := \Phi^B_{\mathbf{t};m}(x)(\Phi^S_{\mathbf{t};m}(y))^T. \tag{4.33}$$

Definition 7.4.1 Let $\mu \in \mathbf{N}$, and let S be a banded spd matrix with row and column indices in K_m. The vector $\Phi^S_{\mathbf{t};m}$ of splines is called an *approximate dual of order* μ of the B-spline basis $\Phi^B_{\mathbf{t};m}$, if the kernel K^S in (4.33) satisfies

$$\sup_{x\in I}\int_I |K^S(x,y)|\,dy < \infty, \tag{4.34}$$

and

$$K(x,y) - K^S(x,y) = \sum_{k,\ell} a_{k,\ell}\frac{d^\mu}{dx^\mu}N^B_{\mathbf{t};m+\mu,k}(x)\frac{d^\mu}{dy^\mu}N^B_{\mathbf{t};m+\mu,\ell}(y), \tag{4.35}$$

where $A = [a_{k,\ell}]_{k,\ell}$ is a real and symmetric matrix and the sum in (4.35) converges almost everywhere.

We give several explanations and examples.

Remark 7.4.2 (a) The integer μ in Definition 7.4.1 describes the order of vanishing moments of the difference $K - K^S$, as specified in (4.35). Since the knots t_k grow at most polynomially, as assumed in (4.31), a consequence of (4.35) is given by

$$\int_I y^\nu\, K^S(x,y)\,dy = x^\nu, \qquad \text{a.e. } x \in I, \quad 0 \le \nu \le \min(m,\mu) - 1. \tag{4.36}$$

In other words, the kernel K^S defines a quasi-interpolation operator that reproduces polynomials of degree $\min(m,\mu) - 1$. Note that the integral in (4.36) is defined for almost all x, since K^S has compact support. In [16, Theorem 2] we show that, for a bounded interval I, the conditions (4.35) and (4.36) are equivalent. Note also, that identity (4.35) is equivalent to the matrix equation

$$\Gamma^{-1} - S = E^B_{\mathbf{t};m,\mu}\; A\; (E^B_{\mathbf{t};m,\mu})^T, \tag{4.37}$$

see (4.15).

(b) The assumption that S be a banded matrix merely means that every $N^S_{\mathbf{t};m,k}$ has compact support. This condition is always satisfied and therefore not meaningful, if $I = [a,b]$. What we have in mind for the construction of approximate duals can be better explained by the term "local support" rather than compact support. We wish to find approximate duals $\Phi^S_{\mathbf{t};m}$ whose matrix S has bandwidth $\le c\mu$, where c is a universal constant (we will later choose $c = 1$). This assures that, regardless of the number of knots in the interval $[a,b]$, the splines N^S are linear combinations of at most $2c\mu - 1$ B-splines. This matches the notion of locality that is usually employed for quasi-interpolants.

(c) If I is a bounded interval $[a,b]$, the condition (4.34) is again not meaningful, since the sum in (4.32) is finite and each B-spline $N^B_{\mathbf{t};m,k}$ has finite L^1-norm. It is not trivial, however, to construct approximate duals such that the uniform boundedness in (4.34) is achieved, where the upper bound can be chosen independently of the interval I and the knot vector $\mathbf{t}$. We will show in 7.4.3, that the construction of an "optimal" approximate dual in this section does achieve this goal.

The following examples present special cases of the general construction principle, that follows in 7.4.2. The first two examples also show how the shift-invariant setting of Sections 2 and 3 appears as a special case.

Example 7.4.1 (a) We first consider the knot vector $\mathbf{t} = \mathbf{Z}$. The L^2-normalized B-splines are $N^B_{\mathbf{t};m,k} = N_{\mathbf{t};m,k} = N_m(\cdot - k)$, where N_m is the cardinal B-spline in Section 2.1. The case $m = \mu = 2$ was already analyzed in Example 1, where the Laurent polynomial VMR-function $S(z) = (8 - z - z^{-1})/6$ gives rise to an approximate dual of N_m. The analogue in the time-domain is defined by the bi-infinite Toeplitz matrix

$$S = \begin{bmatrix} \ddots & \ddots & \ddots & & \\ & -\frac{1}{6} & \frac{4}{3} & -\frac{1}{6} & \\ & & -\frac{1}{6} & \frac{4}{3} & -\frac{1}{6} \\ & & \ddots & \ddots & \ddots \end{bmatrix}.$$

This matrix defines the approximate dual

$$N^S_{\mathbf{t};2,0}(x) = (8N_m(x) - N_m(x-1) - N_m(x+1))/6,$$
$$N^S_{\mathbf{t};2,k}(x) = N^S_{\mathbf{t};2,0}(x-k), \quad k \in \mathbf{Z},$$

of order 2. Identity (4.37) is satisfied, where A is the bi-infinite Toeplitz matrix whose entries $a_{k,\ell} = a_{k-\ell}$ are defined by the Laurent series

$$\frac{\Phi(z)^{-1} - S(z)}{(1-z)^2(1-1/z)^2} = \sum_{k=0}^{\infty} a_k z^k.$$

(b) Let $\mathbf{t}$ be the set of all integers repeated with multiplicity 2. The splines of order 4 with knot vector $\mathbf{t}$ are the cubic splines with double integer knots studied in Example 8. We obtain that $N^B_{\mathbf{t};4,2k+\ell} = \sqrt{2}N_{\mathbf{t};4,2k+\ell} = \sqrt{2}N_{4,2;\ell}(\cdot - k)$, $\ell = 0, 1$. Here, the time-domain analogue of the VMR matrix S in (3.22) is the bi-infinite block Toeplitz matrix

$$S = \frac{1}{5832} \begin{bmatrix} \ddots & -48 & & & & \\ & 1689 & -387 & & & \\ & -5767 & 1689 & -48 & & \\ & 18088 & -9432 & 1689 & -387 & \\ & -9432 & 18088 & -5767 & 1689 & \\ & 1689 & -5767 & 18088 & -9432 & \\ & -387 & 1689 & -9432 & 18088 & \\ & & -48 & 1689 & -5767 & \ddots \\ & & & -387 & 1689 & \\ & & & & -48 & \end{bmatrix}$$

whose main diagonal is constant 18088/5832. It satisfies (4.37) with $\mu = 4$.

(c) The matrices S in examples (a)–(b) remain unchanged, if the knot vector is scaled by a constant $h > 0$. This effect is due to the L^2-normalization of the B-splines $N^B_{\mathbf{t};m,k}$, which incorporates such scaling.

One of the achievements in our joint paper [16] with W. He is the exact computation of the matrix S of minimal bandwidth such that $\Phi^S_{\mathbf{t};m}$ is an approximate dual. The functions F_ν in (4.25) play an important role in this computation. Particular instances of this result appeared in Theorems 7.2.4 and 7.3.2 and were formulated in the Fourier domain. Here, we give a representation that has a similar form as in (2.35) and (3.21), but appears in the time-domain. In doing so, we must substitute the constants u_k in (2.35) and the constant diagonal matrices U_ν in (3.21) by new diagonal matrices U_ν with positive diagonal entries.

Theorem 7.4.2 *For every $1 \le \mu \le m$, there exists a unique spd matrix S with bandwidth μ such that $\Phi^S_{\mathbf{t};m}$ is an approximate dual of order μ for the spline basis $\Phi^B_{\mathbf{t};m}$. Moreover, S has the form*

$$S = I + E^B_{\mathbf{t};m,1} U_{\mathbf{t};m,1} (E^B_{\mathbf{t};m,1})^T + \cdots + E^B_{\mathbf{t};m,\mu-1} U_{\mathbf{t};m,\mu-1} (E^B_{\mathbf{t};m,\mu-1})^T, \tag{4.38}$$

where the matrices $E^B_{\mathbf{t};m,\nu}$ are defined in (4.15) and $U_{\mathbf{t};m,\nu}$ are diagonal matrices with diagonal entries

$$u_k^{(\nu)} := \frac{m!(m-\nu-1)!}{(m+\nu)!(m+\nu-1)!} F_\nu(t_{k+1}, \ldots, t_{k+m+\nu-1}). \tag{4.39}$$

The corresponding kernel K^S has the form

$$\begin{aligned} K^S(x,y) = & \sum_{k \in \mathrm{K}_m} N^B_{\mathbf{t};m,k}(x) N^B_{\mathbf{t};m,k}(y) + \\ & \sum_{\nu=1}^{\mu-1} \sum_{k \in \mathrm{K}_{m+\nu}} u_k^{(\nu)} \frac{d^\nu}{dx^\nu} N^B_{\mathbf{t};m+\nu,k}(x) \frac{d^\nu}{dy^\nu} N^B_{\mathbf{t};m+\nu,k}(y). \end{aligned} \tag{4.40}$$

The positivity of all diagonal elements of the matrices U_ν in (4.38) follows directly from (4.39). Hence, it is trivial to conclude that S is an spd matrix. It is worthwhile to note the following.

Proposition 7.4.1 . *The matrix $S = S(\mathbf{t})$ in (4.38) is invariant under shifts and scaling of the knot vector; that is,*

$$S(\mathbf{t} - c) = S(h\mathbf{t}) = S(\mathbf{t}) \tag{4.41}$$

holds for any $c \in \mathbf{R}$ and $h > 0$, where $\mathbf{t} - c = \{t_k - c,\ k \in \mathrm{K}\}$.

Proof: The functions F_ν in (4.25) satisfy

$$F_\nu(x_1 - c, \ldots, x_r - c) = F_\nu(x_1, \ldots, x_r),$$
$$F_\nu(hx_1, \ldots, hx_r) = h^{2\nu} F_\nu(x_1, \ldots, x_r),$$

while

$$E^B_{\mathbf{t}-c;m,\nu} = E^B_{\mathbf{t};m,\nu},$$
$$E^B_{h\mathbf{t};m,\nu} = h^{-\nu} E^B_{\mathbf{t};m,\nu}.$$

Hence, each summand in (4.38) is invariant under the shift of the knots and the factors $h^{2\nu}$ and $h^{-2\nu}$ in $E^B_{h\mathbf{t};m,\nu} U_{h\mathbf{t};m,\nu} (E^B_{h\mathbf{t};m,\nu})^T$ cancel each other. This confirms the equation (4.41).

Remark 7.4.3 We explain in more detail, how the matrix S in the previous result is related to the VMR Laurent polynomial (matrix) $S(z)$ in (2.35) and (3.21). If $\mathbf{t} = \mathbf{Z}$, the matrix U_ν in (4.38) is a bi-infinite diagonal Toeplitz matrix and S is a bi-infinite banded Toeplitz matrix with bandwidth μ. The symbol of S is the VMR Laurent polynomial $S(z)$ in (2.35). Hence, the positivity of the coefficients $u_k^{(\nu)}$ is equivalent to the positivity of u_k in (2.35). Likewise, if $\mathbf{t}$ is the set of all integers repeated with multiplicity r, S is a bi-infinite block Toeplitz matrix with $r \times r$ blocks, and its symbol is a slight modification of the VMR Laurent polynomial matrix $S(z)$ in (3.21), see also Examples 8 and 9(b). The only difference appears in the use of the L^2-normalized B-splines for the definition (4.38) versus the L^∞-normalized B-splines in (3.21). Once again, the positivity of the diagonal entries of $U_\nu(z)$ in (3.21) can be viewed as a consequence of 7.4.2.

The property (4.34) of boundedness of the kernel K^S in (4.40) is a consequence of the following result that is proven in [16, Section 5.7].

Theorem 7.4.3 *Let $\mathbf{t}$ be a knot vector and $u_k^{(\nu)}$ be the numbers in (4.39), $0 \le \nu \le m-1$. Then the kernel*

$$K^{(\nu)}_{\mathbf{t};m}(x, y) := \frac{\partial^{2\nu}}{\partial x^\nu \partial y^\nu} \sum_{k \in \mathrm{K}_{m+\nu}} u_k^{(\nu)} N^B_{\mathbf{t};m+\nu,k}(x) N^B_{\mathbf{t};m+\nu,k}(y)$$

satisfies

$$\int_I K^{(\nu)}_{\mathbf{t};m}(x, y)\, dy = \delta_{\nu,0}, \qquad x \in I, \tag{4.42}$$

and

$$\int_I |K^{(\nu)}_{\mathbf{t};m}(x, y)|\, dy \le \frac{2^\nu (m + \nu - 1)!}{\nu!(m-1)!}, \qquad x \in I. \tag{4.43}$$

Moreover, $K^{(0)}_{\mathbf{t};m}(x, y) \ge 0$ for all $x, y \in I$.

This result confirms that the kernel K^S of the approximate dual in (4.40) satisfies a similar condition of uniform boundedness for all intervals I and all knot vectors $\mathbf{t}$ as the ortho-kernel K in (4.28), see (4.30). We hope to extend our arguments in the future, in order to give a new and accessible proof to de Boor's conjecture.

A special case of 7.4.2 is the case of Bernstein polynomials, which we describe in the following example.

Example 7.4.2 For $m \in \mathbf{N}$, we let $n := m-1$ and $\mathbf{t} = \{0, \ldots, 0, 1, \ldots, 1\}$ with m-fold knots at 0 and 1 and no interior knots. The Bernstein basis of the space of polynomials of degree n is given by

$$p_{n,k}(x) = \binom{n}{k} x^k (1-x)^{n-k}, \qquad 0 \le k \le n.$$

The L^2-normalized B-splines on $[0,1]$ with respect to this knot vector are $N^B_{\mathbf{t};m,k-m+1} = \sqrt{n+1}p_{n,k}$. More generally, we have

$$N^B_{\mathbf{t};m+\nu,k-m+1} = \sqrt{n+\nu+1}p_{n+\nu,k+\nu} \tag{4.44}$$

for all $0 \le \nu \le n$, $0 \le k \le n-\nu$. In [16], it is shown that the reproducing kernel of the space of polynomials of degree at most n on $[0,1]$ can be written as

$$K(x,y) = (n+1)\sum_{\nu=0}^{n} \frac{1}{(2\nu)!} \sum_{k=0}^{n-\nu} \frac{\binom{k+\nu}{\nu}\binom{n-k}{\nu}}{\binom{n+\nu}{2\nu}\binom{n+\nu}{\nu}} \frac{d^\nu}{dx^\nu} p_{n+\nu,k+\nu}(x) \frac{d^\nu}{dy^\nu} p_{n+\nu,k+\nu}(y). \tag{4.45}$$

The partial sum for $0 \le \nu \le \mu - 1$ provides the kernel K^S that is associated with the approximate dual of order μ in 7.4.2. More precisely, if we make use of (4.44), then the partial sum of (4.45) becomes (4.40), where

$$u^{(\nu)}_{k-m+1} = \frac{m}{m+\nu} \frac{\binom{k+\nu}{\nu}\binom{m-1-k}{\nu}}{\binom{m-1+\nu}{2\nu}\binom{m-1+\nu}{\nu}}, \qquad 0 \le k \le m-1-\nu, \qquad 0 \le \nu \le m-1.$$

This result is related to work by Sablonnière [53] on quasi-interpolants of Bernstein polynomials. A multivariate analogue of the identity (4.45) for the reproducing kernel was recently found in [39]. We also note that the difference of the inverse Gramian of the B-spline basis $\Phi^B_{\mathbf{t};m}$ and the matrix S in (4.38) is given by

$$\begin{aligned}\Gamma^{-1} - S =& \sum_{\nu=\mu}^{m-1} E^B_{\mathbf{t};m,\nu} U_{\mathbf{t};m,\nu} (E^B_{\mathbf{t};m,\nu})^T \\ =& E^B_{\mathbf{t};m,\mu} \left[U_{\mathbf{t};m,\mu} + \sum_{\nu=\mu+1}^{m-1} E^B_{\mathbf{t};m+\mu,\nu-\mu} U_{\mathbf{t};m,\nu} (E^B_{\mathbf{t};m+\mu,\nu-\mu})^T \right] (E^B_{\mathbf{t};m,\mu})^T.\end{aligned}$$

The sum in brackets is an explicit expression of the matrix A in (4.37).

An algorithm for the computation of S for arbitrary (finite) knot vectors and $1 \le \mu \le 4$ is included in MATLAB syntax in the appendix. We make use of this algorithm for the computations in the following examples.

Example 7.4.3 (a) We let $m = 2$, $I = [0, \infty)$ and $\mathbf{t} = \{0, 0, 1, 2, 3, \ldots\}$ with double knot $t_{-1} = t_0 = 0$. The B-spline $N^B_{\mathbf{t};2,-1}(x) = \sqrt{2}(1 - x)\chi_{[}0, 1]$ has a double knot at 0. All other B-splines $N^B_{\mathbf{t};2,k} = N_m(\cdot - k)$, $k \geq 0$, are precisely those of part (a) in Example 9. The matrix

$$S = \begin{bmatrix} \frac{3}{2} & -\frac{1}{2\sqrt{2}} & & & \\ -\frac{1}{2\sqrt{2}} & \frac{17}{12} & -\frac{1}{6} & & \\ & -\frac{1}{6} & \frac{4}{3} & -\frac{1}{6} & \\ & & -\frac{1}{6} & \frac{4}{3} & \ddots \\ & & & \ddots & \ddots \end{bmatrix}$$

defines an approximate dual of order 2 on the interval $[0, \infty)$. The interval $I = [0, n + 1]$ is obtained by assigning knots $t_{n+1} = t_{n+2} = n + 1$. The last B-spline is $N^B_{\mathbf{t};2,n}(x) = N^B_{\mathbf{t};2,-1}(n + 1 - x)$, and the approximate dual is defined by the matrix

$$S = \begin{bmatrix} \frac{3}{2} & -\frac{1}{2\sqrt{2}} & & & & & \\ -\frac{1}{2\sqrt{2}} & \frac{17}{12} & -\frac{1}{6} & & & & \\ & -\frac{1}{6} & \frac{4}{3} & \ddots & & & \\ & & \ddots & \ddots & \ddots & & \\ & & & \ddots & \frac{4}{3} & -\frac{1}{6} & \\ & & & & -\frac{1}{6} & \frac{17}{12} & -\frac{1}{2\sqrt{2}} \\ & & & & & -\frac{1}{2\sqrt{2}} & \frac{3}{2} \end{bmatrix}.$$

Hence, approximate duals in the shift-invariant setting on intervals with bounded endpoints are obtained by a modification of the first and last two rows and columns of S, leaving the main part of the matrix as in Example 9(a).

(b) A modification of 5 columns of the bi-infinite Toeplitz matrix S in Example 9(b) is needed, in order to define an approximate dual for $m = \mu = 4$ with double knots in $\mathbf{N}$ and a quadruple knot 0. The first 6 columns of $5832S$ are

$$\begin{bmatrix} 17496 & -16038 & 2916\sqrt{2} & -729\sqrt{2} & 0 & 0 \\ -16038 & 37098 & -\frac{21897}{2}\sqrt{2} & \frac{6957}{2}\sqrt{2} & -144\sqrt{2} & 0 \\ 2916\sqrt{2} & -\frac{21897}{2}\sqrt{2} & 24237 & -23733/2 & 1881 & -387 \\ -729\sqrt{2} & \frac{6957}{2}\sqrt{2} & -23733/2 & 19187 & -5879 & 1689 \\ 0 & -144\sqrt{2} & 1881 & -5879 & 18104 & -9432 \\ 0 & 0 & -387 & 1689 & -9432 & 18088 \\ 0 & 0 & 0 & -48 & 1689 & -5767 \\ 0 & 0 & 0 & 0 & -387 & 1689 \\ 0 & 0 & 0 & 0 & 0 & -48 \end{bmatrix}.$$

The subsequent columns are those of Example 9(b). The same modification occurs, if a knot of multiplicity 4 is introduced at the right endpoint b. Then the last 6 columns have the form which is obtained by flipping the above matrix up/down and left/right.

As mentioned previously, the positive definiteness of the matrix S in (4.38) is obvious. This property, however, is not sufficient for the construction of spline frames to be presented in the next section. Instead, an analogue of the positivity condition (1.16) must be developed. For this purpose, we let

$$\mathbf{t}_j \subset \mathbf{t}_{j+1}, \qquad j \geq 0,$$

be two knot vectors on the interval I as in (4.17). The knot refinement can be almost arbitrary, if all knots have multiplicity at most m and the number of knots inserted between two adjacent knots of $\mathbf{t}_j$ is bounded by a constant n_j. This condition is quite realistic. It is required by the method of proof in [16], where the refinement matrix $P_{j;m}$ in (4.18) is written as a finite product whose factors are block diagonal matrices, with blocks of the form (4.20). Under these very weak conditions, the following is proven in [16, Sections 5.5–5.6].

Theorem 7.4.4 *Let $1 \leq \mu \leq m$. Let S_j and S_{j+1} denote the matrices in (4.38) for knot vectors $\mathbf{t}_j \subset \mathbf{t}_{j+1}$ that satisfy the aforementioned conditions. Then the matrix*

$$S_{j+1} - P^B_{j,m} S_j (P^B_{j,m})^T \tag{4.46}$$

is positive semi-definite and banded, and there exists a positive semi-definite and banded matrix Z_j, with row and column indices in $\mathrm{K}^{(j+1)}_{m+\mu}$, such that

$$S_{j+1} - P^B_{j,m} S_j (P^B_{j,m})^T = E^B_{j+1;m,\mu} Z_j (E^B_{j+1;m,\mu})^T. \tag{4.47}$$

Moreover, the bandwidth of Z_j equals the bandwidth of the matrix on the left-hand side of (4.47) minus μ.

We will show in the next section, that the matrix $S_{j+1} - P^B_{j;m} S_j (P^B_{j;m})^T$ serves the same purpose for the construction of tight frames as the Laurent polynomial matrix $\mathcal{M}(z)$ in (2.46) and (3.17) does for the shift-invariant setting. The assertion of positive semi-definiteness generalizes the positivity condition (1.16), which was necessary for the construction of tight frame generators with knots at the half-integers. Moreover, the matrix Z_j in the factorization (4.47) is the time-domain analogue of $\mathcal{M}_0(z)$ in Example 7.2.1 and (3.18). The factor $E^B_{j;m,\mu}$ and its transpose are needed, in order to construct spline functions with knots in $\mathbf{t}_{j+1}$, which have μ vanishing moments, see Lemma 7.4.1. Any symmetric factorization

$$S_{j+1} - P^B_{j;m} S_j (P^B_{j;m})^T = \underbrace{E^B_{j+1;m,\mu} R}_{=:\, Q_j}\; R^T (E^B_{j+1;m,\mu})^T$$

can be employed in order to define a vector of spline functions

$$\Psi_j = [\psi_{j,k}]_k := \Phi^B_{j+1;m}\, Q_j, \tag{4.48}$$

where each $\psi_{j,k}$ has μ vanishing moments. This shows that there is a close relation of VMR functions and approximate duals. We will come back to this point in Section 5, where we discuss the construction of a tight frame of $L^2(I)$.

A more precise statement can be made about the sparsity of the matrix $Z := Z_j$ in (4.47). This is important for adaptive refinements of the knot vector, where the number of new knots that are inserted between two adjacent knots of $\mathbf{t}_j$ varies (between 0 and n_j, say). Roughly speaking, Z has zero rows and columns where the left-hand side of (4.47) has zero rows or columns. This situation occurs if there are large regions, where no knots are inserted. Moreover, we can define the lower profile of a matrix $A = [a_{k,\ell}]_{k,\ell \in \mathrm{K}}$ by the sequence

$$\lambda_\ell(A) := \max(\{\ell - 1\} \cup \{k;\ a_{k,\ell} \neq 0\}), \qquad \ell \in \mathrm{K}.$$

If A has bandwidth μ, then $\lambda(\ell) \leq \ell + \mu - 1$, of course. The factorization in (4.47) leads to a symmetric matrix Z, whose lower profile equals, up to an identification of appropriate columns of the two matrices, the lower profile of the matrix in (4.46) reduced by μ. Instead of going into more technical details, we explain the structure of Z by an example.

Example 7.4.4 We let $m = \mu = 4$ and $\mathbf{t}_0 = [0, 0, 0, 0, 1, 2, \ldots, 19, 20, 20, 20, 20]$. The matrix S_0 in (4.38) has dimension 23×23 and bandwidth 4. If we insert simple knots at 3.5 and 4.5 only, the matrix S_1 has dimension 25. The matrix $S_1 - P^B_{0;4} S_0 (P^B_{0;4})^T$ is symmetric, positive semi-definite and has rank 9. The lower profile of its first 13 columns is given by the sequence of row indices (putting $(1,1)$ as the upper left corner) $[5, 7, 8, 9, 10, 11, 12, 12, 13, 13, 13, 13, 13]$. All of the columns $14 - 25$ are zero. Hence, the points of insertion do not affect this region. After performing the factorization in (4.47), we obtain the banded, symmetric and positive semi-definite matrix Z of dimension 21×21, whose columns $10 - 21$ vanish. The lower profile of the first 9 columns of Z is given by $[1, 3, 4, 5, 6, 7, 8, 8, 9]$, and the 9×9 block in the upper left corner has full rank. The Cholesky factorization $Z = R * R^T$ can be computed, where the upper left block of $100 * R$ is

$$\begin{bmatrix}
0.94491 & & & & & & & & \\
 & 3.93398 & 0.39340 & & & & & & \\
 & & 6.24787 & 1.54318 & & & & & \\
 & & & 5.78332 & 2.77857 & & & & \\
 & & & & 4.77005 & 3.36881 & & & \\
 & & & & & 4.97520 & 2.17523 & & \\
 & & & & & & 6.29258 & 0.45862 & \\
 & & & & & & & 5.86692 & \\
 & & & & & & & & 3.69358
\end{bmatrix}.$$

7.5 TIGHT SPLINE FRAMES WITH NON-UNIFORM KNOTS

We show in this section that approximate duals of B-splines appear naturally in the characterization and construction of tight spline frames. In particular, the result of 7.4.4 is an important tool for this construction. Moreover, a characterization of all tight spline frames is given, and three comprehensive examples are provided.

First, we must define the spline MRA of $L^2(I)$ in the setting of splines with nonuniform knots. A nested sequence of knot vectors

$$\mathbf{t}_0 \subset \mathbf{t}_1 \subset \mathbf{t}_2 \subset \dots \tag{5.1}$$

is given, where each knot vector satisfies (4.1) and, if I is unbounded, (4.2) and (4.31) as well. If I has a finite endpoint, then $\mathbf{t}_0$, and therefore all $\mathbf{t}_j$'s, are supposed to have an m-fold knot there. Moreover, there is a bound $n \in \mathbf{N}$ for the maximum number of knots inserted between two knots, namely

$$n := \max_{j \geq 0} \max_k \#([t_k^{(j)}, t_{k+1}^{(j)}] \cap \mathbf{t}_{j+1}) < \infty,$$

and the knot vectors become dense in I, which means that

$$\lim_{j \to \infty} \sup_k (t_{k+1}^{(j)} - t_k^{(j)}) = 0. \tag{5.2}$$

Then it follows from standard arguments in spline approximation, that the spaces

$$V_j := \mathcal{S}(\mathbf{t}_j; m) \cap L^2(I), \qquad j \geq 0,$$

are dense in $L^2(I)$.

We define function families

$$\Psi_j := [\psi_{j,k};\ k \in \mathrm{M}_j] = \Phi_{j+1;m}^B\, Q_j, \qquad j \geq 0, \tag{5.3}$$

where Q_j is a real matrix with row indices in K_{j+1} and column indices in a set denoted by M_j. The localization properties of this family are defined as follows.

Definition 7.5.1 The sequence of families Ψ_j, $j \geq 0$, is called *locally supported* (with respect to the B-spline bases Φ_j^B, $j \geq 1$), if there exist integers n_1, n_2 (not depending on j) such that each $\psi_{j,k} \in \Psi_j$ is a linear combination of at most n_1 consecutive B-splines, and at every point $x \in I$ at most n_2 functions $\psi_{j,\ell} \in \Psi_j$ do not vanish.

In other words, each family Ψ_j is locally finite, and the supports of the functions $\psi_{j,k}$ shrink at the same rate as the supports of the B-splines when j tends to infinity.

Recall from (4.38) the special definition of the approximate dual of the B-spline basis $\Phi_{j;m} := \Phi_{\mathbf{t}_j;m}$, whose matrix $S_j := S(\mathbf{t}_j)$ has bandwidth μ. We define the quadratic form

$$T_j f := \sum_{k \in \mathrm{K}_j} \langle f, N_{\mathbf{t}_j;m,k}^B \rangle\ S_j\ \langle N_{\mathbf{t}_j;m,k}^B, f \rangle, \qquad f \in L^2(I). \tag{5.4}$$

It follows from (5.2) that for every $\epsilon > 0$, there exists $j_0 \in \mathbf{N}$ such that

$$K^{S_j}(x,y) = 0 \quad \text{for all} \quad |x-y| \geq \epsilon \text{ and } j \geq j_0. \tag{5.5}$$

Together with (4.42) and the uniform bounds (4.43), we obtain

$$\lim_{j \to \infty} T_j f = \|f\|^2 \quad \text{for all} \quad f \in L^2(I). \tag{5.6}$$

This is one of the key identities that we employ for the proof of the following result.

Theorem 7.5.2 *Let $m, \mu \in \mathbf{N}$ with $1 \leq \mu \leq m$ and let $\{\mathbf{t}_j\}_{j \geq 0}$ be a sequence of knot vectors as described above. Assume that banded matrices R_j are defined, such that*

$$S_{j+1} - P^B_{j,m} S_j (P^B_{j,m})^T = E^B_{j+1,m,\mu} R_j R_j^T (E^B_{j+1,m,\mu})^T. \tag{5.7}$$

Then the families

$$\Psi_j = [\psi_{j,k}]_{k \in \mathrm{M}_j} := \Phi_{j;m} Q_j, \qquad \textit{with } Q_j := E^B_{j+1,m,\mu} R_j \textit{ and } j \geq 0, \tag{5.8}$$

are locally supported and constitute a tight frame of $L^2(I)$, in the sense that

$$\|f\|^2 = T_0 f + \sum_{j=0}^{\infty} \sum_{k \in \mathrm{M}_j} |\langle f, \psi_{j,k} \rangle|^2 \quad \textit{for all} \quad f \in L^2(I). \tag{5.9}$$

Moreover, all the wavelets $\psi_{j,k}$ in (5.8) have (at least) μ vanishing moments.

The method of proof in [16] is an adaptation of the telescoping argument (2.44). More precisely, the factorization (5.7) allows us to conclude that

$$T_{j+1} f - T_j f = \sum_{k \in \mathrm{M}_j} |\langle f, \psi_{j,k} \rangle|^2, \qquad j \geq 0, \ f \in L^2(I). \tag{5.10}$$

The identity (5.9) follows from (5.6). Moreover, all the functions $\psi_{j,k}$ have μ vanishing moments by Lemma 7.4.1.

7.5.2 shows that the construction of tight frames of splines can be reduced to a simple problem of linear algebra, by means of the appoximate duals of B-splines. For an arbitrary sequence of knot vectors $\mathbf{t}_j$, $j \geq 0$, of at most polynomial growth, we only have to find the factorization (5.7), for each $j \geq 0$, in order to define a tight frame with μ vanishing moments. The theoretical base for this factorization was already given in 7.4.4, where we showed that the matrices $S_{j+1} - P^B_{j,m} S_j (P^B_{j,m})^T$ are positive semidefinite and admit the factorization

$$S_{j+1} - P^B_{j,m} S_j (P^B_{j,m})^T = E^B_{j+1,m,\mu} Z_j (E^B_{j+1,m,\mu})^T$$

with a positive semi-definite and banded matrix Z_j. Therefore, the construction of the matrix Q_j hinges on the existence of the factorization

$$Z_j = R_j R_j^T. \tag{5.11}$$

For finite matrices, this is trivially achieved by the Cholesky factorization of Z_j. For bi-infinite banded Toeplitz matrices, a factorization of Cholesky type, with a banded Toeplitz matrix R_j, is obtained as an application of the Riesz-Fejér Theorem [49, pp. 117-118]. The analogue for bi-infinite banded block Toeplitz matrices was recently obtained in [35]. For other types of infinite matrices, results for the factorization (5.11) with **banded** R_j exist under the additional assumption, that Z_j is strictly positive definite, see [20]. We refer to the survey article of van der Mee et al. [42] for ongoing research in this direction.

Remark 7.5.1 The "coarse scale component" $T_0 f$ in the identity (5.9) is indispensable for the following reason. We only discuss the case where I is a bounded interval. If all the wavelets $\psi_{j,k}$ in (5.9) have μ vanishing moments, then $T_0 p = \|p\|^2$ must hold for all polynomials p of degree $\mu - 1$, since all other summands in (5.9) vanish. Consequently, the kernel K^{S_0} reproduces all polynomials of degree $\mu - 1$. We mentioned in Remark 14(a), that this is equivalent to the fact that S_0 defines an approximate dual of order μ. Hence, the tight frame condition (5.9) is the natural condition to ask, when we define a tight frame based on a spline MRA which starts with a coarsest level V_0.

It is interesting to ask how the property of local support can be made more precise for the wavelets $\psi_{j,k}$. In order to describe the sparsity of the Cholesky decomposition of the matrix $S_{j+1} - P^B_{j,m}S_j(P^B_{j,m})^T$, we need to describe the lower profile of this matrix. For this purpose, we recall the definition of the index sequence $\eta(k)$, $k \in \mathrm{K}_j$, from (4.19) in Section 4.1. This sequence describes the lower profile of $P^B_{j;m}$. A second index sequence $\zeta(i)$, $i \in \mathrm{K}_{j+1}$, is defined that denotes the lower profile of the transpose of $P^B_{j;m}$. In other words, $\eta(k)$ is the largest row index of the nonzero entries in the kth column of $P^B_{j;m}$, and $\zeta(i)$ is the largest column index of the nonzero entries of the ith row of this matrix. It follows by elementary combinatorial arguments, that the lower profile of the matrix $S_{j+1} - P^B_{j,m}S_j(P^B_{j,m})^T$ is given by

$$\ell(i) := \eta(\zeta(i) + \mu - 1).$$

The matrix Q_j in the Cholesky factorization has the same lower profile. Note that no fill-in of nonzero elements occurs, since ℓ is an increasing sequence. If we define $\nu(k)$ to be the number of new knots in $\mathbf{t}_{j+1}$, that lie in the open interval $(t^{(j)}_k, t^{(j)}_{k+m+\mu-1})$, we also obtain that

$$i \le \ell(i) \le i + \nu(\kappa(i)) + \mu - 1.$$

These considerations lead to the following result.

Proposition 7.5.1 *Let S_j be the spd matrix in 7.4.2, and assume that the Cholesky factorization of Z_j in (4.47) exists. Then there exists a factorization*

$$Q_jQ_j^T = S_{j+1} - P^B_{j,m}S_j(P^B_{j,m})^T$$

where Q_j defines the wavelets $\psi_{j,i}$ of a tight frame with μ vanishing moments, and each $\psi_{j,i}$ is a linear combination of at most $\nu(\zeta(i)) + \mu$ consecutive B-splines of the basis $\Phi_{j+1;m}$, starting with $N_{j+1;m,i}$. In particular, the wavelet $\psi_{j,i}$ is a spline in V_{j+1} whose support is contained in $[t^{(j+1)}_i, t^{(j)}_{\zeta(i)+m+\mu-1}]$.

We can compare the previous result with Remark 5, in the case where precisely one knot is inserted between two adjacent knots of $\mathbf{t}_j$. In this particular case, we have $\nu(k)$ is constant $m+\mu-1$. Hence, the number $\nu(\zeta(i))+\mu = 2\mu+m-1$ is the same number that we denoted by n_1 in Remark 5 of Section 2.1. This is the number of nonzero B-splines in the representation of one function of a pair $(\psi^{(1)}, \psi^{(2)})$ of minimally supported tight frame generators. This shows, that there is essentially no difference between the support of wavelets for the shift-invariant setting and for nonuniform knot sequences.

The next theorem shows that approximate duals are essential for the characterization of tight frames even in much more generality. A similar result is given for non-spline frames in [16].

Theorem 7.5.3 *Let $1 \le \mu \le m$, S_0 be an spd banded matrix such that $\Phi_{0,m}S_0$ is an approximate dual of order μ, and let locally supported families $\Psi_j := \Phi_{j;m}Q_j$, $j \ge 0$, be defined, where $Q_j = E^B_{j+1,m,\mu}R_j$ with some banded matrix R_j. Then the functions $\psi_{j,k}$ define a tight frame of $L^2(I)$, in the sense of (5.9), if and only if there exist spd banded matrices S_j, $j \ge 1$, such that the following statements hold:*

(i) $\Phi^B_{j;m}S_j$ is an approximate dual of order μ of the B-spline basis $\Phi^B_{j;m}$;
(ii) $\lim_{j\to\infty} T_jf = \|f\|^2$ for all $f \in L^2(I)$;

(iii) $S_{j+1} - P^B_{j;m} S_j (P^B_{j,m})^T = Q_j Q_j^T$.

We end our discussion by giving several examples that explain the general approach sketched in the results of this section.

7.5.1 Piecewise linear tight frames

We discuss the construction of the wavelets $\psi_{0,k}$, where $\mathbf{t}_0 \subset \mathbf{t}_1$ are nested knot vectors. The families Ψ_j, $j \geq 1$, are constructed analogously. Here we consider piecewise linear B-splines and 2 vanishing moments, hence $m = \mu = 2$. The matrices S_0 and S_1 in (4.38) are tridiagonal of dimensions $N_0 + 2$ and $2N_0 + 3$, respectively. The diagonal matrices U_0 and U_1 in (4.39) have diagonal entries

$$u_k^{(0)} = 1 \quad \text{and} \quad u_k^{(1)} = \frac{(t_{k+2}^{(j)} - t_{k+1}^{(j)})^2}{6},$$

for $j = 0, 1$. We present an explicit construction for the case of a bounded interval $[a, b]$, where all interior knots are simple and one "new" knot of $\mathbf{t}_1 \setminus \mathbf{t}_0$ is placed between two adjacent knots of $\mathbf{t}_0$; in other words, we assume that

$$a = \underbrace{t_{-1}^{(1)}}_{=t_{-1}^{(0)}} = \underbrace{t_0^{(1)}}_{=t_0^{(0)}} < t_1^{(1)} < \underbrace{t_2^{(1)}}_{=t_1^{(0)}} < \cdots < \underbrace{t_{2N_0}^{(1)}}_{=t_{N_0}^{(0)}} < t_{2N_0+1}^{(1)} < \underbrace{t_{2N_0+2}^{(1)}}_{=t_{N_0+1}^{(0)}} = \underbrace{t_{2N_0+3}^{(1)}}_{=t_{N_0+2}^{(0)}} = b.$$

The upper index (1) will be dropped from now on. In this case, the factorization

$$S(\mathbf{t}_1) - P^B_{m,0} S(\mathbf{t}_0)(P^B_{m,0})^T = E^B_{1;m,2} Z_0 (E^B_{1;m,1})^T$$

is obtained where Z_0 is a tridiagonal symmetric matrix of dimension $N_1 := 2N_0 + 1$. Instead of a Cholesky decomposition, we choose a more economical factorization $Z_0 = R_0 R_0^T$, where R_0 has the form

$$R_0 = D_1 \begin{bmatrix} t_3 - a & & & & & \\ t_4 - t_3 & 1\ t_1 - t_0 & & & & \\ & t_5 - t_1 & & & & \\ & t_6 - t_5 & 1\ t_3 - t_2 & & & \\ & & t_7 - t_3 & & & \\ r & & t_8 - t_7 & \ddots & & \\ & & & & 1\ t_{2N_0-1} - t_{2N_0-2} & \\ & & & & & b - t_{2N_0-1} \end{bmatrix} D_2$$

and where D_1 and D_2 are diagonal matrices with diagonal entries (indexed from 1 to $2N_0 + 1$) of the form

$$D_{1;k,k} = \frac{2}{\sqrt{t_{k+2} - t_{k-2}}}, \qquad 1 \leq k \leq 2N_0 + 1,$$

$$D_{2;k,k} = \frac{(t_{k+1} - t_{k-1})\sqrt{(t_{k+3} - t_k)(t_k - t_{k-3})}}{12\sqrt{2(t_{k+3} - t_{k-3})}}, \qquad k = 1, 3, \ldots, 2N_0 + 1,$$

and

$$D_{2;k,k} = \tfrac{1}{12\sqrt{2}}\Big((t_{k+2} - t_{k-1})(t_{k+1} - t_{k-2}) \times$$
$$((t_k - t_{k-1})(t_k - t_{k-2})(t_{k+2} - t_{k+1}) + (t_{k+1} - t_k)(t_{k+2} - t_k)(t_{k-1} - t_{k-2}))\Big)^{1/2}$$

for all $k = 2, 4, \ldots, 2N_0$. Here we let $t_k := t_k^{(1)}$ and $t_{-2} := a$, $t_{2N_0+4} := b$. The wavelet family Ψ_0 is then defined by the coefficient matrix $Q_0 := E_{\mathbf{t}_0;2,2}R_0$. Note that the multiplication by $E_{\mathbf{t}_0;2,2}$ spreads the support of the column vectors of R_0 by 2. Hence, the wavelets $\psi_{0,2k}$, $1 \le k \le N_0$, have a 3-tap coefficient sequence. The support of $\psi_{0,2k}$ is the interval $[t_{k-1}^{(0)}, t_{k+1}^{(0)}]$ and all knots are simple. The wavelets $\psi_{0,2k+1}$, $1 \le k \le N_0 - 1$, have a 5-tap coefficient sequence, support in $[t_{k-1}^{(0)}, t_{k+2}^{(0)}]$, and simple knots. There are only two wavelets, namely $\psi_{0,1}$ and ψ_{0,N_1}, which have a double knot at one of the endpoints of the interval $[a, b]$. These wavelets have a 4-tap coefficient sequence. $\psi_{0,1}$ has the support $[a, t_2^{(0)}]$, and ψ_{0,N_1} has the support $[t_{N_0-1}^{(0)}, b]$. Hence, our construction leads to $2N_0 - 1$ interior wavelets and 2 boundary wavelets. All wavelets have two vanishing moments, that is

$$\int_a^b x^\nu \psi_{0,k}(x)\,dx = 0 \qquad \text{for all } 1 \le k \le N_1, \quad \nu = 0, 1.$$

In the special case, where the knots in $\mathbf{t}_0$ are equidistant (with stepsize h_0) and the new knots in $\mathbf{t}_1$ are placed in the middle of each knot interval, our construction leads to

$$Q_0 = \frac{1}{12}\left[\begin{array}{cccccccccc}
6\sqrt{6} & & & & & & & & & \\
-9\sqrt{3} & 6 & \sqrt{6} & & & & & & & \\
2\sqrt{3} & -12 & 2\sqrt{6} & & & & & & & \\
\sqrt{3} & 6 & -6\sqrt{6} & 6 & \sqrt{6} & & & & & \\
 & & 2\sqrt{6} & -12 & 2\sqrt{6} & & & & & \\
 & & \sqrt{6} & 6 & -6\sqrt{6} & & & & & \\
 & & & & 2\sqrt{6} & \ddots & & & & \\
 & & & & \sqrt{6} & & & & & \\
 & & & & & & 6 & \sqrt{6} & & \\
 & & & & & & -12 & 2\sqrt{6} & & \\
 & & & & & & 6 & -6\sqrt{6} & 6 & \sqrt{3} \\
 & & & & & & & 2\sqrt{6} & -12 & 2\sqrt{3} \\
 & & & & & & & \sqrt{6} & 6 & -9\sqrt{3} \\
 & & & & & & & & & 6\sqrt{6}
\end{array}\right]$$

The interior wavelets (with coefficient sequences in columns 2 to $N_1 - 1$) are shifts of the two generators $\psi_{0,2}$ and $\psi_{0,3}$, namely

$$\psi_{0,2k+2}(x) = \psi_{0,2}(x - kh_0), \quad \psi_{0,2k+3}(x) = \psi_{0,3}(x - kh_0), \qquad 1 \le k \le N_0 - 1.$$

Moreover, all of these interior wavelets are symmetric. If we fix the stepsize $h_0 = 1$, then these generators are identical with the functions ψ^1 and ψ^2 that were constructed in the shift-invariant setting of Example 1. The current construction reveals the necessary adaptation to the bounded interval $[a, b]$ by assigning one boundary wavelet at each end of the interval.

7.5.2 Piecewise cubic tight frames with equidistant simple knots

For simplicity of the presentation, we restrict to the case of simple equidistant interior knots of stepsize $h_0 = 1$ in an interval $[0, N+1]$. The boundary knots at 0 and $N+1$ are assumed to have multiplicity 4, so that

$$t_k^{(0)} = k \text{ for } 1 \le k \le N, \qquad t_k^{(1)} = k/2 \text{ for } 1 \le k \le 2N+1,$$

while $t_k^{(0)} = t_k^{(1)} = 0$ for $k = -3, -2, -1, 0$ and $t_{N+k}^{(0)} = t_{2N+1+k}^{(1)} = N+1$ for $k = 1, 2, 3, 4$. Hence, $\mathbf{t}_1$ is given by inserting knots at the half integers. The generic method described in Section 5, for 4 vanishing moments, employs the Cholesky factorization of the matrix Z_0 in (4.47), which has dimension $N_1 := 2N + 1$. This leads to the definition of N_1 non-symmetric wavelets with 4 vanishing moments. We construct another factorization in [16], which reflects the shift-invariant structure of the interior wavelets that was also observed in Section 5.1 for the case of equidistant knots. For this purpose, we choose a larger number of wavelets, namely $3N - 14$ interior wavelets and 6 boundary wavelets for each endpoint, giving a total of $3N - 2$. The gain, by increasing the number of wavelets by roughly 3/2, is that all the interior wavelets (i.e., those with support strictly inside $(0, N+1)$ or with a simple knot at one of the endpoints) are symmetric, have 4 vanishing moments, and are shifts of three "mother" wavelets. Moreover, the construction is scale-invariant, such that the same coefficient sequences (for interior and boundary wavelets) can be employed for all scales V_j, if uniform refinement of the knot vector is used across all scales.

The matrix Z_0 of size $N_1 \times N_1$ in (4.47) is positive definite and has bandwidth 7. In order to simplify the factorization, two symmetric reductions

$$Z_1 := (I - K_2)(I - K_1)Z_0(I - K_1^T)(I - K_2^T)$$

are performed in [16], which yield a matrix Z_1 of bandwidth 3. A factorization $Z_1 = BB^T$ is then defined, where the "interior" part B_i of B is given by

$$B_i = \begin{bmatrix}
0 &&&&&&&&& \\
0 &&&&&&&&& \\
a &&&&&&&&& \\
b && d &&&&&&& \\
a & c & e & a &&&&&& \\
&& d & b && d &&&& \\
&&& a & c & e &&&& \\
&&&&& d &&&& \\
&&&&&& \ddots &&& \\
&&&&&&&& d & \\
&&&&&&& c & e & a \\
&&&&&&&& d & b \\
&&&&&&&&& a \\
&&&&&&&&& 0 \\
&&&&&&&&& 0
\end{bmatrix}.$$

This ansatz exhibits the symmetry and shift-invariance of the resulting wavelets. Then 6 columns are added to the left and right of B_i, that define the coefficient sequences of 6 boundary wavelets. All these computations can be done analytically. The numerical values for the coefficient sequences of the wavelets that result from this construction are given in two tables. Table 7.3 gives the coefficients $\hat{q}_k^{(i)}$ of the generators $\psi^{(i)}$ for the interior wavelets, if they are written in the form

$$\psi^{(i)}(x) = \sum_{k=-3}^{N_1-4} \hat{q}_k^{(i)} \frac{d^4}{dx^4} N_{\mathbf{t}_1;8,k}, \quad i = 1, 2, 3. \tag{5.12}$$

For convenience, we employ the L^∞-normalization of the splines in (5.12). The supports of these functions are

$$\operatorname{supp}\psi^{(1)} = [0,6], \quad \operatorname{supp}\psi^{(2)} = [1,6], \quad \operatorname{supp}\psi^{(3)} = [0,7].$$

All of the $3N-14$ interior wavelets are given by

$$\psi^{(i)}(\cdot - k), \qquad i = 1,2,3, \quad 0 \le k \le N-6, \qquad \psi^{(1)}(\cdot - N + 5).$$

The graphs of $\psi^{(i)}$, $i = 1,2,3$, are shown in Figure 10. Table 7.4 gives the coefficients of the 6 boundary wavelets for the left endpoint of the interval, using the same representation as in (5.12). The first three of these functions have a knot of multiplicity 4 at zero and supports $[0, 2.5]$, $[0,3]$, $[0,4]$, respectively. The fourth boundary wavelet has a triple knot at 0 and support $[0,5]$. The last two boundary wavelets have a double knot at 0 and support $[0,5]$, $[0,6]$, respectively. The reflection of these functions yields the 6 boundary wavelets at the other endpoint $N+1$. The graphs of the boundary wavelets for the left endpoint are shown in Figure 11.

Table 7.3. Coefficients ($\times 100$) of the generators $\psi^{(i)}$ as in expansion (5.12)

i	$\hat{q}_0^{(i)}$	$\hat{q}_1^{(i)}$	$\hat{q}_2^{(i)}$	$\hat{q}_3^{(i)}$	$\hat{q}_4^{(i)}$	$\hat{q}_5^{(i)}$	$\hat{q}_6^{(i)}$
1	0.171217	1.369738	3.091033	1.369738	0.171217		
2			0.267942	2.143537	0.267942		
3	0.112045	0.896364	2.883961	4.248047	2.883961	0.896364	0.112045

Table 7.4. Coefficients ($\times 100$) of the 6 boundary wavelets as in expansion (5.12)

i	$\hat{q}_{-3}^{(i)}$	$\hat{q}_{-2}^{(i)}$	$\hat{q}_{-1}^{(i)}$	$\hat{q}_0^{(i)}$	$\hat{q}_1^{(i)}$	$\hat{q}_2^{(i)}$	$\hat{q}_3^{(i)}$	$\hat{q}_4^{(i)}$
1	0.468951							
2	0.208884	1.193513						
3	0.046733	0.588400	2.238939	0.279867				
4		0.217826	1.577574	3.111291	1.110131	0.138766		
5			0.051599	0.511502	2.242497	0.280312		
6			0.245479	1.950235	3.573393	2.594618	0.818264	0.102283

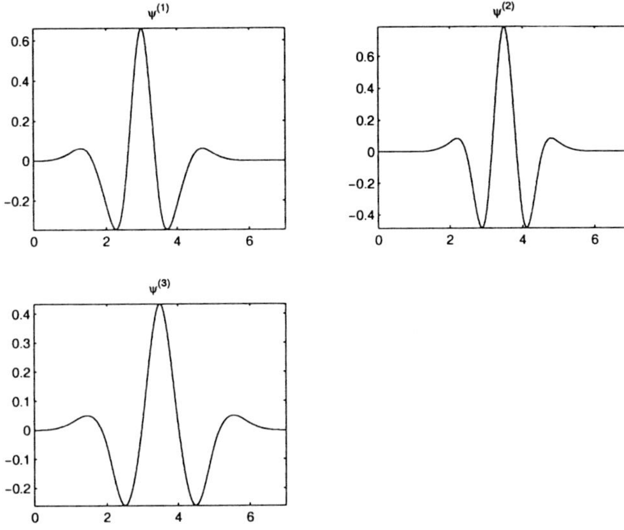

Figure 7.10. Generators of tight frame of cubic splines with 4 vanishing moments

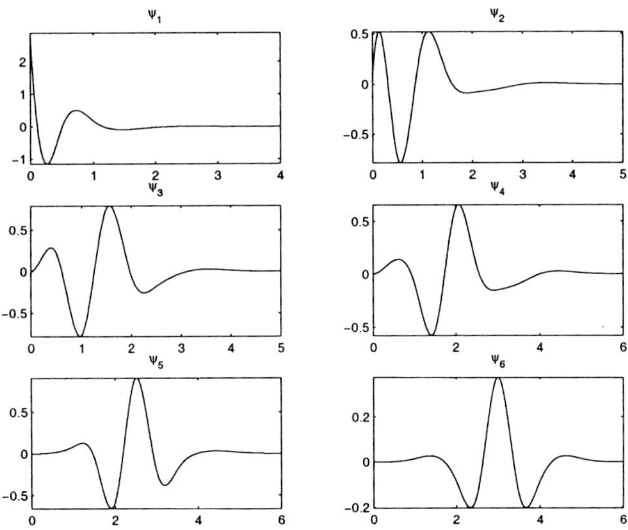

Figure 7.11. Boundary wavelets of tight frame of cubic splines with 4 vanishing moments

Remark 7.5.2 The three generators $\psi^{(i)}$, $i = 1, 2, 3$, in the previous example also constitute a tight frame Ψ of $L^2(\mathbf{R})$. This construction yields three symmetric generators with 4 vanishing moments and coefficient sequences (in terms of the B-spline basis $\Phi_{\mathbf{t}_1;4}$) of 7, 9, and 11 nonzero coefficients, respectively. This underlines the fact that our general method is also useful in order to find generators of tight frames in the shift-invariant setting discussed in Section 2.

7.5.3 Tight frames of cubic splines with equidistant knots of multiplicity 2

The consideration of splines with double knots at the integers leads to an MRA that is generated by two functions $\phi^{(1)}$ and $\phi^{(2)}$, in the case the shifts extend over $\mathbf{Z}$. If we restrict our consideration to the interval $[0, N+1]$, as in Section 5.2, the knot vector for cubic splines ($m = 4$) has the form

$$\mathbf{t}_0 = \{0,0,0,0,1,1,2,2,\ldots,N,N,N+1,N+1,N+1,N+1\}.$$

The refined knot vectors $\mathbf{t}_j$, $j \geq 1$, are obtained by inserting double knots at the midpoints of each knot interval. For $\mathbf{t}_1$, for example, we insert double knots at $1/2+k$, $0 \leq k \leq N$. The dimension of the spline space V_0 is $2N+4$, that of V_1 is $4N+6$. This setting is comparable to that of Section 3.

Remark 7.5.3 As mentioned in Section 3, no analogue for the construction of tight frames from this type of MRA has yet been developed based on the Fourier-domain approach. Our time-domain construction, however, makes no assumptions concerning the number of generators of the MRA. In fact, the absence of techniques based on any sort of shift-invariance makes our new technique versatile for much more general settings, where multiplicities of knots can vary and adaptive refinements are allowed. Therefore, our new results provide, at least for spline spaces, a unified treatment of several types of MRA. The current example shall serve as a simple illustration.

For the construction of a tight frame for the given MRA, we proceed as sketched out before. The approximate dual S_0 of order $\mu = 4$ was already computed in Example 9(b) and Example 11(b). The matrix Z_0 in (4.47) has dimension $4N+2$, is positive definite and has bandwidth 8. Instead of its Cholesky factorization, we now wish to find another factorization that defines interior wavelets that have four vanishing moments, are symmetric or anti-symmetric and are translates of only a small number of generators $\psi^{(i)} \in V_1$ (it turns out 5 generators are enough). At both endpoints of the interval we require several boundary wavelets which have also 4 vanishing moments.

Table 7.5. Coefficients ($\times 1000$) of the generators $\psi^{(i)}$ as in expansion (5.12)

i	$\hat{q}_0^{(i)}$	$\hat{q}_1^{(i)}$	$\hat{q}_2^{(i)}$	$\hat{q}_3^{(i)}$	$\hat{q}_4^{(i)}$	$\hat{q}_5^{(i)}$	$\hat{q}_6^{(i)}$	$\hat{q}_7^{(i)}$
1	0.092642	0.370569	1.852847	0.989527	−0.989527	−1.852847	−0.370569	−0.092642
2	0.126349	0.505395	2.526977	3.156191	3.156191	2.526977	0.505395	0.126349

i	$\hat{q}_3^{(i)}$	$\hat{q}_4^{(i)}$	$\hat{q}_5^{(i)}$	$\hat{q}_6^{(i)}$	$\hat{q}_7^{(i)}$	$\hat{q}_8^{(i)}$
3	0.526730	1.601752	0.086252	−0.086252	−1.601752	−0.526730
4	0.580480	2.180883	1.757771	1.757771	2.180883	0.580480
5		0.869741	3.478964	3.478964	0.869741	

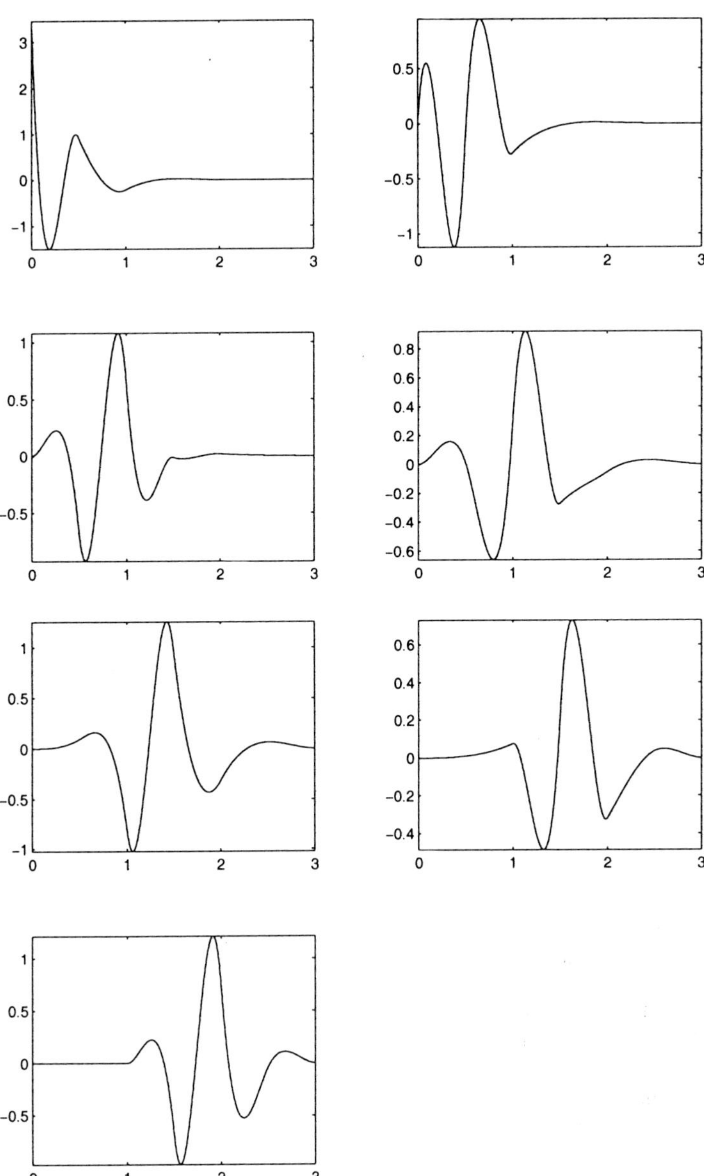

Figure 7.12. Boundary wavelets of tight frame of cubic splines with double knots and 4 vanishing moments

Similar to the simple knot case, three symmetric reductions

$$Z_1 = (I - K_3)(I - K_2)(I - K_1)Z_0(I - K_1^T)(I - K_2^T)(I - K_3^T)$$

Table 7.6. Coefficients ($\times 1000$) of the 7 boundary wavelets as in expansion (5.12)

i	$\hat{q}_{-3}^{(i)}$	$\hat{q}_{-2}^{(i)}$	$\hat{q}_{-1}^{(i)}$	$\hat{q}_{0}^{(i)}$	$\hat{q}_{1}^{(i)}$	$\hat{q}_{2}^{(i)}$	$\hat{q}_{3}^{(i)}$	$\hat{q}_{4}^{(i)}$
1	1.030983	1.417601	0.644364	0.096655				
2		1.964342	1.523281	0.719836	0.300617	0.060123	0.015031	
3			2.170762	1.104518	0.574380	0.134319	0.038137	0.001519
4			0.909528	3.566099	2.804337	1.352000	0.523422	0.061807
5				0.987016	3.948064	3.102908	1.320567	0.181613
6				0.100948	0.403790	2.018952	1.126572	0.207278
7							2.193554	0.731185

(with tridiagonal matrices $I-K_i$) lead to a matrix Z_1 with bandwidth 4. The factorization of Z_1 leads to the definition of 7 boundary wavelets at both endpoints and 5 generators $\psi^{(i)} \in V_1$ of the interior wavelets, such that

$$\psi^{(i)}(\cdot - k), \quad 1 \le i \le 5, \ 0 \le k \le N-4, \qquad \psi^{(i)}(\cdot - N + 3), \quad i = 1, 2, \tag{5.13}$$

represent all interior wavelets. The coefficients of these generators, with respect to the representation (5.12), are given in Table 7.5. Note that the same functions were defined in Section 3, Example 8. Their graphs are also shown in Section 3, Figure 9. The coefficients in Table 7.5 differ from those in Example 8 only by a constant factor, as we chose a different normalization in (5.12). The following symmetry properties are revealed by a simple analysis: $\psi^{(2)}$, $\psi^{(4)}$, and $\psi^{(5)}$ are symmetric, while $\psi^{(1)}$ and $\psi^{(3)}$ are anti-symmetric. Their supports are given by

$$\operatorname{supp}\psi^{(1)} = \operatorname{supp}\psi^{(2)} = [0,4], \quad \operatorname{supp}\psi^{(3)} = \operatorname{supp}\psi^{(4)} = \operatorname{supp}\psi^{(5)} = [1,4].$$

The splines $\psi^{(1)}$, $\psi^{(2)}$ have simple knots 0 and 4 and double knots at $.5, 1, \ldots, 3.5$, while $\psi^{(3)}$ and $\psi^{(4)}$ have double knots at $1, 1.5, \ldots, 4$. The spline $\psi^{(5)}$ has simple knots $1, 4$ and double knots $1.5, 2, \ldots, 3.5$. The coefficients of the 7 boundary wavelets for the left endpoint are given in Table 7.6, and their graphs are shown in Figure 12. The boundary wavelets for the right endpoint are obtained as the mirror images. Together with the interior wavelets, we have a total of $5N+1$ functions. This number is roughly $5/4$ times the number that we obtain by the Cholesky factorization of the matrix Z_0. Therefore, at a small expense we gain symmetry of the wavelets while keeping the supports rather small.

APPENDIX

The computation of the psd matrix S that defines an approximate dual for the B-spline basis on the bounded interval $[a,b]$ is given in MATLAB syntax. The vector `knots` is the knot vector (with multiplicity m of the boundary knots and all other knots of multiplicity $\le m$), `m` is the order of the B-spline basis, and `mu` is the order of the approximate dual.

```
function S = make_S(knots,m,mu)
```

```
   % compute approximate dual of order mu
   % for B-spline basis of order m
   % use Horner-like scheme for S
   S = make_U(knots,m,mu-1);
           % produce the diagonal matrix U_{mu-1}
   for nu=mu-2:-1:0
      EB = make_EB(knots,m+nu);
           % produce the matrix E^B_{knots;m+nu,1}
      S  = EB*S*EB' + make_U(knots,m,nu);
   end

function U = make_U(knots,m,mu)
   % compute F_{2 mu} by means of centered moments
   % and normalize to give U_{mu}
   N = length(knots)-m;
           % dimension of spline space
   temp_knots = knots(2:end - 1);
   switch mu
    case 0,
        udiag = ones(1,N);
    case 1,
        a=make_moment(temp_knots,2,m+mu-1);
        udiag = (m*a)/((m+1)*(m-1));
    case 2,
        a=make_moment(temp_knots,2,m+mu-1);
        b=make_moment(temp_knots,4,m+mu-1);
        udiag = ((m^2 - m +1)*a.^2 - m*b)/(2*(m+2)*m*(m-1)*(m-2));
    case 3,
        a=make_moment(temp_knots,2,m+mu-1);
        b=make_moment(temp_knots,3,m+mu-1);
        c=make_moment(temp_knots,4,m+mu-1);
        d=make_moment(temp_knots,6,m+mu-1);
        c1 = (m^2-3*m+5)*(m+2)/(6*(m+3)*(m+1)^2*(m-1)*(m-2)*(m-3));
        c2 = -(m^2-m+4)/(2*(m+3)*(m+1)^2*(m-1)*(m-2)*(m-3));
        c3 = -(3*m^2-3*m+2)/(3*(m+3)*(m+1)^2*m*(m-1)*(m-2)*(m-3));
        c4 = 1/(3*(m+3)*(m+1)*(m-1)*(m-2)*(m-3));
        udiag = c1*a.^3 + c2*a.*c + c3*b.^2 + c4*d;
     end
   U=spdiags(udiag(:),[0],length(udiag),length(udiag));

function a = make_moment(knots,nu,k)
   % compute centered moments of degree nu
   % for all sets of k consecutive knots
   t=knots(:);
   lt=length(t);
   tmp=repmat(t,1,k);
   tmp=tmp(:);
   trep=zeros(lt+1,k);
   trep(:)=[tmp;zeros(k,1)];
```

```
  trep=trep';
     % now contains, in each column, k consecutive knots
  tstar=sum(trep)/k;
     % the mean value
  a=sum((trep-repmat(tstar,k,1)).^nu)/k;
     % the centered moment of degree nu
  a=a(1:lt-k+1);

function EB = make_EB(knots,m)
  % compute the matrix E^B_{knots;m,1}
  a=m./(knots(m+1:end)-knots(1:end-m));
  b=(m+1)./(knots(m+2:end)-knots(1:end-m-1));
  a=sqrt(a(:));
  b=sqrt(b(:));
  EB=spdiags([[a(1:end-1).*b;0], -[a(2:end).*b;0] ],[0,-1],...
             length(a),length(a)-1);
```

ACKNOWLEDGEMENT

Both authors were supported by NSF grant # CCR–0098331, and the first author was also supported by NSF grant # CCR–9988289 and ARO grant # DAAD–19–00–1–0512.

REFERENCES

[1] A. Aldroubi, Portraits of frames, Proc. Amer. Math. Soc. **123**(6) (1995), 1661–1668.

[2] J. J. Benedetto and S. Li, The theory of multiresolution analysis frames and applications to filter banks, Appl. Comp. Harmonic Anal. **5**(4) (1998), 389–427.

[3] L. Beutel, K. Y. Lee, and J. Stöckler, Practical algorithms for tight spline frames on intervals, manuscript in preparation.

[4] C. de Boor, "A Practical Guide to Splines, Revised Edition," Springer-Verlag, Berlin, 2002.

[5] C. de Boor, R. DeVore, and A. Ron, Approximation from shift-invariant subspaces of $L_2(\mathbf{R}^d)$, Trans. Amer. Math. Soc. **341** (1994), 787–806.

[6] C. de Boor, R. DeVore, and A. Ron, On the construction of multivariate (pre)wavelets, Constr. Approximation **9** (1993), 123–166.

[7] C. de Boor, R. DeVore, and A. Ron, The structure of finitely generated shift-invariant spaces in $L_2(\mathbf{R}^d)$, J. Funct. Anal. **119** (1994), 37–78.

[8] C. de Boor, R. DeVore, and A. Ron, Approximation orders of FSI spaces in $L_2(\mathbf{R}^d)$, Constr. Approximation **14** (1998), 411-427 and 631–652.

[9] C. de Boor and G. J. Fix, Spline approximation by quasiinterpolants, J. Approx. Theory **8** (1973), 19–45.

[10] C.K. Chui, "An Introduction to Wavelets," Academic Press, Boston, 1992.

[11] C. K. Chui, "Wavelets: A Mathematical Tool for Signal Analysis," SIAM, Philadelphia, 1997.

[12] C. K. Chui and J. M. De Villiers, Spline-wavelets with arbitrary knots on a bounded interval: Orthogonal decomposition and computational algorithms, Comm. Appl. Anal. **2**(4) (1998), 457–486.

[13] C. K. Chui and W. He, Compactly supported tight frames associated with refinable functions, Appl. Comp. Harmonic Anal. **8** (2000), 293–319.

[14] C. K. Chui, W. He, and J. Stöckler, Compactly supported tight and sibling frames with maximum vanishing moments, Appl. Comp. Harmonic Anal., to appear.

[15] C. K. Chui, W. He, and J. Stöckler, Tight frames with maximum vanishing moments and minimum support, in "Approximation Theory X: Wavelets, Splines, and Applications," C. K. Chui, L. L. Schumaker, and J. Stöckler (eds.), Vanderbilt University Press, 2001, 187–206.

[16] C. K. Chui, W. He, and J. Stöckler, Tight spline frames with arbitrary knot sequences and desirable order of vanishing moments, preprint.

[17] C. K. Chui, W. He, J. Stöckler, and Q. Sun, Compactly supported tight affine frames with integer dilations and maximum vanishing moments, Advances in Comp. Math., to appear.

[18] C. K. Chui and J.-A. Lian, Nonstationary wavelets and refinement sequences of nonuniform B-splines, in "Approximation Theory X: Wavelets, Splines, and Applications," C. K. Chui, L. L. Schumaker, and J. Stöckler (eds.), Vanderbilt University Press, 2001, 207–229.

[19] C. K. Chui and X. L. Shi, Bessel sequences and affine frames, Appl. Comp. Harmonic Anal. **1**(1) (1993), 29–49.

[20] C. K. Chui, P. W. Smith, and J. D. Ward, Cholesky factorization of positive definite bi-infinite matrices, Numer. Funct. Anal. Optimization **5** (1982), 1–20.

[21] A. Cohen, I. Daubechies, and P. Vial, Wavelets on the interval and fast wavelet transforms, Appl. Comp. Harmonic Anal. **1**(1) (1993), 54–81.

[22] I. Daubechies, Orthonormal bases of compactly supported wavelets, Comm. Pure Appl. Math. **41** (1988), 909-996.

[23] I. Daubechies, "Ten Lectures on Wavelets," CBMS-NSF Reg. Conf. Series in Applied Mathematics #61, SIAM, Philadelphia, 1992.

[24] I. Daubechies and B. Han, Pairs of dual wavelet frames from any two refinable functions, preprint.

[25] I. Daubechies, B. Han, A. Ron, and Z. W. Shen, Framelets: MRA-based constructions of wavelet frames, Appl. Comp. Harmonic Anal., to appear.

[26] I. Daubechies, I. Guskov, and W. Sweldens, Commutation for irregular subdivision, Constr. Approximation **17** (2001), 479–514.

[27] S. Demko, Inverses of band matrices and local convergence of spline projections, SIAM J. Numer. Anal. **14** (1977), 616–619.

[28] R. A. DeVore and G. G. Lorentz, "Constructive Approximation," Springer-Verlag, New York, 1993.

[29] P. Dierckx, "Curve and surface fitting with splines," Clarendon Press, Oxford, 1995.

[30] M. Frazier, G. Garrigós, K. Wang, and G. Weiss, A characterization of functions that generate wavelet and related expansion. Proceedings of the conference dedicated to Professor Miguel de Guzmán (El Escorial, 1996). J. Fourier Anal. Appl. **3** (1997), 883–906.

[31] T. N. T. Goodman, S. L. Lee, and W. S. Tang, Wavelets in wandering subspaces, Trans. Am. Math. Soc. **338** (1993), 639–654.

[32] B. Han, On dual wavelet tight frames, Appl. Comp. Harmonic Anal. **4**(4) (1997), 380–413.

[33] B. Han and Q. Mo, Tight wavelet frames generated by three symmetric B-spline functions with high vanishing moments, preprint.

[34] B. Han and Q. Mo, Multiwavelet frames from refinable function vectors, preprint.

the action on the functions ψ of a very special subset of this affine group, the elements of the form $D_{2^{-j}}\,T_{-k}$, $j,k \in \mathbb{Z}$. That is, $\psi_{j,k} = D_{2^{-j}}\,T_{-k}\,\psi$; observe that the translations by $k \in \mathbb{Z}$ are applied first to ψ and, then, the dilations $D_{2^{-j}}$ are applied to $T_{-k}\,\psi$. This set of operators $\{D_{2^{-j}}\,T_{-k} : j,k \in \mathbb{Z}\}$ is not a subgroup of the affine group.

If $\|\psi\|_2 \geq 1$, the fact that $\{\psi_{j,k}\}$, $j,k \in \mathbb{Z}$, is an orthonormal basis is equivalent to the reproducing formula, valid for all $f \in L^2(\mathbb{R})$,

$$f = \sum_{j\in\mathbb{Z}} \sum_{k\in\mathbb{Z}} \langle f, \psi_{j,k}\rangle\, \psi_{j,k}, \tag{1.1}$$

with convergence in $L^2(\mathbb{R})$. Moreover, (1.1) is equivalent to

$$\|f\|_2^2 = \sum_{j\in\mathbb{Z}} \sum_{k\in\mathbb{Z}} |\langle f, \psi_{j,k}\rangle|^2, \tag{1.2}$$

for all $f \in L^2(\mathbb{R})$. If either (1.1) or (1.2) hold and $\|\psi\|_2 \geq 1$, then, clearly, we must have $\|\psi\|_2 = 1$; if $0 < \|\psi\|_2 < 1$, then either of these two equalities assert that the system $\{\psi_{j,k}\}$ is a *normalized tight frame* (a tight frame with constant 1). See Chapter 7 of [2] for proofs of these claims and related questions.

It is natural to ask if other subsets of the affine group can be used to obtain similar normalized tight frames (or, more generally, frames) for $L^2(\mathbb{R})$. A significant discovery in this direction was made by A.Ron and Z.Shen [4]. They showed that, if the system $\{D_{2^{-j}}\,T_{-k}\,\psi : j,k \in \mathbb{Z}\}$ is changed so that, for $j > 0$, $D_{2^{-j}}\,T_{-k}\,\psi$ is replaced by $2^{-j/2}\,T_{-k}\,D_{2^{-j}}$ for all $k \in \mathbb{Z}$, we do obtain a normalized tight frame whenever the original system has this property (and vice versa). In fact, this equivalence is true more broadly in the sense that one of these systems is a frame for $L^2(\mathbb{R})$ if and only if the same is true for the other. One of the goals of this paper is to gain a better understanding of this and related matters.

Most of what we consider applies to more general situations: higher dimensions, "multi-systems" that are obtained by applying these translations and dilations to a finite family $\Psi = \{\psi^1, \ldots, \psi^L\} \subset L^2(\mathbb{R}^n)$, and "dual" systems in which one family, $\Psi = \{\psi^1, \ldots, \psi^L\}$, is used to "analyze" a function and another, $\Phi = \{\phi^1, \ldots, \phi^L\}$, to "synthesize", or reproduce, the given function from the data obtained from the "analysis". That is, (1.1) is extended to equalities of the form

$$f = \sum_{\ell=1}^{L} \sum_{j\in\mathbb{Z}} \sum_{k\in\mathbb{Z}^n} \langle f, \psi^\ell_{j,k}\rangle\, \phi^\ell_{j,k}. \tag{1.3}$$

In order to focus on the properties of appropriate subsets of the affine group and the systems they generate in the manner described above, however, we will restrict our attention to the 1−dimensional case and to systems obtained from a single function. We break this resolve, however, in the following few paragraphs devoted to the description of continuous wavelets. It is our opinion that the more general setting will present a better perspective of the questions we are addressing.

Let D be a closed subgroup of $GL(x, \mathbb{R})$, the general linear group acting on $\mathbb{R}^n$. Let us form the semi-direct product

$$G = D \times \mathbb{R}^n \equiv \{g = (a,b) : a \in D, b \in \mathbb{R}^n\}$$

and endow this set with the product

$$g_1 \circ g_2 = (a_1, b_1) \circ (a_2, b_2) = (a_1 a_2, a_2^{-1} b_1 + b_2). \tag{1.4}$$

This operation corresponds to the action $(a, b)(x) = a(x + b)$ on the points of $\mathbb{R}^n$. A simple calculation shows that $d\lambda(a, b) = d\mu(a)\, db$ is a left Haar measure on D. Moreover, we have $(a, b)^{-1} = (a^{-1}, -ab)$, so that the action, T, of G on functions $\psi \in L^2(\mathbb{R}^n)$ defined by

$$(T_g\, \psi)(x) = |\det a|^{-1/2}\, \psi(g^{-1}(x)) = |\det a|^{-1/2}\, \psi(a^{-1}x - b),$$

$g = (a, b) \in G$, produces a "continuous" system that is a natural analog to the discrete system $\{\psi_{j,k}\}$ we introduced in the first paragraph of this section. We write

$$\psi_{a,b}(x) = |\det a|^{-1/2}\, \psi(a^{-1}x - b), \tag{1.5}$$

for $(a, b) \in G$, in order to complete this analogy.

It is natural to find a condition on ψ that guarantees the reproducing property

$$\|f\|_2^2 = \int_G |\langle f, \psi_{a,b}\rangle|^2\, d\lambda(a, b) = \int_D \Big(\int_{\mathbb{R}^n} |\langle f, \psi_{a,b}\rangle|^2\, db \Big) d\mu(a) \tag{1.6}$$

for all $f \in L^2(\mathbb{R}^n)$, which is clearly an analog of (1.2). The following extension of the "Calderón condition" provides us with a characterization of those ψ for which (1.6) is true.

Theorem 8.1.1 *Equality (1.6) is valid for all $f \in L^2(\mathbb{R}^n)$ if and only if for a.e. $\xi \in \mathbb{R}$*

$$\Delta_\psi(\xi) = \int_D |\hat{\psi}(\xi a)|^2\, d\mu(a) = 1. \tag{1.7}$$

In [6] one can find a rather complete discussion of equality (1.7) and its relation to the original Calderón condition; in particular, a proof of this theorem is presented in the cited article (see Theorem (2.1)). We shall refer to the functions ψ satisfying (1.7) (or, equivalently, (1.6)) as the *continuous wavelet on $\mathbb{R}^n$ associated with the dilation group* D. Let us examine, in view of the result of Ron and Shen that involved the interchange of the order in which translations and dilations are applied to ψ, what happens in the case of continuous wavelets when this order is interchanged.

Perhaps a good way of seeing the effect of such interchange is to endow the set $\{(a, b) : a \in D, b \in \mathbb{R}^n\}$ with the product

$$(a_1, b_1) \circ (a_2, b_2) = (a_1 a_2, b_1 + a_1 b_2) \tag{1.8}$$

that corresponds to the action $(a, b)(x) = ax + b$ on the points of $\mathbb{R}^n$. Let us denote the group having this operation by G^*. We distinguish G and G^* by calling G the *affine group* and G^* the *co-affine group* (associated with the dilation group D). The system $\{\psi_{a,b}\}$ will be referred to as the *affine system* and

$$\psi^*_{a,b}(x) = |\det a|^{-1/2}\, \psi(a^{-1}(x - b)) \tag{1.9}$$

is defined to be the corresponding (continuous) *co-affine system* [1]. We say that ψ is a

[1] With respect to the operation(1.8), $(a, b)^{-1} = (a^{-1}, -a^{-1}b)$. Thus $\psi^*_{a,b}(x) = |\det a|^{-1/2}\, \psi((a, b)^{-1}(x))$.

Beyond Wavelets
G. V. Welland (Editor)
© 2003 Elsevier Science (USA) All rights reserved

9

SPARSITY VS. STATISTICAL INDEPENDENCE IN ADAPTIVE SIGNAL REPRESENTATIONS: A CASE STUDY OF THE SPIKE PROCESS

BERTRAND BÉNICHOU AND NAOKI SAITO

Ecole Nationale Supérieure des Télécommunications
46, rue Barrault 75634 Paris cedex 13 France
benichou@email.enst.fr

Department of Mathematics University of California, Davis
One Shields Avenue Davis, CA 95616
saito@math.ucdavis.edu

Abstract

Finding a basis/coordinate system that can efficiently represent an input data stream by viewing them as realizations of a stochastic process is of tremendous importance in many fields including data compression and computational neuroscience. Two popular measures of such efficiency of a basis are sparsity (measured by the expected ℓ^p norm, $0 < p \leq 1$) and statistical independence (measured by the mutual information). Gaining deeper understanding of their intricate relationship, however, remains elusive. Therefore, we choose to study a simple synthetic stochastic process called the "spike process", which puts a unit impulse at a random location in an otherwise zero vector of length n in each realization.

For this process, we prove the following results: 1) The standard basis is the best in terms of sparsity for all $n \geq 2$ among all possible orthonormal bases in $\mathbf{R}^n$ or all possible invertible linear transformations in $\mathbf{R}^n$ with a fixed determinant value; 2) The standard basis is again the best in terms of statistical independence if $n \geq 5$ and the search of basis is restricted within all possible orthonormal bases in $\mathbf{R}^n$; if $2 \leq n \leq 4$, then the standard basis is *not* the best orthonormal basis in statistical independence; 3) If we extend our basis search to all possible linear invertible transformations in $\mathbf{R}^n$, then the best basis in statistical independence is *not* the standard basis for any $n \geq 2$; 4) The best basis in statistical independence is not unique in general, and there even exist those which turn input spikes into completely dense vectors; 5) There is no linear invertible transformation that achieves the true statistical independence for $n > 2$.

9.1 INTRODUCTION

What is a good coordinate system/basis to efficiently represent a given set of images? We view images as realizations of a certain complicated stochastic process whose probability density function (pdf) is not known a priori. *Sparsity* is important here since this is a measure of how well one can compress the data. A coordinate system producing a few large coefficients and many small coefficients has high sparsity for that data. The sparsity of images relative to a coordinate system is often measured by the expected ℓ^p norm of the coefficients where $0 < p \leq 1$. *Statistical independence* is also important since statistically independent coordinates do not interfere with each other (no crosstalk, no error propagation among them). The amount of statistical dependence of input images relative to a coordinate system is often measured by the so-called mutual information, which is a statistical distance between the true pdf and the product of the one-dimensional marginal pdfs.

Neuroscientists have become interested in efficient representations of images, in particular, images of natural scenes such as trees, rivers, mountains, etc., since mammalian visual systems effortlessly reduce the amount of visual input data without losing the essential information contained in them. Therefore, if we can find what type of basis functions are sparsifying the input images or are providing us with the statistically independent representation of the inputs, then that may shed light on the mechanisms of our visual system. Olshausen and Field [18], [19] pioneered such studies using computational experiments emphasizing the sparsity. Immediately after their experiments, Bell and Sejnowski [1], van Hateren and van der Schaaf [24] conducted similar studies using the statistical independence criterion. Surprisingly, these results suggest that both sparsity and independence criteria tend to produce basis functions efficient to capture and represent edges of various scales, orientations, and positions, which are similar to the receptive field profiles of the neurons in our primary visual cortex. (Note the criticism raised by Donoho and Flesia [9] about the trend of referring to these functions as "Gabor"-like functions; therefore, we just call them "edge-detecting" basis functions in this paper.) However, the relationship between these two criteria has not been understood completely.

These experiments and observations inspired our study in this paper. Our goal here, however, is more modest in that we only study the "spike" process, a simple synthetic stochastic process which puts a unit impulse at a random location in an otherwise zero vector of length n in each realization. It is important to use a simple stochastic process first since we can gain insights and make precise statements in terms of theorems. By these theorems, we now understand what are the precise conditions for the sparsity and statistical independence criteria to select the same basis for the spike process. In fact, we prove the following facts.

- The standard basis is the best in terms of sparsity for all $n \geq 2$ among all possible orthonormal bases in $\mathbf{R}^n$ or all possible invertible linear transformations in $\mathbf{R}^n$ with a fixed determinant value;

- The standard basis is again the best in terms of statistical independence if $n \geq 5$ and the search of basis is restricted within all possible orthonormal bases in $\mathbf{R}^n$; if $2 \leq n \leq 4$, then the standard basis is *not* the best orthonormal basis in statistical independence;
- If we extend our basis search to all possible linear invertible transformations in $\mathbf{R}^n$, then the best basis in statistical independence is *not* the standard basis for any $n \geq 2$;
- The best basis in statistical independence is not unique in general, and there even exist those which turn input spikes into completely dense vectors;
- There is no linear invertible transformation that achieves the true statistical independence for $n > 2$.

These results and observations hopefully lead to deeper understanding of the efficient representations of more complicated stochastic processes such as natural scene images.

Additionally, a very important by-product of this paper is that this simple process can be used for validating any independent component analysis (ICA) software that uses mutual information as a measure of statistical dependence, and any sparse component analysis (SCA) software that uses ℓ^p-norm ($0 < p \leq 1$) as a measure of sparsity. Actual outputs of the software can be compared with the true solutions obtained by our theorems. For example, the ICA software using mutual information of the inputs should not converge for the spike process unless there is some constraint on the basis search (e.g., search within all possible orthonormal bases). Considering the recent popularity of such software ([14], [2], [17]), it is a good thing to have such a simple example that can be generated and tested easily on computers.

Our investigations of other stochastic processes in terms of sparsity and statistical independence, such as the "generalized spike process" (which puts an impulse whose amplitude is sampled randomly from the standard normal distribution $N(0, 1)$ in each realization) and "ramp" process (another simple yet important stochastic process), can be found in Saito [22] and Saito et al. [23], respectively. The latter also contains our numerical experiments on natural scene images.

The organization of this paper is as follows. The next section specifies notation and terminology. Section 3 defines how to quantitatively measure the sparsity and statistical dependence of a stochastic process relative to a given basis. Using a very simple example, Section 4 demonstrates that the sparsity and statistical independence are two clearly different concepts. Section 5 presents our main results. We prove these theorems in Section 6 and Appendices. Finally, we discuss the implications of our results and further research directions in Section 7.

9.2 NOTATION AND TERMINOLOGY

Let us first set our notation and the terminology of basis dictionaries and best bases. Let $\boldsymbol{X} \in \mathbf{R}^n$ be a random vector with some unknown pdf $f_{\boldsymbol{X}}$. Let us assume that the available data $T = \{\boldsymbol{x}_1, \ldots, \boldsymbol{x}_N\}$ were independently generated from this probability model. The set T is often called the training dataset. Let $B = (\boldsymbol{w}_1, \ldots, \boldsymbol{w}_n) \in \mathrm{O}(n)$ (the group of orthonormal transformations in $\mathbf{R}^n$) or $\mathrm{SL}^{\pm}(n, \mathbf{R})$ (the group of invertible volume-preserving transformations in $\mathbf{R}^n$, i.e., their determinants are ± 1). The best-basis paradigm [4], [26], [20], is to find a basis B or a subset of basis vectors such that the features (expansion coefficients) $\boldsymbol{Y} = B^{-1}\boldsymbol{X}$ are useful for the problem at hand (e.g., compression, modeling, discrimination, regression, segmentation) in a computationally fast manner. Let $C(B \mid T)$ be a numerical measure of *deficiency* or *cost* of the basis B given the training dataset T for the problem at hand. For very high-dimensional problems, we often restrict our search within the basis dictionary $D \subset \mathrm{SL}^{\pm}(n, \mathbf{R})$, such as the orthonormal or biorthogonal wavelet packet dictionaries or local cosine or Fourier dictionaries where we never need to compute the full matrix-vector product or the matrix inverse for analysis and synthesis. Under this setting, $B_\star = \arg\min_{B \in D} C(B \mid T)$ is called the *best basis* relative to the cost C and the training dataset T.

We also note that log in this paper implies $\log_2$, unless stated otherwise. The $n \times n$ identity matrix is denoted by I_n, and the $n \times 1$ column vector whose entries are all ones, i.e., $(1, 1, \ldots, 1)^T$, is denoted by $\mathbf{1}_n$.

9.3 SPARSITY VS. STATISTICAL INDEPENDENCE

The concept of sparsity and that of statistical independence are intrinsically different. Sparsity emphasizes the issue of compression directly, whereas statistical independence concerns the relationship among the coordinates. Yet, for certain stochastic processes, these two are intimately related, and often confusing. For example, Olshausen and Field [18], [19] emphasized the sparsity as the basis selection criterion, but they also assumed the statistical independence of the coordinates. Bell and Sejnowski [1] used the statistical independence criterion and obtained the basis functions similar to those of Olshausen and Field. They claimed that they did not impose the sparsity explicitly and such sparsity *emerged* by minimizing the statistical dependence among the coordinates. These motivated us to study these two criteria.

First let us define the measure of sparsity and that of statistical independence in our context.

9.3.1 Sparsity

Sparsity is a key property for compression. The true sparsity measure for a given vector $\boldsymbol{x} \in \mathbf{R}^n$ is the so-called ℓ^0 quasi-norm which is defined as

$$\|\boldsymbol{x}\|_0 \triangleq \#\{i \in [1, n] : x_i \neq 0\},$$

i.e., the number of nonzero components in $\boldsymbol{x}$. This measure is, however, very unstable for even small geometric perturbations of the components in a vector. Therefore, a better measure is the ℓ^p norm:

$$\|\boldsymbol{x}\|_p \triangleq \left(\sum_{i=1}^{n} |x_i|^p \right)^{1/p}, \quad 0 < p \leq 1.$$

In fact, this is a quasi-norm for $0 < p < 1$ since this does not satisfy the triangle inequality, but only satisfies weaker conditions: $\|\boldsymbol{x}+\boldsymbol{y}\|_p \leq 2^{-1/p'}(\|\boldsymbol{x}\|_p + \|\boldsymbol{y}\|_p)$ where $p' = p/(p-1)$ is the conjugate exponent of p; and $\|\boldsymbol{x}+\boldsymbol{y}\|_p^p \leq \|\boldsymbol{x}\|_p^p + \|\boldsymbol{y}\|_p^p$. It is easy to show that $\lim_{p \downarrow 0} \|\boldsymbol{x}\|_p^p = \|\boldsymbol{x}\|_0$. See [6], [7], [8] for the details of the ℓ^p norm properties.

Thus, we use the expected ℓ^p norm minimization as a criterion to find the best basis for a given stochastic process in terms of sparsity:

$$C_p(B \mid \boldsymbol{X}) = E\|B^{-1}\boldsymbol{X}\|_p^p, \tag{3.1}$$

The sample estimate of this cost given the training dataset T is

$$C_p(B \mid T) = \frac{1}{N} \sum_{k=1}^{N} \|\boldsymbol{y}_k\|_p^p = \frac{1}{N} \sum_{k=1}^{N} \sum_{i=1}^{n} |y_{i,k}|^p, \tag{3.2}$$

where $\boldsymbol{y}_k = (y_{1,k}, \ldots, y_{n,k})^T = B^{-1}\boldsymbol{x}_k$ and $\boldsymbol{x}_k$ is the kth sample (or realization) in T. We propose to minimize this cost in order to select the *best sparsifying basis* (BSB):

$$B_p = B_p(T, D) = \arg\min_{B \in D} C_p(B \mid T).$$

Remark 9.3.1 It should be noted that *the minimization of the ℓ^p norm can also be achieved for each realization.* Without taking the average in k in (3.2), one can select the BSB $B_p = B_p(\{\boldsymbol{x}_k\}, D)$ for each realization $\boldsymbol{x}_k \in T$. We can guarantee that

$$\min_{B \in D} C_p(B \mid \{\boldsymbol{x}_k\}) \leq \min_{B \in D} C_p(B \mid T) \leq \max_{B \in D} C_p(B \mid \{\boldsymbol{x}_k\}).$$

For highly variable or erratic stochastic processes, however, $B_p(\{\boldsymbol{x}_k\}, D)$ may significantly change for each k. Thus if we adopt this strategy to compress an entire training dataset consisting of N realizations, we need to store additional information in order to describe a set of N bases. Whether we should adapt a basis per realization or on the average is still an open issue. See Saito et al. [23] for more details.

9.3.2 Statistical Independence

The statistical independence of the coordinates of $\boldsymbol{Y} \in \mathbf{R}^n$ means

$$f_{\boldsymbol{Y}}(\boldsymbol{y}) = f_{Y_1}(y_1) f_{Y_2}(y_2) \cdots f_{Y_n}(y_n),$$

where $f_{Y_k}(y_k)$ is a one-dimensional marginal pdf of $f_{\boldsymbol{Y}}$. Statistical independence is a key property for compressing and modeling a stochastic process because: 1) an n-dimensional stochastic process of interest can be modeled as a set of one-dimensional processes; and 2) damage of one coordinate does not propagate to the others. Of course, in general, it is difficult to find a truly statistically independent coordinate system for a given stochastic process. Such a coordinate system may not even exist for a given stochastic process. Therefore, the next best thing we can do is to find the least statistically-dependent coordinate system within a basis dictionary. Naturally, then, we need to measure the "closeness" of a coordinate system (or random variables) $Y_1, \ldots, Y_n$ to the statistical independence. This can be measured by *mutual information* or *relative entropy* between the true pdf $f_{\boldsymbol{Y}}$ and the product of its marginal pdfs:

$$I(\boldsymbol{Y}) \triangleq \int f_{\boldsymbol{Y}}(\boldsymbol{y}) \log \frac{f_{\boldsymbol{Y}}(\boldsymbol{y})}{\prod_{i=1}^n f_{Y_i}(y_i)} \, \mathrm{d}\boldsymbol{y} = -H(\boldsymbol{Y}) + \sum_{i=1}^n H(Y_i),$$

where $H(\boldsymbol{Y})$ and $H(Y_i)$ are the differential entropy of $\boldsymbol{Y}$ and Y_i respectively:

$$H(\boldsymbol{Y}) = -\int f_{\boldsymbol{Y}}(\boldsymbol{y}) \log f_{\boldsymbol{Y}}(\boldsymbol{y}) \, \mathrm{d}\boldsymbol{y}, \quad H(Y_i) = -\int f_{Y_i}(y_i) \log f_{Y_i}(y_i) \, \mathrm{d}y_i.$$

We note that $I(\boldsymbol{Y}) \geq 0$, and $I(\boldsymbol{Y}) = 0$ if and only if the components of $\boldsymbol{Y}$ are mutually independent. See [5] for more details of the mutual information.

Suppose $\boldsymbol{Y} = B^{-1}\boldsymbol{X}$ and $B \in \mathrm{GL}(n, \mathbf{R})$ with $\det(B) = \pm 1$. We denote this set of matrices by $\mathrm{SL}^{\pm}(n, \mathbf{R})$. Note that the usual $\mathrm{SL}(n, \mathbf{R})$ is a subset of $\mathrm{SL}^{\pm}(n, \mathbf{R})$. Then, we have

$$I(\boldsymbol{Y}) = -H(\boldsymbol{Y}) + \sum_{i=1}^n H(Y_i) = -H(\boldsymbol{X}) + \sum_{i=1}^n H(Y_i),$$

since the differential entropy is *invariant* under such an invertible volume-preserving linear transformation, i.e.,

$$H(B^{-1}\boldsymbol{X}) = H(\boldsymbol{X}) + \log|\det(B^{-1})| = H(\boldsymbol{X}),$$

because $|\det(B^{-1})| = 1$. Based on this fact, we proposed the minimization of the following cost function as the criterion to select the so-called *least statistically-dependent basis* (LSDB) in [21]:

$$C_H(B\,|\,\boldsymbol{X}) = \sum_{i=1}^{n} H\left((B^{-1}\boldsymbol{X})_i\right) = \sum_{i=1}^{n} H(Y_i). \tag{3.3}$$

The sample estimate of this cost given the training dataset T is

$$C_H(B\,|\,T) = -\frac{1}{N}\sum_{k=1}^{N}\sum_{i=1}^{n} \log \widehat{f}_{Y_i}(y_{i,k}),$$

where $\widehat{f}_{Y_i}(y_{i,k})$ is an empirical pdf of the coordinate Y_i, which must be estimated by an algorithm such as the histogram-based estimator with optimal bin-width search of Hall and Morton [11]. Now, we can define the LSDB as

$$B_{LSDB} = B_{LSDB}(T, D) = \arg\min_{B\in D} C_H(B\,|\,T). \tag{3.4}$$

We note that the differences between this strategy and the standard independent component analysis (ICA) algorithms are: 1) restriction of the search in the basis dictionary D; and 2) approximation of the coordinate-wise entropy. For more details, we refer the reader to [21] for the former and [3] for the latter.

We now demonstrate the fact that the sparsity and the statistical independence are two intrinsically different concepts using a simple example.

9.4 TWO-DIMENSIONAL COUNTEREXAMPLE

Let us consider a simple process $\boldsymbol{X} = (X_1, X_2)^T$ where X_1 and X_2 are independently and identically distributed as the uniform random variable on the interval $[-1, 1]$. Thus, the realizations of this process are distributed as the right-hand side of Figure 9.1. Let us consider all possible rotations around the origin as a basis dictionary, i.e., $D = \mathrm{SO}(2, \mathbf{R}) \subset \mathrm{O}(2)$. Then, the sparsity and independence criteria select completely different bases as shown in Figure 9.1. Note that the data points under the BSB coordinates (45 degree rotation) concentrate more around the origin than the LSDB coordinates (with no rotation) and this rotation makes the data representation sparser. This example clearly demonstrates that the BSB and the LSDB are different in general. One can also generalize this example to higher dimensions.

9.5 THE SPIKE PROCESS

An n-dimensional *spike process* simply generates the standard basis vectors $\{\boldsymbol{e}_j\}_{j=1}^{n} \subset \mathbf{R}^n$ in a random order, where $\boldsymbol{e}_j$ has one at the jth entry and all the other entries are zero. One can view this process as a unit impulse located at a random position between 1 and n as shown in Figure 9.2.

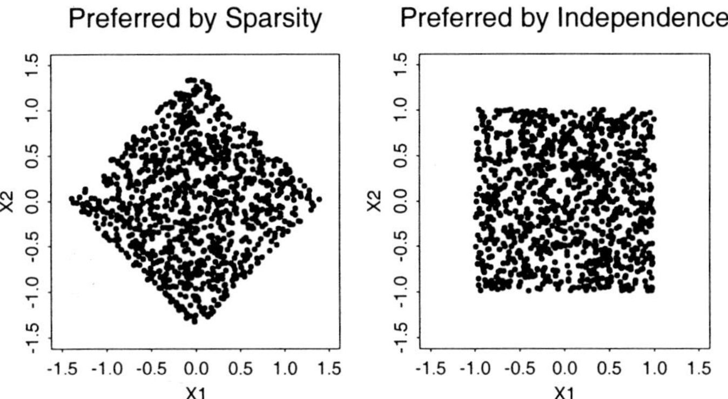

Figure 9.1. Sparsity and statistical independence prefer the different coordinates

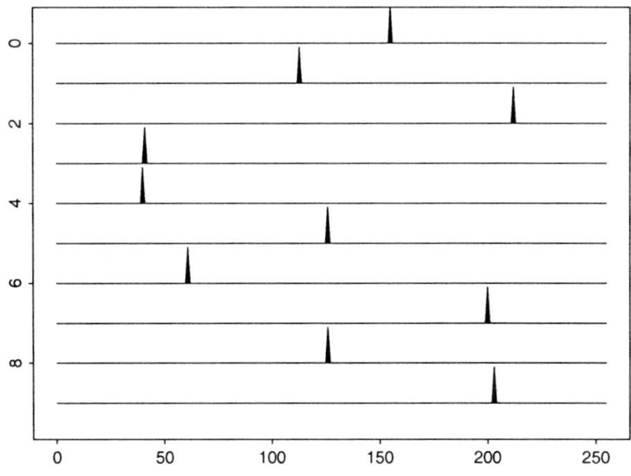

Figure 9.2. Ten realizations of the spike process ($n = 256$)

9.5.1 The Karhunen-Loève Basis

Let us first consider the Karhunen-Loève basis of this process from which we can learn a few things.

Proposition 9.5.1 *The Karhunen-Loève basis for the spike process is any orthonormal basis in* $\mathbf{R}^n$ *containing the "DC" vector* $\mathbf{1}_n = (1, 1, \ldots, 1)^T$.

This means that the KLB is not useful for this process. This is because the spike process is highly non-Gaussian.

9.5.2 The Best Sparsifying Basis

It seems obvious that the standard basis is the BSB among $\mathrm{O}(n)$ by construction; an expansion of a realization of this process into any other basis simply increases the number of nonzero coefficients. However, we still need to verify this. In fact, we have the following theorem.

Theorem 9.5.1 *The BSB for the spike process is the standard basis if* $D = \mathrm{O}(n)$ *or* $\mathrm{SL}^{\pm}(n, \mathbf{R})$.

Remark 9.5.1 It is not meaningful to consider the group $\mathrm{GL}(n, \mathbf{R})$ as a basis dictionary to find the BSB since one can always find an invertible matrix B whose inverse B^{-1} consists of infinitesimally small entries so that the cost $C_p(B^{-1}\boldsymbol{X})$ is close to zero. However, we can consider the subset $\mathrm{GL}_a(n, \mathbf{R}) \subset \mathrm{GL}(n, \mathbf{R})$, which consists of all invertible matrices whose determinant is $a > 0$. (Note that this set $\mathrm{GL}_a(n, \mathbf{R})$ is generally not a group since it does not contain the inverse matrices of its members. Of course $\mathrm{GL}_1(n, \mathbf{R}) = \mathrm{SL}(n, \mathbf{R})$ is a special case.) The following corollary is a minor modification of Theorem 9.5.1 by recognizing that $\mathrm{GL}_a(n, \mathbf{R}) = a^{1/n} \cdot \mathrm{SL}(n, \mathbf{R})$:

Corollary 9.5.2 *If* $D = \mathrm{GL}_a(n, \mathbf{R})$ *with* $a > 0$*, then the BSB must be the scalar multiple of the identity matrix,* $a^{1/n} I_n$.

Remark 9.5.2 Note that when we say the basis is a matrix such as $a^{1/n} I_n$, we really mean that the column vectors of that matrix form the basis. This also means that any permuted and/or sign-flipped (i.e., multiplied by -1) versions of those column vectors also form the basis. Therefore, when we say the basis is a matrix A, we mean not only A but also its permuted and sign-flipped versions of A. This remark also applies to all the propositions, lemmas, and theorems below, unless stated otherwise.

9.5.3 Statistical Dependence and Entropy of the Spike Process

Before considering the LSDB of this process, let us note a few specifics about the spike process. First, although the standard basis is the BSB for this process, it clearly does not provide the statistically independent coordinates. The existence of a single spike at one location prohibits spike generation at other locations. This implies that these coordinates are highly statistically dependent.

Second, we can compute the true entropy $H(\boldsymbol{X})$ for the spike process unlike other complicated stochastic processes. Since the spike process selects one possible vector from the standard basis of $\mathbf{R}^n$ with uniform probability $1/n$, the true entropy $H(\boldsymbol{X})$ is clearly $\log n$. This is one of the rare cases where we know the true high-dimensional entropy of the process.

9.5.4 The LSDB among $\mathrm{O}(n)$

Let us now consider $\mathrm{O}(n)$, the set of all possible orthonormal bases in $\mathbf{R}^n$, as our basis dictionary. Then, we have the following theorem.

Theorem 9.5.3 *The LSDB among* $\mathrm{O}(n)$ *is the following:*

- *for* $n \geq 5$, *either the standard basis or the basis whose matrix representation is*

$$\frac{1}{n}\begin{bmatrix} n-2 & -2 & \cdots & -2 & -2 \\ -2 & n-2 & \ddots & & -2 \\ \vdots & \ddots & \ddots & \ddots & \vdots \\ -2 & & \ddots & n-2 & -2 \\ -2 & -2 & \cdots & -2 & n-2 \end{bmatrix}; \tag{5.1}$$

- *for* $n = 4$, *the Walsh basis, i.e.,* $\frac{1}{2}\begin{bmatrix} 1 & 1 & 1 & 1 \\ 1 & 1 & -1 & -1 \\ 1 & -1 & 1 & -1 \\ 1 & -1 & -1 & 1 \end{bmatrix}$;

- *for* $n = 3$, $\begin{bmatrix} \frac{1}{\sqrt{3}} & \frac{1}{\sqrt{6}} & \frac{1}{\sqrt{2}} \\ \frac{1}{\sqrt{3}} & \frac{1}{\sqrt{6}} & \frac{-1}{\sqrt{2}} \\ \frac{1}{\sqrt{3}} & \frac{-2}{\sqrt{6}} & 0 \end{bmatrix}$; *and*

- *for* $n = 2$, $\frac{1}{\sqrt{2}}\begin{bmatrix} 1 & 1 \\ 1 & -1 \end{bmatrix}$, *and this is the only case where the true independence is achieved.*

Remark 9.5.3 There is an important geometric interpretation of (5.1). This matrix can also be written as:

$$B_{HR(n)} \triangleq I_n - 2\frac{\mathbf{1}_n}{\sqrt{n}}\frac{\mathbf{1}_n^T}{\sqrt{n}}.$$

In other words, this matrix represents the *Householder reflection* with respect to the hyperplane $\{\boldsymbol{y} \in \mathbf{R}^n \mid \sum_{i=0}^{n} y_i = 0\}$ whose unit normal vector is $\mathbf{1}_n/\sqrt{n}$.

Below, we use the notation $B_{\mathrm{O}(n)}$ for the LSDB among $\mathrm{O}(n)$ to distinguish it from the LSDB among $\mathrm{GL}(n, \mathbf{R})$, which is denoted by $B_{\mathrm{GL}(n)}$. So, for example, for $n \geq 5$, $B_{\mathrm{O}(n)} = I_n$ or $B_{HR(n)}$.

9.5.5 The LSDB among $\mathrm{GL}(n, \mathbf{R})$

Before discussing the LSDB among a larger dictionary of bases, let us remark an important specifics for a discrete stochastic process.

Let $\boldsymbol{X}$ be a random vector obeying a discrete stochastic process with a probability mass function (pmf) $f_{\boldsymbol{X}}$. This means that there are only finite number of possible values (or states) $\boldsymbol{X}$ can take. Clearly the spike process is a discrete process since the only possible values are $\{\boldsymbol{e}_1, \ldots, \boldsymbol{e}_n\}$, the standard basis vectors. Then, for any invertible transformation $B \in \mathrm{GL}(n, \mathbf{R})$ with $\boldsymbol{Y} = B^{-1}\boldsymbol{X}$, be it orthonormal or not, the total entropy of the process before and after the transformation is exactly the same. Indeed, in the definition of discrete Shannon entropy, $-\sum_j p_j \log p_j$, the values that the random variable takes are of no importance; only the number of possible values the random variable can take and its pmf matter. In our case, it is clear that the events $\{\boldsymbol{X} = \boldsymbol{a}_i\}$ and $\{\boldsymbol{Y} = \boldsymbol{b}_i\}$ where $\boldsymbol{b}_i = B^{-1}\boldsymbol{a}_i$ are equivalent; otherwise the transformation would not be invertible. This implies that the corresponding probabilities are equal:

$$\Pr\{\boldsymbol{X} = \boldsymbol{a}_i\} = \Pr\{\boldsymbol{Y} = \boldsymbol{b}_i\}.$$

Therefore, considering the expression of the discrete Shannon entropy, this proves that

$$H(\boldsymbol{Y}) = H(\boldsymbol{X}),$$

as long as the transformation matrix belongs to $\mathrm{GL}(n, \mathbf{R})$. Note that for the continuous case, this is only true if $B \in \mathrm{SL}^{\pm}(n, \mathbf{R})$. Therefore, for a discrete stochastic process like the spike process, the LSDB among $\mathrm{GL}(n, \mathbf{R})$ can be selected by just minimizing the sum of the coordinate-wise entropy as (3.4) as if $D = \mathrm{SL}^{\pm}(n, \mathbf{R})$. In other words, there is no important distinction in the LSDB selection from $\mathrm{GL}(n, \mathbf{R})$ and from $\mathrm{SL}^{\pm}(n, \mathbf{R})$ for discrete stochastic processes. Therefore, we do not have to treat these two cases separately. Note that the case of the BSB is a different story as we already mentioned in Remark 2.

Now we have the following theorem:

Theorem 9.5.4 *The LSDB among* $\mathrm{GL}(n, \mathbf{R})$ *with* $n \geq 2$ *is the following basis pair (for analysis and synthesis respectively):*

$$B_{\mathrm{GL}(n)}^{-1} = \begin{bmatrix} a & a & \cdots & \cdots & \cdots & \cdots & a \\ b_2 & c_2 & b_2 & \cdots & \cdots & \cdots & b_2 \\ b_3 & b_3 & c_3 & b_3 & \cdots & \cdots & b_3 \\ \vdots & \vdots & & \ddots & & & \vdots \\ \vdots & \vdots & & & \ddots & & \vdots \\ b_{n-1} & \cdots & \cdots & \cdots & b_{n-1} & c_{n-1} & b_{n-1} \\ b_n & \cdots & \cdots & \cdots & \cdots & b_n & c_n \end{bmatrix}, \tag{5.2}$$

$$B_{\mathrm{GL}(n)} = \begin{bmatrix} \left(1 + \sum_{k=2}^{n} b_k d_k\right)/a & -d_2 & -d_3 & \cdots & -d_n \\ -b_2 d_2/a & d_2 & 0 & \cdots & 0 \\ -b_3 d_3/a & 0 & d_3 & \ddots & \vdots \\ \vdots & \vdots & \ddots & \ddots & 0 \\ -b_n d_n/a & 0 & \cdots & 0 & d_n \end{bmatrix}, \tag{5.3}$$

where a, b_k, c_k *are arbitrary real-valued constants satisfying* $a \neq 0$, $b_k \neq c_k$, *and* $d_k = 1/(c_k - b_k)$, $k = 2, \ldots, n$.

If we restrict ourselves to $D = \mathrm{SL}^{\pm}(n, \mathbf{R})$, *then the parameter* a *must satisfy:*

$$a = \pm \prod_{k=2}^{n} (c_k - b_k)^{-1}.$$

Remark 9.5.4 The LSDB such as (5.1) and the LSDB pair (5.2), (5.3) provide us with further insight into the difference between sparsity and statistical independence. In the case of (5.1), this is the LSDB, yet it does not sparsify the spike process at all. In fact, these coordinates are completely dense, i.e., $C_0 = n$. We can also show that the sparsity measure C_p gets worse as $n \to \infty$. More precisely, we have the following proposition.

Proposition 9.5.2

$$\lim_{n\to\infty} C_p \left(B_{HR(n)} \,|\, \boldsymbol{X} \right) = \begin{cases} \infty & \text{if } 0 \le p < 1; \\ 3 & \text{if } p = 1. \end{cases}$$

It is interesting to note that this LSDB approaches to the standard basis as $n \to \infty$. This also implies that

$$\lim_{n\to\infty} C_p \left(B_{HR(n)} \,|\, \boldsymbol{X} \right) \ne C_p \left(\lim_{n\to\infty} B_{HR(n)} \,|\, \boldsymbol{X} \right).$$

As for the analysis LSDB (5.2), the ability to sparsify the spike process depends on the values of b_k and c_k. Since the parameters a, b_k and c_k are arbitrary as long as $a \ne 0$ and $b_k \ne c_k$, let us put $a = 1$, $b_k = 0$, $c_k = 1$, for $k = 2, \ldots, n$. Then we get the following specific LSDB pair:

$$B_{\mathrm{GL}(n)}^{-1} = \begin{bmatrix} 1 & 1 & \cdots & 1 \\ 0 & & & \\ \vdots & & I_{n-1} & \\ 0 & & & \end{bmatrix}, \quad B_{\mathrm{GL}(n)} = \begin{bmatrix} 1 & -1 & \cdots & -1 \\ 0 & & & \\ \vdots & & I_{n-1} & \\ 0 & & & \end{bmatrix}.$$

This analysis LSDB provides us with a sparse representation for the spike process (though this is clearly not better than the standard basis). For $\boldsymbol{Y} = B_{\mathrm{GL}(n)}^{-1} \boldsymbol{X}$,

$$C_p = E\left[\|\boldsymbol{Y}\|_p^p \right] = \frac{1}{n} \times 1 + \frac{n-1}{n} \times 2 = 2 - \frac{1}{n}, \quad 0 \le p \le 1.$$

Now, let us take $a = 1$, $b_k = 1$, $c_k = 2$ for $k = 2, \ldots, n$ in (5.2) and (5.3). Then we get

$$B_{\mathrm{GL}(n)}^{-1} = \begin{bmatrix} 1 & 1 & \cdots & 1 \\ 1 & 2 & \ddots & \vdots \\ \vdots & \ddots & \ddots & 1 \\ 1 & \cdots & 1 & 2 \end{bmatrix}, \quad B_{\mathrm{GL}(n)} = \begin{bmatrix} n & -1 & \cdots & -1 \\ -1 & & & \\ \vdots & & I_{n-1} & \\ -1 & & & \end{bmatrix}. \tag{5.4}$$

The sparsity measure of this process is:

$$C_p = \frac{1}{n} \times n + \frac{n-1}{n} \times \{(n-1) + 2^p\} = n + (2^p - 1)\left(1 - \frac{1}{n}\right), \quad 0 \le p \le 1.$$

Therefore, the spike process under this analysis basis is completely dense, i.e., $C_p \ge n$ for $0 \le p \le 1$ and the equality holds if and only if $p = 0$. Yet this is still the LSDB.

Finally, from Theorems 9.5.3 and 9.5.4, we can prove the following corollary:

Corollary 9.5.5 *There is no invertible linear transformation providing the statistically independent coordinates for the spike process for $n > 2$. In fact, the mutual information $I\left(B_{O(n)}^T \boldsymbol{X}\right)$ and $I\left(B_{GL(n)}^{-1} \boldsymbol{X}\right)$ are monotonically increasing as a function of n, and both approaches to $\log e \approx 1.4427$ as $n \to \infty$.*

Remark 9.5.5 Although the spike process is very simple, we have the following interpretation. Consider a stochastic process generating a basis vector randomly at a time selected from some orthonormal basis. Then, that basis itself is both the BSB and the LSDB among $O(n)$. Theorem 9.5.3 claims that once we transform the data to the spikes, one cannot do any better than that both in sparsity and independence within $O(n)$ with $n \ge 5$. Of course, if one extends the search to nonlinear transformations, then it becomes a different story. We refer the reader to our recent articles Lin et al. [15], [16], for the details of a nonlinear algorithm.

9.6 PROOFS OF PROPOSITIONS AND THEOREMS

9.6.1 Proof of Proposition

9.5.1 Let $\boldsymbol{X} = (X_1, X_2, \ldots, X_n)^T$ be a random vector generated by this process. For each of its realizations, a randomly chosen coordinate among these n positions takes the value 1, while the others take the value 0. Hence each X_i, $i = 1, \ldots, n$, takes the values 1 with probability $1/n$ and the value 0 with probability $1 - 1/n$. Let us calculate the covariance of these variables. First, we have:

$$E(X_i) = \frac{1}{n} \times 1 + \left(1 - \frac{1}{n}\right) \times 0 = \frac{1}{n} \quad \text{for } i = 1, \ldots, n$$

$$E(X_i X_j) = \begin{cases} E(X_i^2) = E(X_i) & \text{if } i = j; \\ 0 & \text{if } i \ne j, \end{cases}$$

since one of these two variables will always take the value 0. Let $R = (R_{ij})$ be the covariance matrix of this process. Then, we have:

$$R_{ij} = E(X_i X_j) - E(X_i)E(X_j) = \frac{1}{n}\delta_{ij} - \frac{1}{n^2}$$

We know that a basis is a Karhunen-Loève basis if and only if it is orthonormal and diagonalizes the covariance matrix. Thus, we will now calculate the eigenvalue decomposition of the covariance matrix $R = \frac{1}{n} I_n - \frac{1}{n^2} \mathbf{1}_n \mathbf{1}_n^T$.

We now need to calculate the determinant:

$$P_R(\lambda) \triangleq \det(\lambda I_n - R) = \begin{vmatrix} \lambda - \frac{1}{n} + \frac{1}{n^2} & \frac{1}{n^2} & \cdots & \frac{1}{n^2} \\ \frac{1}{n^2} & \ddots & \ddots & \vdots \\ \vdots & \ddots & \ddots & \frac{1}{n^2} \\ \frac{1}{n^2} & \cdots & \frac{1}{n^2} & \lambda - \frac{1}{n} + \frac{1}{n^2} \end{vmatrix},$$

which is of the generic form:

$$\Delta(a,b) \triangleq \begin{vmatrix} a+b & b & \cdots & b \\ b & a+b & \ddots & \vdots \\ \vdots & \ddots & \ddots & b \\ b & \cdots & b & a+b \end{vmatrix},$$

with the values $a = \lambda - 1/n$ and $b = 1/n^2$. We can easily evaluate this determinant by subtracting the last row from all the others followed by adding all $n-1$ columns to the last column:

$$\Delta(a,b) = \begin{vmatrix} a & 0 & \cdots & 0 & 0 \\ 0 & a & \ddots & \vdots & \vdots \\ \vdots & \ddots & \ddots & 0 & \vdots \\ 0 & \cdots & 0 & a & 0 \\ b & \cdots & \cdots & b & a+nb \end{vmatrix} = a^{n-1}(a+nb). \tag{6.1}$$

Putting $a = \lambda - 1/n$ and $b = 1/n^2$, we have the characteristic polynomial P_R of R as $P_R(\lambda) = \lambda(\lambda - 1/n)^{n-1}$. Hence, the eigenvalues of R are $\lambda = 0$ or $1/n$.

It is now obvious that the vector $\mathbf{1}_n$ is an eigenvector for R associated with the eigenvalue 0, i.e., $\mathbf{1}_n \in \ker R$. Indeed, we have

$$R\mathbf{1}_n = \left(\frac{1}{n}I_n - \frac{1}{n^2}\mathbf{1}_n\mathbf{1}_n^T\right)\mathbf{1}_n = \frac{1}{n}\,\mathbf{1}_n - \frac{1}{n^2}\,n\mathbf{1}_n = 0.$$

Since $\dim\ker R = 1$, $\ker R$ is a one-dimensional subspace spanned by $\mathbf{1}_n$. Considering that R is symmetric and only has two distinct eigenvalues, we know that the eigenspace associated to the eigenvalue $1/n$ is orthogonal to $\ker R$, which is the hyperplane $\{\boldsymbol{y} \in \mathbf{R}^n \mid \sum_{i=1}^n y_i = 0\}$. Therefore, the orthogonal bases that diagonalize R are the bases formed by the adjunction of $\mathbf{1}_n$ to any orthogonal basis of $\ker R^\perp$. The Walsh basis, which consists of oscillating square waves, is such a basis, although it is just one among many. □

9.6.2 Proof of Theorem 9.5.1

We first prove the case $D = \mathrm{SL}^\pm(n,\mathbf{R})$. Then, the case of $D = \mathrm{O}(n)$ is automatic since this is just a special case of $\mathrm{SL}^\pm(n,\mathbf{R})$.

Let B be any matrix in $\mathrm{SL}^\pm(n,\mathbf{R})$, and let $\boldsymbol{b}_j$ be its jth column vector. Let us first write the cost function (3.1) for the spike process in terms of the matrix elements of B:

$$C_p(B \mid \boldsymbol{X}) = \frac{1}{n} \sum_{i=1}^{n} \sum_{j=1}^{n} |b_{ij}|^p = \frac{1}{n} \sum_{j=1}^{n} \|\boldsymbol{b}_j\|_p^p . \tag{6.2}$$

It is a well-known fact [13, p.112] that for any nonsingular real-valued matrix B (i.e., $\in \mathrm{GL}(n, \mathbf{R})$), there exists a unique QR factorization

$$B = Q\,R, \tag{6.3}$$

where $Q \in \mathrm{O}(n)$ and R is an $n \times n$ upper triangular matrix,

$$R = \begin{bmatrix} r_{11} & r_{12} & \cdots & r_{1n} \\ 0 & \ddots & & \vdots \\ \vdots & \ddots & \ddots & \vdots \\ 0 & \cdots & 0 & r_{nn} \end{bmatrix}, \tag{6.4}$$

with $r_{jj} > 0$, $j = 1, \ldots, n$. Furthermore, since $\det(B) = \pm 1$, we can assume $\prod_{j=1}^{n} r_{jj} = 1$. Let $\boldsymbol{q}_j$ be the jth column vector of Q. Then, from (6.3) and (6.4), we have

$$\boldsymbol{b}_j = r_{j1}\boldsymbol{q}_1 + \cdots + r_{jj}\boldsymbol{q}_j, \quad j = 1, \ldots, n.$$

Now, the cost function (6.2) can be written as:

$$\begin{aligned} n \cdot C_p(B \mid \boldsymbol{X}) &= \sum_{j=1}^{n} \|\boldsymbol{b}_j\|_p^p \\ &= \|r_{11}\boldsymbol{q}_1\|_p^p + \|r_{12}\boldsymbol{q}_1 + r_{22}\boldsymbol{q}_2\|_p^p + \cdots + \|r_{1n}\boldsymbol{q}_1 + \cdots + r_{nn}\boldsymbol{q}_n\|_p^p . \end{aligned} \tag{6.5}$$

$$n \cdot C_p(B \mid \boldsymbol{X}) = \|r_{11}\boldsymbol{q}_1\|_p^p + \|r_{12}\boldsymbol{q}_1 + r_{22}\boldsymbol{q}_2\|_p^p + \cdots + \|r_{1n}\boldsymbol{q}_1 + \cdots + r_{nn}\boldsymbol{q}_n\|_p^p . \tag{6.6}$$

Let us now consider this last expression term by term. The first term is bounded from below as follows:

$$\|r_{11}\boldsymbol{q}_1\|_p^p = r_{11}^p \|\boldsymbol{q}_1\|_p^p \geq r_{11}^p,$$

since $\boldsymbol{q}_1$ is an ℓ^2-unit vector in $\mathbf{R}^n$ (i.e., $\|\boldsymbol{q}_1\|_2 = 1$), and its ℓ^p norm ($0 \leq p \leq 1$) is clearly minimized if $\boldsymbol{q}_1 = \boldsymbol{e}_k$ or $-\boldsymbol{e}_k$ for some $k \in \{1, \ldots, n\}$. The second term of (6.5) is bounded from below as:

$$\|r_{12}\boldsymbol{q}_1 + r_{22}\boldsymbol{q}_2\|_p^p \geq (r_{12}^2 + r_{22}^2)^{p/2},$$

since the vector $r_{12}\boldsymbol{q}_1 + r_{22}\boldsymbol{q}_2$ has the ℓ^2 length $(r_{12}^2 + r_{22}^2)^{1/2}$, and among any vector of that length in $\mathbf{R}^n$, the minimum ℓ^p norm is attained if $r_{12}\boldsymbol{q}_1 + r_{22}\boldsymbol{q}_2 = \pm(r_{12}^2 + r_{22}^2)^{1/2}\boldsymbol{e}_k$ for some $k \in \{1, \ldots, n\}$, i.e., if it is aligned along one of the standard basis vectors. We can repeat this argument and finally we have:

$$n \cdot C_p(B \mid \boldsymbol{X}) \geq r_{11}^p + (r_{12}^2 + r_{22}^2)^{p/2} + \cdots + (r_{1n}^2 + \cdots + r_{nn}^2)^{p/2}.$$

Let $g(R)$ denote the righthand side of this inequality. Since all the diagonal elements of R are positive, this is further bounded from below by

$$g(R) \geq r_{11}^p + r_{22}^p + \cdots + r_{nn}^p,$$

by setting all the nondiagonal elements of R to zero. This is again bounded from below by

$$n \cdot r_{kk}^p,$$

where $r_{kk} = \min(r_{11}, \dots, r_{nn})$. Combining this with the fact that $\prod_{j=1}^n r_{jj} = 1$ and $r_{jj} \in \mathbf{R}$, we must have $r_{jj} = 1$ for $j = 1, \dots, n$, i.e., $R = I_n$ as the minimizer of the function $g(R)$. That is,

$$\min g(R) = g(I_n) = n.$$

Coming back to the matrix $B = QR$, the minimizer of $C_p(B \mid \boldsymbol{X})$ must satisfy $B = Q$, and furthermore

$$\begin{aligned} n \cdot C_p(Q \mid \boldsymbol{X}) &= \|\boldsymbol{q}_1\|_p^p + \cdots + \|\boldsymbol{q}_n\|_p^p \\ &\geq n, \end{aligned}$$

where the equality holds if and only if Q is a permutation matrix or the sign flipped version of such a matrix, by the same argument of the minimization of the ℓ^p norm of an ℓ^2-unit vector. This implies that B must be the identity matrix modulo permutations and sign flips. □

9.6.3 Coordinate-wise Entropy of the Spike Process

Before proceeding to the proof of Theorems 9.5.3 and 9.5.4, let us consider coordinate-wise entropy of the spike process and define some convenient quantities for characterizing a basis in $\mathrm{O}(n)$ or $\mathrm{GL}(n, \mathbf{R})$.

Let us consider an invertible matrix $U = (u_{ij})_{i,j=1,\dots,n} = B^{-1} \in \mathrm{GL}(n, \mathbf{R})$, and the vector $\boldsymbol{Y} = U\boldsymbol{X}$. Let us consider the ith coordinate of $\boldsymbol{Y}$, $Y_i = \sum_{j=1}^n u_{ij} X_j$. For each realization of the spike process $\boldsymbol{X}$, Y_i takes one of the values $\{u_{ij}, j = 1, \dots, n\}$. More precisely, we have $\Pr\{X_j = 1\} = 1/n$ and $\Pr\{X_j = 0\} = 1 - 1/n$, for $j = 1, \dots, n$. Thus, if all $\{u_{ij}, j = 1, \dots, n\}$ were distinct, Y_i would take these values with a uniform pmf. But there is no particular reason that allows us to think $\{u_{ij}, j = 1, \dots, n\}$ are mutually distinct. Therefore, we shall group these values in "classes" of equality. Let us introduce, for each $i \in \{1, \dots, n\}$, an integer $k(i)$ equal to the number of distinct values in the ith row vector $\{u_{ij}, j = 1, \dots, n\}$, and the vector $c(i) = (\alpha_1(i), \dots, \alpha_{k(i)}(i)) \in \mathbf{N}^{k(i)}$, where each component counts the number of occurrences of each distinct value in the ith row vector. We will call $k(i)$ the *class* of the ith row and $c(i)$ the *index* of that row. Clearly, we have

$$1 \leq k(i) \leq n \quad \text{and} \quad \sum_{\ell=1}^{k(i)} \alpha_\ell(i) = n.$$

For example, with $n = 3$, if we had

$$\begin{cases} Y_1 = X_1 + X_2 + X_3 \\ Y_2 = 5X_1 + 2X_2 + 2X_3 \\ Y_3 = -X_1 + X_2 \end{cases},$$

then we would get

$$\begin{cases} k(1) = 1,\ c(1) = (3) \\ k(2) = 2,\ c(2) = (2, 1) \\ k(3) = 3,\ c(3) = (1, 1, 1) \end{cases},$$

since $\{u_{1j}\} = \{1, 1, 1\}$ in which we find three 1's, $\{u_{2j}\} = \{5, 2, 2\}$ in which we find two 2's, one 5, and $\{u_{3j}\} = \{-1, 1, 0\}$ in which we find one -1, one 1, and one 0.

Let us now examine the coordinate-wise entropy in terms of the quantities we have just defined. Suppose the value u appears $\alpha_\ell(i)$ times in $\{u_{ij}, j = 1, \ldots, n\}$. Then the probability of the event $\{Y_i = u\}$ is $\alpha_\ell(i)/n$. Therefore, we have

$$H(Y_i) = -\sum_{\ell=1}^{k(i)} \frac{\alpha_\ell(i)}{n} \log \frac{\alpha_\ell(i)}{n}.$$

We shall now describe the different values that this coordinate-wise entropy takes as the number of distinct values and their occurrences vary. Because the entropy is a measure of uncertainty, we can intuitively guess that a coordinate with a small class number generates small entropy.

$k(i) = 1$: This necessarily means that $c(i) = (n)$, i.e., all the $\{u_{ij}, j = 1, \ldots, n\}$ are identical. Since there is no uncertainty about this coordinate, its entropy should be 0. Indeed, $H(Y_i) = -\sum_{k=1}^{1} \frac{n}{n} \log \frac{n}{n} = 0$.

$k(i) = 2$: Let us consider the link between the uncertainty and the index $c(i)$. $k(i) = 2$ means that Y_i can take only two distinct values. The least scattered distribution of these two values corresponds to the case $c(i) = (1, n-1)$. This is also the distribution closest to the certain case $k(i) = 1$ and $c(i) = (n)$. We now show that the case $c(i) = (1, n-1)$ generates the smallest entropy. Suppose that Y_i can take two distinct values with index (α_1, α_2), $\alpha_1 + \alpha_2 = n$. In other words, Y_i takes these two values with probability α_1/n and $\alpha_2/n = 1 - \alpha_1/n$, respectively. Without loss of generality, we can assume $\alpha_1 \le \alpha_2$. Then, the entropy of the coordinate Y_i is

$$\begin{aligned} H(Y_i) &= -\left[\frac{\alpha_1}{n} \log \frac{\alpha_1}{n} + \frac{\alpha_2}{n} \log \frac{\alpha_2}{n}\right] \\ &= -\left[\frac{\alpha_1}{n} \log \frac{\alpha_1}{n} + \left(1 - \frac{\alpha_1}{n}\right) \log \left(1 - \frac{\alpha_1}{n}\right)\right] \\ &= f\left(\frac{\alpha_1}{n}\right), \end{aligned}$$

where the function f is defined as

$$f(x) \triangleq -[x \log(x) + (1-x)\log(1-x)], \quad 0 \le x \le 1, \tag{6.7}$$

which is displayed in Figure 9.3. The following properties of this function f are basic and will be used repeatedly in this paper:

- For all $x \in [0, 1]$, $f(x) \ge 0$ and $f(x) = 0$ if and only if $x = 0$ or $x = 1$;
- For all $x \in [0, 1]$, $f(x) = f(1 - x)$;
- f is increasing on $[0, 1/2]$, and decreasing on $[1/2, 1]$;
- f is concave on $[0, 1]$.

Since $\alpha_1 \le \alpha_2$, it suffices to consider α_1 with $1 \le \alpha_1 \le n/2$. So, we have $1/n \le \alpha_1/n \le 1/2$, and in this interval, $f(\alpha_1/n)$ is strictly increasing. In other words,

$$f\left(\frac{1}{n}\right) \le f\left(\frac{\alpha_1}{n}\right) \le f\left(\frac{1}{2}\right) = 1.$$

Therefore, the entropy is minimal when $\alpha_1 = 1$ and $\alpha_2 = n - 1$. For $\alpha_1 \ge 2$, we clearly have $H(Y_i) \ge f(2/n)$.

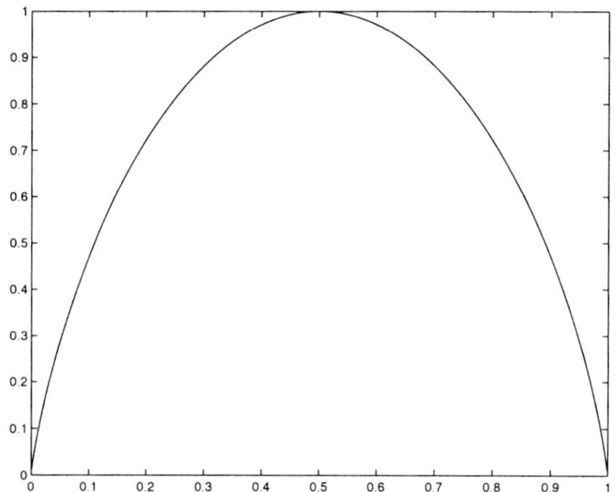

Figure 9.3. A plot of $f : x \to -[x \log x + (1-x)\log(1-x)]$

$k(i) \geq 3$: To find a lower bound of $H(Y_i) = -\sum_{\ell=1}^{k(i)} \frac{\alpha_\ell(i)}{n} \log \frac{\alpha_\ell(i)}{n}$, we need the following lemma:

Lemma 9.6.1 *Let $k \geq 3$ be an integer, and let $(\alpha_1, \ldots, \alpha_k)$ be a set of strictly positive integers with $\sum_{j=1}^{k} \alpha_j = n$. Then,*

$$\sum_{j=1}^{k} \frac{\alpha_j}{n} \log \frac{\alpha_j}{n} \leq -\left(1 + \frac{2(k-2)}{n}\right) f\left(\frac{1}{n}\right).$$

See 9.8.1 for the proof of this lemma.
Lemma 9.6.1 implies that

$$H(Y_i) \geq \left(1 + \frac{2(k-2)}{n}\right) f\left(\frac{1}{n}\right) \geq \left(1 + \frac{2}{n}\right) f\left(\frac{1}{n}\right).$$

We can now summarize these results as the following lemma:

Lemma 9.6.2 *The coordinate-wise entropy of the spike process after transformed by a basis in* $\mathrm{GL}(n, \mathbf{R})$ *can be computed or bounded as follows:*

$$\text{if } k(i) = 1, \text{ then } H(Y_i) = 0; \tag{6.8}$$

$$\text{if } k(i) = 2, \text{ then } H(Y_i) \begin{cases} = f(1/n) & \text{if } \alpha_1(i) = 1; \\ \geq f(2/n) & \text{if } 2 \leq \alpha_1(i) \leq n/2; \end{cases} \tag{6.9}$$

$$\text{if } k(i) \geq 3, \text{ then } H(Y_i) \geq \left(1 + \frac{2}{n}\right) f\left(\frac{1}{n}\right). \tag{6.10}$$

Let us now come back to our invertible transformation U; we are searching for the LSDB among $\mathrm{O}(n)$ or $\mathrm{GL}(n, \mathbf{R})$. This means that the cost of the LSDB, i.e., the sum of the coordinate-wise entropy of the LSDB coordinates, cannot be larger than that of the standard basis. Therefore we will always keep the standard basis in mind as a reference basis with which we shall compare the performance of all other bases.

The standard basis corresponds to $U = I_n$. Every row of the standard basis has index $k(i) = 2$ and $c(i) = (1, n-1)$. Hence the entropy cost of the standard basis is

$$C_H(I_n \mid \boldsymbol{X}) = n \times f(1/n) = n \log n - (n-1)\log(n-1). \tag{6.11}$$

We saw that, assuming $k(i) > 1$, $H(Y_i) \geq f(1/n)$, with equality if and only if $k(i) = 2$ and $c(i) = (1, n-1)$. Therefore a basis with $k(i) > 1$ for every $i \in \{1, \ldots, n\}$ has no chance to win over the standard basis, and the best thing one can do with such a basis is to match the entropy with that of the standard basis, i.e., a basis with $k(i) = 2$ and $c(i) = (1, n-1)$ for every i.

So, the only chance to beat the standard basis is to have some "class 1" rows (i.e., $k(i) = 1$) in a basis. However, we will never find an invertible matrix with multiple one class 1 rows. Indeed, a class 1 row is necessarily proportional to $\mathbf{1}_n^T = (1, 1, \ldots, 1)$, and it is evident that no more than one class 1 row can exist in any invertible matrix.

9.6.4 Proof of Theorem 9.5.3

Let us start with a simple remark. If we assume that B is an orthonormal basis, then $U = B^{-1} = B^T$. Hence the rows of U are in fact the basis vectors of this basis. In the case of an orthonormal matrix, the presence of one row of class 1 imposes a constraint on the other rows, since these rows must form an orthonormal basis. The following lemma describes one of these constraints.

Lemma 9.6.3 *If $k(1) = 1$, then it is impossible to have two class 2 rows with index $(1, n-1)$ in a matrix $U \in \mathrm{O}(n)$. In other words, If $k(1) = 1$, then there do not exist $i_1, i_2 \in \{1, \ldots, n\}$ such that $i_1 \neq i_2$ and $c(i_1) = c(i_2) = (1, n-1)$.*

The proof of this lemma can be found in 9.8.2.

Hence, assuming that $k(1) = 1$, we can have at most one row of class 2 with index $(1, n-1)$. All the other rows will be of either class $k(i) > 2$ or class $k(i) = 2$ with index $(\alpha_1, n-\alpha_1)$, $1 < \alpha_1 \leq n/2$. Considering the minimization of the sum of the coordinate-wise entropy, we must have one row of class 1 and one row of class 2 with index $(1, n-1)$. All the other cases always increase the entropy, i.e., dependency. From (6.9) and (6.10), the entropy of a row with either $k(i) > 2$ or $k(i) = 2$ with index $(\alpha_1, n-\alpha_1)$, $1 < \alpha_1 \leq n/2$ is bounded from below as

$$\begin{aligned} H(Y_i) &\geq \min\left(\left(1 + \frac{2}{n}\right) f\left(\frac{1}{n}\right), f\left(\frac{2}{n}\right)\right) \\ &= f\left(\frac{1}{n}\right) + \min\left(\frac{2}{n} f\left(\frac{1}{n}\right), f\left(\frac{2}{n}\right) - f\left(\frac{1}{n}\right)\right). \end{aligned}$$

Therefore, combining this with (6.8) for $k(1) = 1$ and (6.9) for $\alpha_1 = 1$, we have

$$\sum_{i=1}^{n} H(Y_i) \geq 0 + f\left(\frac{1}{n}\right) + (n-2)\left[f\left(\frac{1}{n}\right) + \min\left(\frac{2}{n} f\left(\frac{1}{n}\right), f\left(\frac{2}{n}\right) - f\left(\frac{1}{n}\right)\right)\right]. \tag{6.12}$$

We now use the following lemma:

Lemma 9.6.4 *For $n \geq 6$,*

$$\min\left(\frac{2}{n}f\left(\frac{1}{n}\right), f\left(\frac{2}{n}\right) - f\left(\frac{1}{n}\right)\right) = \frac{2}{n}f\left(\frac{1}{n}\right).$$

Proof Let us define a function: $r(x) \triangleq x\left[\frac{2}{x}f\left(\frac{1}{x}\right) - \left(f\left(\frac{2}{x}\right) - f\left(\frac{1}{x}\right)\right)\right]$ for $x \geq 2$, where f is defined in (6.7). This is a continuous and monotonically-decreasing function for $x \geq 2$, since

$$r'(x) = -\frac{2}{x^2}\log(x-1) + \log\frac{x-2}{x-1} < 0 \quad \text{for } x \geq 2.$$

Moreover, we have $r(5) \approx 0.199$ and $r(6) \approx -0.310$, and we can find a zero of $r(x)$ numerically, i.e., $r(x^*) = 0$ where $x^* \approx 5.3623$. These prove that this function is negative if $x \geq x^*$. Therefore, for each integer $n \geq 6$, $r(n) < 0$, i.e.,

$$\frac{2}{n}f\left(\frac{1}{n}\right) < f\left(\frac{2}{n}\right) - f\left(\frac{1}{n}\right).$$

□

Using this lemma for $n \geq 6$, (6.12) can be written as

$$\sum_{i=1}^{n} H(Y_i) \geq f\left(\frac{1}{n}\right) + (n-2)\left[f\left(\frac{1}{n}\right) + \frac{2}{n}f\left(\frac{1}{n}\right)\right] = \left[\frac{2(n-2)}{n} + n - 1\right] f\left(\frac{1}{n}\right).$$

Therefore, if we compare the mutual information of the new coordinates to that of the standard basis, we have

$$I(\boldsymbol{Y}) - I(\boldsymbol{X}) \geq \left[\frac{2(n-2)}{n} + n - 1\right] f\left(\frac{1}{n}\right) - nf\left(\frac{1}{n}\right) = \left[\frac{2(n-2)}{n} - 1\right] f\left(\frac{1}{n}\right),$$

That is,

$$I(\boldsymbol{Y}) - I(\boldsymbol{X}) \geq \frac{n-4}{n} f\left(\frac{1}{n}\right) > 0.$$

Thus, $B = U^{-1} = U^T$ is not the LSDB. We have therefore proved that any orthonormal basis yields a larger mutual information than the standard basis for the spike process for $n \geq 6$.

We can summarize our results so far.

- For $n \geq 6$, the standard basis is the LSDB among $\mathrm{O}(n)$.
- Any basis that yields the same mutual information as the standard basis necessarily consists of only class 2 rows with index $(1, n-1)$.

Now the question is whether there is any other basis except the standard basis satisfying this condition. The following lemma concludes the proof of Theorem 9.5.3 for $n \geq 6$.

Lemma 9.6.5 *For $n \geq 2$, an orthonormal basis consisting of class 2 rows with index $(1, n-1)$ other than the standard basis is uniquely (modulo permutations and sign flips as described in Remark 2) determined as* (5.1), *i.e.,*

$$B_{HR(n)} = \frac{1}{n} \begin{bmatrix} n-2 & -2 & \cdots & -2 \\ -2 & n-2 & \ddots & \vdots \\ \vdots & \ddots & \ddots & -2 \\ -2 & \cdots & -2 & n-2 \end{bmatrix}.$$

The proof of this lemma can be found in 9.8.3. Note that this matrix becomes a permuted and sign-flipped version of I_2 when $n = 2$, and approaches to the identity matrix as $n \to \infty$.

We now prove the particular cases, $n = 2, 3, 4, 5$ in Theorem 9.5.3. For these small values of n, we cannot use Lemma 9.6.4 anymore since we have

$$\min\left(\frac{2}{n} f\left(\frac{1}{n}\right), f\left(\frac{2}{n}\right) - f\left(\frac{1}{n}\right)\right) = f\left(\frac{2}{n}\right) - f\left(\frac{1}{n}\right).$$

Therefore, we prove these cases by examining exhaustively all possible indexes and the coordinate-wise entropy they generate.

$\underline{n = 2:}$ The only possible classes of rows in this case are class 1 with index (2) and class 2 with index (1, 1), which generate the following entropy values (in bits):

$$\begin{aligned} (2) : H(Y_i) &= 0; \\ (1,1) : H(Y_i) &= 2 \times \left(-\frac{1}{2}\log\frac{1}{2}\right) = \log 2 = 1. \end{aligned}$$

The rows of the standard basis are of class 2 with index (1, 1). Therefore, a basis with one class 1 row and one class 2 row generates lower entropy than the standard basis. Because of the orthonormality condition, it is easy to show that it must be

$$U^T = B = \frac{1}{\sqrt{2}} \begin{bmatrix} 1 & 1 \\ 1 & -1 \end{bmatrix},$$

or its permuted and sign flipped versions. In this case, the total coordinate-wise entropy is $0 + 1 = 1$ bit whereas the true joint entropy $H(\boldsymbol{X})$ is also $\log 2 = 1$. Therefore, the mutual information is 0, i.e., this basis provides the true statistically independent coordinates. The fact that this is the only case when the statistical independence is achieved if the basis search is restricted to $O(n)$ will become evident when one goes through the cases of $n = 3, 4, 5$ below.

$\underline{n = 3:}$ The possible indexes are (3), (1, 2) and (1, 1, 1), which generate the following entropy values (in bits):

$$\begin{aligned} (3) : H(Y_i) &= 0; \\ (1,2) : H(Y_i) &= f\left(\frac{1}{3}\right) = -\frac{1}{3}\log\frac{1}{3} - \frac{2}{3}\log\frac{2}{3} = \log 3 - \frac{2}{3}; \\ (1,1,1) : H(Y_i) &= 3 \times \left(-\frac{1}{3}\log\frac{1}{3}\right) = \log 3. \end{aligned}$$

Once again, the only possibility for a basis to generate lower entropy than the standard basis is to include a class 1 row with index (3). But here we still cannot have two class 2 rows of index (1, 2) on top of the class 1 row since Lemma 9.6.3 still holds for $n = 3$.

Therefore, the best combination is to have one row for each possible class, which leads to the following global coordinate-wise entropy:

$$0 + \log 3 - \frac{2}{3} + \log 3 \simeq 2.50 < 3 \log 3 - 2 \log 2 \simeq 2.75,$$

that is, this best possible basis is better than the standard basis. Therefore, the LSDB is a basis including a vector of each class. Considering the orthonormality of the basis, we can only have the following basis or its permuted or sign-flipped versions for $n = 3$:

$$U^T = B = \begin{bmatrix} \frac{1}{\sqrt{3}} & \frac{1}{\sqrt{6}} & \frac{1}{\sqrt{2}} \\ \frac{1}{\sqrt{3}} & \frac{1}{\sqrt{6}} & \frac{-1}{\sqrt{2}} \\ \frac{1}{\sqrt{3}} & \frac{-2}{\sqrt{6}} & 0 \end{bmatrix}.$$

$\underline{n = 4}$: The possible indexes are: (4), $(1,3)$, $(2,2)$, $(1,1,2)$, and $(1,1,1,1)$, which generate the following entropy values (in bits):

$$\begin{aligned} (4) : H(Y_i) &= 0; \\ (1,3) : H(Y_i) &= f\left(\frac{1}{4}\right) = -\frac{1}{4}\log\frac{1}{4} - \frac{3}{4}\log\frac{3}{4} \simeq 0.811; \\ (2,2) : H(Y_i) &= f\left(\frac{1}{2}\right) = 1; \\ (1,1,2) : H(Y_i) &= -\frac{1}{4}\log\frac{1}{4} - \frac{1}{4}\log\frac{1}{4} - \frac{1}{2}\log\frac{1}{2} = 1.5; \\ (1,1,1,1) : H(Y_i) &= 4 \times \left(-\frac{1}{4}\log\frac{1}{4}\right) = 2. \end{aligned}$$

The total coordinate-wise entropy of the standard basis is $4\log 4 - 3\log 3 \approx 3.245$ bits. Note that all the rows of the standard basis is of class 2 with index $(1,3)$. Let U be an orthonormal basis, and let $\{\boldsymbol{b}_i^T, i = 1,\ldots,4\}$ be its rows. If U generates smaller entropy than the standard basis, it necessarily includes one class 1 row. There is no other choice. Without loss of generality, let us assume that $\boldsymbol{b}_1^T$ is of class 1, i.e., $c(1) = (4)$. We now prove that we cannot have a class 2 row with index $(1,3)$ in such a U if the total coordinate-wise entropy of U is smaller than that of the standard basis. Suppose that $\boldsymbol{b}_2^T$ of class 2 with index $(1,3)$, i.e., $c(2) = (1,3)$. If so, we cannot have any more class 2 row with index $(1,3)$ in U by Lemma 9.6.3. Now, U cannot include a class 4 row vector of index $(1,1,1,1)$. If so, these three rows (i.e., rows of class 1, 2, and 4) would generate the entropy $0 + 0.811 + 2 = 2.811$ bits. Hence, as we can easily see from the bit counts of the class indexes above, any other admissible choice for the remaining row would generate larger total coordinate-wise entropy than the standard basis does. Therefore we can discard these combinations immediately, and the indexes of $\boldsymbol{b}_3^T$ and $\boldsymbol{b}_4^T$ must be chosen from $(2,2)$ and $(1,1,2)$. Since $\boldsymbol{b}_2^T$ is of the form (a,a,a,b) its orthogonality with $\boldsymbol{b}_1^T$ implies that $\boldsymbol{b}_2^T$ is proportional to the vector $(1,1,1,-3)$. If $\boldsymbol{b}_3^T$ were of index $(2,2)$, it would be of the form (c,c,d,d) and its orthogonality with $\boldsymbol{b}_1^T$ implies that $\boldsymbol{b}_3^T$ is proportional to $(c,c,-c,-c)$. On the other hand, the orthogonality with $\boldsymbol{b}_2^T$ implies $c + c - c + 3c = 0$, i.e., $c = 0$, which is impossible. Therefore the only possibility for $\boldsymbol{b}_3^T$ and $\boldsymbol{b}_4^T$ would be of class 3 rows with index $(1,1,2)$. Such a row generates the coordinate-wise entropy 1.5 bits. The total coordinate-wise entropy generated by such a basis U is therefore at least $0 + 0.811 + 2 \times 1.5 = 3.811$ bits, which is larger than that of the standard

$$\begin{aligned} I(\boldsymbol{Y}) &= -H(\boldsymbol{X}) + \sum_{i=1}^{n} H(Y_i) \\ &= -\log n + (n-1) f\left(\frac{1}{n}\right) \\ &= -\log n + (n-1)\left[\log n - \frac{n-1}{n}\log(n-1)\right] \\ &= (n-2)\log n - \frac{(n-1)^2}{n}\log(n-1). \end{aligned} \tag{6.13}$$

Let $h(n)$ denote the last expression in (6.13). Note that $h(2) = 0$, i.e., we can achieve the true independence for $n = 2$. If $n > 2$, this function is strictly positive and monotonically increasing. By expanding the natural logarithm version of $h(x)$, we have

$$\begin{aligned} \ln 2 \times h(x) &= (x-2)\ln x - \frac{(x-1)^2}{x}\ln(x-1) \\ &= (x-2)\ln x - \left(x - 2 + \frac{1}{x}\right)\left(\ln x + \ln\left(1 - \frac{1}{x}\right)\right) \\ &= (x-2)\ln x - \left(x - 2 + \frac{1}{x}\right)\left(\ln x - \frac{1}{x} - \frac{1}{2x^2} + o\left(\frac{1}{x^2}\right)\right) \\ &= -\frac{\ln x}{x} + \left(x - 2 + \frac{1}{x}\right)\left(\frac{1}{x} + \frac{1}{2x^2} + o\left(\frac{1}{x^2}\right)\right) \\ &= 1 - \frac{\ln x}{x} - \frac{3}{2x} + o\left(\frac{1}{x}\right) \end{aligned}$$

In other words, we have established

$$I(\boldsymbol{Y}) \sim \frac{1}{\ln 2}\left(1 - \frac{\ln n}{n}\right) \qquad \text{as } n \to \infty.$$

Hence we have

$$\lim_{n\to\infty} I\left(B_{\mathrm{GL}(n)}^{-1}\boldsymbol{X}\right) = \frac{1}{\ln 2} = \log \mathrm{e} \approx 1.4427.$$

Therefore, for $n > 2$, there is no invertible linear transformation that gives truly independent coordinates for the spike process.

As for the orthonormal case, using (6.11), we have

$$I\left(B_{\mathrm{O}(n)}^{T}\boldsymbol{X}\right) = n\log n - (n-1)\log(n-1) - \log n = \log\left(1 + \frac{1}{n-1}\right)^{n-1}.$$

Now, it is easy to see

$$\lim_{n\to\infty} I\left(B_{\mathrm{O}(n)}^{T}\boldsymbol{X}\right) = \log \mathrm{e}.$$

This completes the proof of Corollary 9.5.5. □

9.7 DISCUSSION

In general, sparsity and statistical independence are two completely different concepts as an adaptive basis selection criterion, as demonstrated by the rotations of the 2D uniform distribution in Section 9.4 For the spike process, however, we showed that the BSB and the LSDB *can* coincide (i.e., the standard basis) if we restrict our basis search to $\mathrm{O}(n)$

with $n \geq 5$. However, we also showed that the standard basis is not the only LSDB in this case. To our surprise, there exists another orthonormal basis (5.1) representing the Householder reflector, which attains exactly the same level of the statistical dependence as the standard basis, if the statistical dependence is quantified by the mutual information or equivalently by the total coordinate-wise entropy C_H defined in (3.3). Yet this LSDB does not sparsify the process at all if we measure the sparsity by the expected ℓ^p norm C_p defined in (3.1) where $0 \leq p \leq 1$. It is also interesting to note that this Householder reflector approaches to the standard basis as $n \to \infty$. Furthermore, if we extend our basis search to $\mathrm{SL}^{\pm}(n, \mathbf{R})$ or $\mathrm{GL}(n, \mathbf{R})$, then the LSDB and the BSB *cannot* coincide.

What do these results and the effort to prove these theorems suggest? First, it is clear that proving theorems on the LSDB and computing it for more complicated stochastic processes would be much more difficult than the BSB. To deal with statistical dependency, we need to consider the probability law of the underlying process (e.g., entropy or the marginal pdfs) explicitly. On the other hand, the sparsity criterion does not require such explicit information. In fact, one can even find the BSB for each realization rather than for the whole realizations, which is impossible for the LSDB. see Saito et al. [23], [22] for further information about this issue.

Second, it is now clear that both criteria prefer sharply concentrated (i.e., peaky) marginal distributions. There is, however, a fundamental difference: the sensitivity on the location (mean) of the marginal pdfs. The Shannon entropy is location invariant, i.e., its value does not change regardless of where the mean of the distribution is located, whereas the expected ℓ^p norm is very sensitive to the mean. This is one of the reasons why the LSDB is non-unique and different from the BSB as shown in Theorems 9.5.3 and 9.5.4.

Third, the LSDB unfortunately cannot tell how close it is to the true statistical independence; it can only tell that it is the best one (i.e., the closest one to the statistical independence) among the given set of possible bases. In order to quantify the absolute statistical dependence, we need to estimate the true high-dimensional entropy of the original process, $H(\boldsymbol{X})$, which is an extremely difficult task in general. We would like to note, however, a recent attempt to estimate the high-dimensional entropy of the process by Hero and Michel [12], which uses the minimum spanning trees of the input data and does not require us to estimate the pdf of the process. We feel that this type of techniques will help assessing the absolute statistical dependence of the process under the LSDB coordinates.

Then, why the sparse basis of Olshausen and Field and the ICA basis of Bell and Sejnowski were more or less the same? Our interpretation to this phenomenon is the following. First of all, both teams applied their algorithms to the natural scene image patches after essentially centering and sphering them. Hence there is no location sensitivity problem of the BSB and the LSDB as we described above (although Olshausen and Field used the cost $\sum_{i=1}^{n} E(1+Y_i^2)$ instead of $\sum_{i=1}^{n} E|Y_i|^p$ and Bell and Sejnowski used their "infomax" algorithm rather than directly minimizing the mutual information). This implies that these two algorithms both prefer the basis that makes the input image patches sharply concentrated around the origin. Second, the "edge-detecting" basis functions they obtained essentially convert an input image patch to a spike or spike-like image. In other words, in our opinion, the image patch size such as 16×16 pixels were crucial in their experiments, as Donoho and Flesia also observed [9]. Since those image patches are of small size, they tend to have simpler image contents such as simple oriented edges. It seems to us that if their algorithms were computationally feasible to accept image patches of larger size such as 64×64 or 128×128, both the BSB and the LSDB would be very different from such simple "edge-detecting" basis functions. These large size image patches (due to rich scene variations and contents in the patches of these sizes) cannot be converted to spikes by those simple basis functions. See also Remark 5 about this viewpoint.

These observations, therefore, suggest that the pursuit of sparse representations should be encouraged rather than that of statistically independent representations, if we believe that mammalian vision systems were evolved and developed by the principle of data compression. This is also the viewpoint indicated by Donoho [8]. However, this does not mean to downgrade the importance of the statistical independence altogether. If we want to separate mixed signals or to build empirical models of stochastic processes for simulation purposes, then pursuing the statistical independence should be encouraged, and we expect to see further interplay between these two criteria.

Finally, there are a few interesting generalizations of the spike process, which need to be addressed in the near future. One is the spike process with varying amplitude. The spike process whose amplitude obeys the normal distribution was treated by Donoho et al. [10] to demonstrate the superiority of the non-Gaussian coding using spike location information over the Gaussian-KLB coding (see also a recent follow-up article by Weidmann and Vetterli [25]). We have started investigating this "generalized spike process" and have succeeded in obtaining the same result for the BSB as the simple spike process dealt in this paper, but the different results for the KLB and the LSDB, which will be reported elsewhere [22]. The other generalization is to randomly throw in multiple spikes to a single realization. If one throws in more and more spikes to one realization, the standard basis is getting worse in terms of sparsity. It will be an interesting exercise to consider the BSB and the LSDB for such situations.

Except in very special circumstances, it would be extremely difficult to find the BSB of a complicated stochastic process (e.g., natural scene images) that truly converts its realizations to the simple spike process. More likely, a theoretically and computationally feasible basis that sparsifies the realizations of a complicated process well (e.g., curvelets for the natural scene images [9]) may generate expansion coefficients that can be viewed as an amplitude-varying multiple spike process. In order to tackle this scenario, we certainly need to: 1) develop such a basis adapted to a specific stochastic process; and 2) deepen our understanding of the amplitude-varying multiple spike process. There is no doubt that these pursuits force us to explore the territory "beyond wavelets".

ACKNOWLEDGEMENT

The second author (N.S.) would like to thank Dr. Jean-Marie Aubry (Université Paris XII) for his checking the proof of Theorem 9.5.1.

This research was partially supported by NSF DMS-99-73032, DMS-99-78321, and ONR YIP N00014-00-1-0469.

9.8 APPENDICES

9.8.1 Appendix A: Proof of Lemma 9.6.1

First we need to show another lemma as follows:

Lemma 9.8.1 *Let $p_2 \geq p_1 \geq 1$ be positive integers such that $p_1 + p_2 \leq n$. Then*

$$\frac{p_1}{n}\log\frac{p_1}{n} + \frac{p_2}{n}\log\frac{p_2}{n} \leq \frac{p_1+p_2}{n}\log\frac{p_1+p_2}{n} - \frac{2}{n}f\left(\frac{1}{n}\right),$$

where f is defined in (6.7).

Proof The left-hand side of the inequality can be written as

$$\frac{p_1}{n}\log\frac{p_1}{n} + \frac{p_2}{n}\log\frac{p_2}{n} = \left(\frac{p_1+p_2}{n}\right)\left[\frac{p_1}{p_1+p_2}\log\frac{p_1}{n} + \frac{p_2}{p_1+p_2}\log\frac{p_2}{n}\right]$$
$$= \left(\frac{p_1+p_2}{n}\right)\left[\log\frac{p_1+p_2}{n} + \frac{p_1}{p_1+p_2}\log\frac{p_1}{p_1+p_2} + \frac{p_2}{p_1+p_2}\log\frac{p_2}{p_1+p_2}\right]$$
$$= \left(\frac{p_1+p_2}{n}\right)\log\left(\frac{p_1+p_2}{n}\right) + \left(\frac{p_1+p_2}{n}\right)\left[-f\left(\frac{p_1}{p_1+p_2}\right)\right] \tag{8.1}$$

However, it is clear that

$$\frac{1}{2} \geq \frac{p_1}{p_1+p_2} \geq \frac{1}{p_1+p_2} \geq \frac{1}{n}.$$

From the monotonicity of $f(x)$ for $x \in [0, 1/2]$, we deduce

$$1 = f\left(\frac{1}{2}\right) \geq f\left(\frac{p_1}{p_1+p_2}\right) \geq f\left(\frac{1}{n}\right),$$

which we can rewrite as

$$-1 \leq -f\left(\frac{p_1}{p_1+p_2}\right) \leq -f\left(\frac{1}{n}\right).$$

This inequality, nonnegativity of f, and the assumption of this lemma yields

$$\left(\frac{p_1+p_2}{n}\right)\left[-f\left(\frac{p_1}{p_1+p_2}\right)\right] \leq -\frac{2}{n}f\left(\frac{1}{n}\right).$$

This inequality combined with (8.1) completes the proof of Lemma 9.8.1. □

Coming back to the proof of Lemma 9.6.1, we now use induction as follows.

<u>$k = 3$:</u> Since $\alpha_1 + \alpha_2 < n$, we can use Lemma 9.8.1 to assert

$$\frac{\alpha_1}{n}\log\frac{\alpha_1}{n} + \frac{\alpha_2}{n}\log\frac{\alpha_2}{n} \leq \frac{\alpha_1+\alpha_2}{n}\log\frac{\alpha_1+\alpha_2}{n} - \frac{2}{n}f\left(\frac{1}{n}\right).$$

Therefore,

$$\sum_{j=1}^{3}\frac{\alpha_j}{n}\log\frac{\alpha_j}{n} \leq \frac{\alpha_3}{n}\log\frac{\alpha_3}{n} + \frac{\alpha_1+\alpha_2}{n}\log\frac{\alpha_1+\alpha_2}{n} - \frac{2}{n}f\left(\frac{1}{n}\right)$$
$$= \frac{\alpha_3}{n}\log\frac{\alpha_3}{n} + \left(1-\frac{\alpha_3}{n}\right)\log\left(1-\frac{\alpha_3}{n}\right) - \frac{2}{n}f\left(\frac{1}{n}\right)$$
$$= -f\left(\frac{\alpha_3}{n}\right) - \frac{2}{n}f\left(\frac{2}{n}\right).$$

We used the fact $\sum_{j=1}^{3}\alpha_j = n$ to derive the equality in the second line of the above expression. Since $\alpha_j \geq 1$ for $j = 1, 2, 3$, we must have $(n-1)/n > \alpha_3/n \geq 1/n$. Considering the symmetry of $f(x)$ around $x = 1/2$ and its behavior, we can deduce that

$$\sum_{j=1}^{3} \frac{\alpha_j}{n} \log \frac{\alpha_j}{n} \leq -f\left(\frac{1}{n}\right) - \frac{2}{n} f\left(\frac{2}{n}\right) \leq -\left(1 + \frac{2}{n}\right) f\left(\frac{1}{n}\right).$$

This nails down the case $k = 3$.

$\underline{k \Rightarrow k+1}$: Let us demonstrate that, assuming that the formula is true for $k \geq 3$, it is still true for $k+1$. We can decompose the sum $\sum_{j=1}^{k+1} \frac{\alpha_j}{n} \log \frac{\alpha_j}{n}$ in the following way:

$$\sum_{j=1}^{k+1} \frac{\alpha_j}{n} \log \frac{\alpha_j}{n} = \frac{\alpha_{k+1}}{n} \log \frac{\alpha_{k+1}}{n} + \frac{\alpha_k}{n} \log \frac{\alpha_k}{n} + \sum_{j=1}^{k-1} \frac{\alpha_j}{n} \log \frac{\alpha_j}{n}. \tag{8.2}$$

But once again, since $\alpha_k + \alpha_{k+1} < n$, we can use Lemma 9.8.1 to reach

$$\frac{\alpha_{k+1}}{n} \log \frac{\alpha_{k+1}}{n} + \frac{\alpha_k}{n} \log \frac{\alpha_k}{n} \leq \frac{\alpha_{k+1} + \alpha_k}{n} \log \frac{\alpha_{k+1} + \alpha_k}{n} - \frac{2}{n} f\left(\frac{1}{n}\right).$$

Let us rename a sequence $\{\alpha_j\}$ as follows:

$$\beta_j = \begin{cases} \alpha_{j+1} + \alpha_j & \text{if } j = k; \\ \alpha_j & \text{if } j = 1, \ldots, k-1. \end{cases}$$

Then, using the induction assumption, the lefthand side of (8.2) can be bounded from above as follows:

$$\sum_{j=1}^{k+1} \frac{\alpha_j}{n} \log \frac{\alpha_j}{n} \leq \frac{\beta_k}{n} \log \frac{\beta_k}{n} + \sum_{j=1}^{k-1} \frac{\beta_j}{n} \log \frac{\beta_j}{n} - \frac{2}{n} f\left(\frac{1}{n}\right).$$

Since $\sum_{j=1}^{k} \beta_j = \sum_{j=1}^{k+1} \alpha_j = n$, we can state that

$$\begin{aligned} \sum_{j=1}^{k+1} \frac{\alpha_j}{n} \log \frac{\alpha_j}{n} &\leq \sum_{j=1}^{k} \frac{\beta_j}{n} \log \frac{\beta_j}{n} - \frac{2}{n} f\left(\frac{1}{n}\right) \\ &\leq -\left(1 + \frac{2(k-2)}{n}\right) f\left(\frac{1}{n}\right) - \frac{2}{n} f\left(\frac{1}{n}\right) \\ &= -\left(1 + \frac{2(k-1)}{n}\right) f\left(\frac{1}{n}\right). \end{aligned}$$

This concludes the proof of Lemma 9.6.1.

□

9.8.2 Appendix B: Proof of Lemma 9.6.3

Let us prove this lemma by *reductio ad absurdum*. Let us assume that, for example, $c(2) = c(3) = (1, n-1)$. Since the first row of U is proportional to $(1, 1, \ldots, 1)$, all the other rows must satisfy $\sum_{j=1}^{n} u_{ij} = 0$ for $i = 2, \ldots, n$ because of the orthonormality condition. Let us now consider the second row $(u_{21}, \ldots, u_{2n})$. Since $c(2) = (1, n-1)$, let us assume $u_{21} = a$ and $u_{2j} = b$, $j = 2, \ldots, n$ for some $a, b \in \mathbf{R}$. Then the orthonormality condition implies $a + (n-1)b = 0$. Since the norm of this row vector has to be one, we also have $a^2 + (n-1)b^2 = 1$. From these two constraints, we have $(n-1)^2 b^2 + (n-1)b^2 = 1$. This implies $a = \pm\sqrt{\frac{n-1}{n}}$ and $b = \mp\frac{1}{\sqrt{n(n-1)}}$.

As the second and third rows of U must be linearly independent, we need to assume that the third row is $(c, d, c, \ldots, c)$ for some $c, d \in \mathbf{R}$. (We cannot assume $(d, c, \ldots, c)$ for the third row since its inner product with the second row gives $ad + (n-1)bc = 0$, which leads to $c = d$ using the values of a and b obtained above.) Then, similarly to the second row, we also get $d + (n-1)c = 0$, $d^2 + (n-1)c^2 = 1$. Thus, we have $d = \pm a$ and $c = \pm b$. Then, regardless of the choice of the signs for a, b, c, d, the orthogonality of the second and third rows yields

$$0 = (n-2)b^2 + 2ab = (n-2) \cdot \frac{1}{n(n-1)} - 2 \cdot \frac{1}{n}.$$

This leads to $2 = \frac{n-2}{n-1}$, i.e., $2n - 2 = n - 2$, and finally $n = 0$. This contradiction implies that the assumption made is impossible, and proves the lemma. □

9.8.3 Appendix C: Proof of Lemma 9.6.5

Our strategy of proving this lemma is the following. First we will show that the LSDB selected from $\mathrm{O}(n)$, which consists of only class 2 row vectors with index $(1, n-1)$, must be of the form (modulo permutations and sign flips):

$$\begin{bmatrix} a_1 & b_1 & \cdots & \cdots & \cdots & b_1 \\ b_2 & a_2 & b_2 & \cdots & \cdots & b_2 \\ \vdots & & \ddots & & & \vdots \\ \vdots & & & \ddots & & \vdots \\ b_{n-1} & \cdots & \cdots & b_{n-1} & a_{n-1} & b_{n-1} \\ b_n & \cdots & \cdots & \cdots & b_n & a_n \end{bmatrix}. \tag{8.3}$$

where $a_k^2 + (n-1)b_k^2 = 1$ for $k = 1, \ldots, n$. We then derive the final form (5.1) using the orthonormality of the row vectors of this matrix (8.3).

Since each row is of class 2 with index $(1, n-1)$, only one the entry in a row must be different from all the other $n-1$ entries. Therefore, without loss of generality, in the kth row, let a_k be such a distinguishing entry and b_k be the other $n-1$ entries. Let $B = U^T$ be the LSDB under consideration. Suppose U has the ith and jth rows in which the locations of a_i and a_j coincide. Without loss of generality (modulo row and column permutations), we can assume that U is of the following form.

$$\begin{bmatrix} a_1 & b_1 & b_1 & \cdots & b_1 \\ a_2 & b_2 & b_2 & \cdots & b_2 \\ b_3 & a_3 & b_3 & \cdots & b_3 \\ \vdots & & \ddots & \ddots & \vdots \\ b_n & b_n & \cdots & a_n & b_n \end{bmatrix}. \begin{bmatrix} a_1 & b_1 & \cdots & \cdots & \cdots & b_1 \\ a_2 & b_2 & \cdots & \cdots & \cdots & b_2 \\ b_3 & a_3 & b_3 & \cdots & \cdots & b_3 \\ \vdots & & \ddots & & & \vdots \\ b_{n-1} & \cdots & \cdots & a_{n-1} & b_{n-1} & b_{n-1} \\ b_n & \cdots & \cdots & b_n & a_n & b_n \end{bmatrix}. \tag{8.4}$$

From the normalization condition, we must have:

$$a_k^2 + (n-1)b_k^2 = 1 \quad \text{for } k = 1, \ldots, n. \tag{8.5}$$

From the orthonormality condition, $U^T U = I_n$, the diagonal entries of $U^T U$ are:

$$\begin{aligned}
(U^T U)_{1,1} &= 1 = a_1^2 + a_2^2 + \sum_{j=3}^{n} b_j^2, \\
(U^T U)_{k,k} &= 1 = a_k^2 + \sum_{j=1, j\neq k}^{n} b_j^2, \quad 2 \leq k < n, \\
(U^T U)_{n,n} &= 1 = \sum_{j=1}^{n} b_j^2.
\end{aligned}$$

These imply that $a_k^2 = b_k^2$ for $k \geq 3$. Inserting this to (8.5) and noting that we must have $a_k \neq b_k$ because of the class 2 condition, we obtain:

$$a_k = \pm 1/\sqrt{n},\ b_k = \mp 1/\sqrt{n}, \quad \text{for } k \geq 3. \tag{8.6}$$

Consider now the off-diagonal entry of $U^T U$, for example,

$$\begin{aligned}
(U^T U)_{1,2} &= 0 = a_1 b_1 + a_2 b_2 + a_3 b_3 + b_4^2 + \cdots + b_n^2, \\
(U^T U)_{1,n} &= 0 = a_1 b_1 + a_2 b_2 + b_3^2 + b_4^2 + \cdots + b_n^2
\end{aligned}$$

Inserting (8.6) into these, we get

$$\begin{aligned}
a_1 b_1 + a_2 b_2 - \frac{1}{n} + \frac{n-3}{n} &= 0 \\
a_1 b_1 + a_2 b_2 + \frac{n-2}{n} &= 0.
\end{aligned}$$

This is a contradiction (i.e., $a_1 b_1 + a_2 b_2$ cannot have two different values). Therefore U cannot have two rows where the distinguishing entries a_i, a_j share the same column index as (8.4). It is clear that we cannot have more than two such rows. Therefore, U must be of the form (8.3).

Now, let us compute the entries of (8.3). The normalization condition (8.5) still holds. Computing the diagonal entries of $U^T U = I_n$, we have

$$(U^T U)_{k,k} = 1 = a_k^2 + \sum_{j=1, j\neq k}^{n} b_j^2 \quad \text{for } k = 1, \ldots, n. \tag{8.7}$$

Combining (8.5) and (8.7), we have:

$$n b_k^2 = \sum_{j=1}^{n} b_j^2 \quad \text{for } k = 1, \ldots, n.$$

This implies that $b_1^2 = \cdots = b_n^2$. Then, from the normalization condition (8.5), we must have $a_1^2 = \cdots = a_n^2$ also. Consider now the off-diagonal entry of $U^T U$:

$$(U^T U)_{1,2} = 0 = a_1 b_1 + a_2 b_2 + (n-2) b_1^2.$$

Now, we must have $b_2 = b_1$ or $b_2 = -b_1$. So, the above equation can be written as

$$(U^T U)_{1,2} = 0 = a_1 b_1 \pm a_2 b_1 + (n-2) b_1^2.$$

This implies that either $b_1 = 0$ or $a_1 \pm a_2 + (n-2)b_1 = 0$. $b_1 = 0$ leads to $b_k = 0$ and $a_k = \pm 1$ for $k = 1, \ldots, n$, i.e., the standard basis. Let us consider now the other case, i.e., $a_1 \pm a_2 + (n-2)b_1 = 0$. Since $a_2 = a_1$ or $a_2 = -a_1$, these lead to either $b_1 = 0$ or $2a_1 + (n-2)b_1 = 0$. The former case has been already treated. Thus, let us proceed the latter case. From this, we have

$$a_1 = \left(1 - \frac{n}{2}\right) b_1. \tag{8.8}$$

Inserting this into (8.5), we have

$$b_1^2 = \frac{4}{n^2}.$$

Consequently,

$$a_1^2 = 1 - (n-1) \cdot \frac{4}{n^2} = \left(\frac{n-2}{n}\right)^2.$$

Because (8.8) is also true for all k, i.e., $a_k = (1 - n/2)b_k$, $k = 1, \ldots, n$, we have:

$$a_k = \pm \frac{n-2}{n}, \; b_k = \mp \frac{2}{n}, \quad \text{for } k = 1, \ldots, n. \tag{8.9}$$

This means that the matrix U must be of the following form or its permuted and sign-flipped versions:

$$B_{O(n)} = \frac{1}{n} \begin{bmatrix} n-2 & -2 & \cdots & \cdots & -2 \\ -2 & n-2 & \ddots & & \vdots \\ \vdots & \ddots & \ddots & \ddots & \vdots \\ \vdots & & \ddots & n-2 & -2 \\ -2 & \cdots & \cdots & -2 & n-2 \end{bmatrix} . U = \frac{1}{n} \begin{bmatrix} n-2 & -2 & \cdots & -2 \\ -2 & n-2 & \ddots & \vdots \\ \vdots & \ddots & \ddots & -2 \\ -2 & \cdots & -2 & n-2 \end{bmatrix}.$$

It turns out that this is symmetric, so we have $B = U$. This completes the proof of Lemma 9.6.5. □

REFERENCES

[1] A. J. Bell and T. J. Sejnowski. The 'independent components' of natural scenes are edge filters. *Vision Research*, 37:3327–3338, 1997.

[2] J. F. Cardoso. An efficient batch algorithm: JADE. `http://sig.enst.fr/ cardoso/guidesepsou.html`. See also `http://tsi.enst.fr/ cardoso/icacentral/index.html` for collections of contributed ICA software.

[3] J.-F. Cardoso. High-order contrasts for independent component analysis. *Neural Computation*, 11:157–192, 1999.

[4] R. R. Coifman and M. V. Wickerhauser. Entropy-based algorithms for best basis selection. *IEEE Trans. Inform. Theory*, 38(2):713–719, Mar. 1992.

[5] T. M. Cover and J. A. Thomas. *Elements of Information Theory.* Wiley Interscience, New York, 1991.

[6] M. M. Day. The spaces L^p with $0 < p < 1$. *Bull. Amer. Math. Soc.*, 46:816–823, 1940.

[7] D. L. Donoho. On minimum entropy segmentation. In C. K. Chui, L. Montefusco, and L. Puccio, editors, *Wavelets: Theory, Algorithms, and Applications*, pages 233–269. Academic Press, San Diego, 1994.

[8] D. L. Donoho. Sparse components analysis and optimal atomic decomposition. *Constructive Approximation*, 17:353–382, 2001.

[9] D. L. Donoho and A. G. Flesia. Can recent innovations in harmonic analysis 'explain' key findings in natural image statistics? *Network: Comput. Neural Syst.*, 12(3):371–393, 2001.

[10] D. L. Donoho, M. Vetterli, R. A. DeVore, and I. Daubechies. Data compression and harmonic analysis. *IEEE Trans. Inform. Theory*, 44(6):2435–2476, 1998. Invited paper.

[11] P. Hall and S. C. Morton. On the estimation of entropy. *Ann. Inst. Statist. Math.*, 45(1):69–88, 1993.

[12] A. O. Hero and O. J. J. Michel. Asymptotic theory of greedy approximations to minimal k-point random graphs. *IEEE Trans. Inform. Theory*, 45(6):1921–1938, 1999.

[13] R. A. Horn and C. R. Johnson. *Matrix Analysis*. Cambridge Univ. Press, 1985.

[14] A. Hyvärinen. The FastICA package for MATLAB. `http://www.cis.hut.fi/projects/ica/fastica/`.

[15] J.-J. Lin, N. Saito, and R. A. Levine. An iterative nonlinear Gaussianization algorithm for resampling dependent components. In P. Pajunen and J. Karhunen, editors, *Proc. 2nd International Workshop on Independent Component Analysis and Blind Signal Separation*, pages 245–250. IEEE, 2000. June 19–22, 2000, Helsinki, Finland.

[16] J.-J. Lin, N. Saito, and R. A. Levine. An iterative nonlinear Gaussianization algorithm for image simulation and synthesis. Technical report, Dept. Math., Univ. California, Davis, 2001. submitted for publication.

[17] B. A. Olshausen. Sparse coding simulation software. `http://redwood.ucdavis.edu/bruno/sparsenet.html`.

[18] B. A. Olshausen and D. J. Field. Emergence of simple-cell receptive field properties by learning a sparse code for natural images. *Nature*, 381:607–609, 1996.

[19] B. A. Olshausen and D. J. Field. Sparse coding with an overcomplete basis set: A strategy employed by V1? *Vision Research*, 37:3311–3325, 1997.

[20] N. Saito. Local feature extraction and its applications using a library of bases. In R. Coifman, editor, *Topics in Analysis and Its Applications: Selected Theses*, pages 269–451. World Scientific Pub. Co., Singapore, 2000.

[21] N. Saito. Image approximation and modeling via least statistically dependent bases. *Pattern Recognition*, 34:1765–1784, 2001.

[22] N. Saito. The generalized spike process, sparsity, and statistical independence. In D. Rockmore and D. Healy, Jr., editors, *Modern Signal Processing*, MSRI Publications, Cambridge University Press, 2003. To appear.

[23] N. Saito, B. M. Larson, and B. Bénichou. Sparsity and statistical independence from a best-basis viewpoint. In A. Aldroubi, A. F. Laine, and M. A. Unser, editors, *Wavelet Applications in Signal and Image Processing VIII*, volume Proc. SPIE 4119, pages 474–486, 2000. Invited paper.

[24] J. H. van Hateren and A. van der Schaaf. Independent component filters of natural images compared with simple cells in primary visual cortex. *Proc. Royal Soc. London, Ser. B*, 265:359–366, 1998.

[25] C. Weidmann and M. Vetterli. Rate distortion behavior of sparse sources. Submitted to IEEE Trans. Info. Theory, Oct. 2001.

[26] M. V. Wickerhauser. *Adapted Wavelet Analysis from Theory to Software.* A K Peters, Ltd., Wellesley, MA, 1994. with diskette.

Beyond Wavelets
G. V. Welland (Editor)
© 2003 Elsevier Science (USA) All rights reserved

10

NONUNIFORM FILTER BANKS: NEW RESULTS AND OPEN PROBLEMS

SONY AKKARAKARAN AND P.P. VAIDYANATHAN

Department of Electrical Engineering 136-93
California Institute of Technology, Pasadena, CA 91125
sony@systems.caltech.edu ppvnath@systems.caltech.edu

Abstract

A nonuniform filter bank (FB) is one whose channel decimation rates need not all be equal. While the theory and design of uniform FBs is a very well developed subject, there are several interesting open issues in the area of nonuniform FBs. Most nonuniform FB designs either result in approximate or *near-perfect* reconstruction, or involve cascading uniform FBs in tree structures. This leaves unanswered many important theoretical issues involved in obtaining perfect reconstruction (PR) in nonuniform FBs. The purpose of this paper is to address these issues. We only study FBs with *integer* decimation rates, as FBs with rational decimators can also be shown to be transformable to them. The central problem of interest is as follows: Let S be a set of positive integers obeying maximal decimation (i.e., with reciprocals summing to unity). Find necessary and sufficient conditions on S for existence of a PRFB belonging to some FB class C and using S as its set of decimators. The class C is defined by some constraint on the filters of its constituent FBs; examples of interest are the class of all rational FBs (FBs with rational filters), FIR FBs, orthonormal FBs, etc. A condition that immediately suggests itself is the one stating that the integers be arrangeable in a tree so that the required PRFB can be built by cascading uniform PRFBs in a tree structure. However, this condition, while clearly sufficient, is *not necessary* for many classes C of interest. In fact there are sets violating it which can be used to build *delay-chain* PRFBs (in which all filters are delays). Many of our new results focus on the class of rational FBs. We strengthen considerably the known necessary conditions in this case, and provide new ones. The basic problem remains unresolved — necessary and sufficient conditions are still unknown, however we believe our

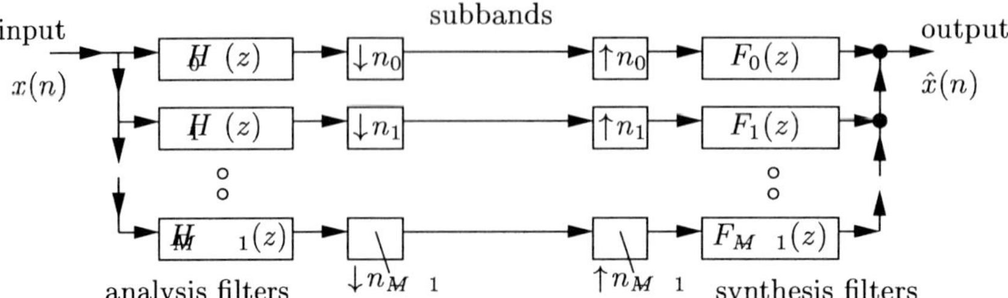

Figure 10.1. Nonuniform filter bank

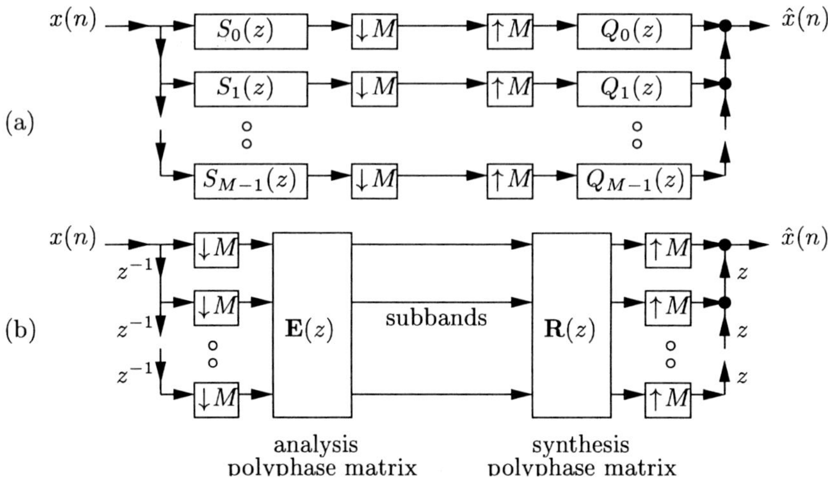

Figure 10.2. Uniform maximally decimated filter bank. (a) Showing analysis and synthesis filters. (b) Polyphase representation

work is an important step towards a full solution. We conclude by listing all known conditions, studying their inter-relationship, and pointing out several open problems.

10.1 INTRODUCTION

Figure 1 shows an M-channel nonuniform filter bank (FB). The FB is said to be maximally decimated if the channel decimation rates n_i are integers satisfying

$$\sum_{i=0}^{M-1} \frac{1}{n_i} = 1. \qquad \text{(maximal decimation condition)} \tag{1.1}$$

Figure 2a shows a maximally decimated *uniform* FB, which is a special case of Fig. 1 where $n_i = M$ for all i. For this case, the system can be equivalently redrawn using the analysis and synthesis polyphase matrices $\mathbf{E}(z)$ and $\mathbf{R}(z)$, as shown in Fig. 2b. The condition for perfect reconstruction (PR) is then easily expressed as $\mathbf{R}(z) = \mathbf{E}^{-1}(z)$. Due to this, the theory and design of uniform PRFBs is an extremely well developed subject. Numerous parameterization results list all possible M-channel uniform PRFBs with various sets of properties such as paraunitariness, FIR filters, linear phase filters, etc.

In contrast, several issues involved in achieving PR in nonuniform FBs remain unresolved. For example, given a general set of positive integers n_i obeying maximal decimation (1.1), how do we determine whether or not there exists a rational PRFB (i.e., one with rational filters) using the n_i as decimators? If the n_i are all equal, clearly such a FB exists (as it is then uniform). Similarly, it also exists if the n_i are arrangeable in a tree so that such a PRFB can be built by cascading uniform PRFBs in a tree structure (Section 4.1). This is the most common approach to achieving PR in nonuniform FBs. In particular, it is used to build the FBs that implement the dyadic wavelet transforms [11], [12]: Such a FB has a *dyadic* decimator–set, i.e., one of form $\{2, 2^2, \ldots, 2^{r-1}, 2^r, 2^r\}$ for some integer $r \geq 1$, and is built using a dyadic tree (i.e., one built from a cascade of r 2-channel FBs). However, there are sets of decimators n_i that cannot be arranged in a tree as described above, and yet permit existence of rational PRFBs in which in fact all filters are *delays*. Further, even if the decimators are arrangeable in a tree, it is possible that there are PRFBs using those decimators that cannot be realized using the tree. These facts will be discussed in detail with examples in Section 4.2. Thus *tree structures of uniform PRFBs are far from being a full solution* to the PR problem for nonuniform FBs.

Derivability of decimators from a tree (as described above) is a sufficient condition for existence of rational PRFBs using the decimators. There are certain other conditions that are known to be necessary, e.g., there are no rational PRFBs using the decimator–set $\{2, 3, 6\}$ because no two decimators of such a FB can be coprime (Section 6.1, [4]). However, a *condition that is both necessary and sufficient remains unknown.* The present work studies this and related problems. An important part of our study is to significantly improve upon the known conditions, i.e., to derive new ones, strengthen necessary conditions and weaken sufficient ones. Another contribution is to study the conditions for reducibility of PRFBs to tree structures. For example, it has been shown [3], [10] that all rational PRFBs with dyadic decimator–sets *must* be derivable from dyadic trees. In Section 7, we will considerably generalize this result. Although these problems in their full generality remain unresolved, we believe the present work to be an important step towards a complete understanding of this subject — an area so rich in open problems even after over two decades of filter bank research.

10.1.1 Relevant earlier work

Trees of uniform FBs, and near–PR designs: A very common approach to nonuniform PRFB design is to cascade uniform PRFBs in a tree-structure, e.g., as is done to implement dyadic wavelet transforms [11], [12]. However, as stated earlier, there are nonuniform PRFBs that cannot be built in this manner. Many works deal with *approximate reconstruction* (or 'near–PR') nonuniform FBs, e.g., the frequency domain approaches of Li et al. [7], the time domain methods of Nayebi et al. [8], and other references therein. These are very useful from a practical standpoint, giving FBs with excellent filter responses and low aliasing distortions. However, they do not address the many theoretical issues involved in obtaining *exact* reconstruction.

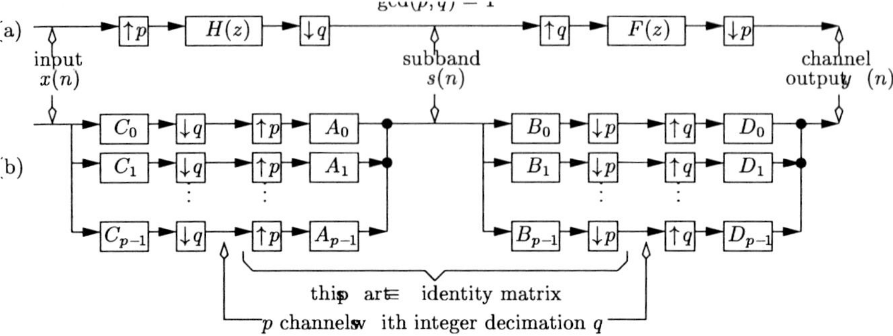

Figure 10.3. FB with rational decimators. (a) Single channel with decimator q/p. (b) Equivalent system of p channels with decimator q. (c) A possible set of filter choices ensuring the equivalence

FBs with fractional decimators: Kovačević and Vetterli have studied a more general system [6] where each channel of the FB has a decimation rate that is fractional, i.e., of form q/p where p, q are coprime positive integers. Such a channel, shown in Fig. 3a, is completely equivalent to the system of Fig. 3b. By this we mean that given any one of these systems, we can choose the filters in the other so that the same input $x(n)$ for both systems always produces the same signals $s(n)$ and $y(n)$ as shown. A choice ensuring this is shown in Fig. 3c (polyphase vectors are defined in Section 1.3). The equivalence under this choice is provable using the discussion on fractional decimation in [11, Section 4.3.3]. If the $A_i(z)$ differ from the special choice of Fig. 3c, we can replace them by this choice and modify the $C_i(z)$ so that the signal $s(n)$ is unaffected. This is done by performing a p-th order polyphase decomposition of the $A_i(z)$, using the fact that p, v are coprime, and moving the resulting polyphase matrix to the left. A similar comment holds for the $B_i(z)$.

From the equivalence shown in Fig. 3, we conclude that the PR problems for integer-decimated and rationally decimated FBs are fully equivalent. Another concern besides PR in rationally decimated FBs is the nature of their spectral analysis: Does a subband represent a contiguous portion of the input spectrum, or do the decimators and expanders in Fig. 3b cause it to contain separate parts, possibly mirrored and shuffled in order? This issue is studied in [6].[1] However, as far as the PR problem is concerned, it is enough to study FBs with integer decimators, and that is the approach we shall use.

Other more general multirate structures: As we will see in Section 2.2, nonuniform PRFBs are hard to design because of certain structural constraints that their associated polyphase matrices must obey. This is the origin of the central problem studied in our work: These structural constraints cannot be obeyed by rational FBs unless their decimators satisfy various conditions, which we aim to characterize. However, the constraints vanish if we

[1] It becomes less serious if we allow modulators at appropriate points within the FB.

use more general systems in the channels of the FB, e.g., if the filters are allowed to be periodically time–varying (Section 2.3). Chen and Qiu [2] and Shenoy [9] have studied multirate and FB design using such more general structures. The PR design then allows as much or even more freedom than that in the well–studied PR designs for the traditional *uniform* FB of Fig. 2. Our work is restricted to the usual nonuniform FB structure of Fig. 1 that does not use such generalized multirate structures.

PR conditions on decimators, and reducibility to tree structures: A necessary condition on the (integer) decimators for PR with rational FBs was first stated in [5]. Called the compatibility condition, it was generalized by Djokovic and Vaidyanathan [4], who also pointed out another such condition (pairwise noncoprimeness). We will considerably generalize these conditions. Another related work has involved showing derivability of FBs using dyadic decimator–sets from dyadic trees [10], [3], as explained earlier. These results too will be significantly strengthened. Among various more general situations studied include certain non–dyadic sets, unconstrained FBs, and tree structures whose constituent FBs need not be uniform.

10.1.2 Outline

Section 2 reviews the PR conditions on the filters of uniform FBs, and their generalization to nonuniform FBs, derivable using a transformation of nonuniform FBs to equivalent uniform ones. It shows how in spite of this transform, the nonuniform PRFB design does not reduce to a uniform PR design, unless the filters of the nonuniform FB are allowed to be *time varying.* In Section 3 we formally state the central problem, and study its solution for classes of unconstrained FBs (where the filters of the FB have no constraints such as rationality). Section 4 analyzes the role of tree structures in the study of the main problem. It shows how tree structures of uniform PRFBs do not provide a full solution (Section 4.2), and how trees can be used to improve upon known PR conditions on the decimators (Section 4.3). Section 5 solves the central problem of the paper for the class of *delay-chains* (FBs in which all filters are delays): It states the necessary and sufficient condition for a set of decimators to be usable to build a PR delay-chain, and presents algorithms to test the condition. Subsequent sections focus mainly on the class of rational FBs. Section 6 states the earlier known necessary conditions on decimators of rational PRFBs, and generalizes them in several ways. Section 7 generalizes [10], [3] by finding weaker conditions on decimators under which all PRFBs using them can be derived from certain tree structures. Section 8 summarizes all known necessary PR conditions on the decimators, and studies their inter-relationships. We conclude by noting many open problems in the area.

10.1.3 Notations, definitions and assumptions

Standard notation: Superscripts (*) and (^{T}) denote the complex conjugate and matrix (or vector) transpose respectively. We use boldface letters for matrices and vectors. We use lowercase letters for discrete sequences and uppercase letters for Fourier and z–transforms. Sometimes lowercase boldface letters are used for vector z–transforms. For sequences $\mathbf{h}(n)$ without z–transforms that are rational functions of z, the notation $\mathbf{H}(z)$ is an abbreviation for the Fourier transform $\mathbf{H}(e^{j\omega})$. For LTI transfer matrices $\mathbf{H}(z)$, the 'paraconjugate' $\mathbf{H}^{*T}(1/z^*)$ is denoted by $\widetilde{\mathbf{H}}(z)$. The L-th root of unity, $e^{-j2\pi/L}$ is denoted by W_L, or by W if the subscript value L is understood. The Kronecker delta function is denoted by δ ($\delta(0) = 1$ and $\delta(x) = 0$ if $x \neq 0$).

Polyphase concepts [11]: The M–fold decimator and expander are represented by $\downarrow_M$ and $\uparrow_M$ respectively, as in Fig. 1. Given a sequence $\mathbf{h}(n)$ with z–transform $\mathbf{H}(z)$, its M–fold decimated version is the sequence $\mathbf{g}(n) = \mathbf{h}(Mn)$, with z–transform denoted by $(\mathbf{H}(z)) \downarrow_M$. Likewise, the M–fold expanded version of $\mathbf{h}(n)$ is

$$\mathbf{f}(n) = \begin{cases} \mathbf{h}(n/M) & \text{if } n/M \text{ is an integer} \\ \mathbf{0} & \text{otherwise} \end{cases}$$

with z–transform denoted by $(\mathbf{H}(z)) \uparrow_M$. With $W = e^{-j2\pi/M}$, we have

$$(\mathbf{H}(z)) \downarrow_M = \frac{1}{M} \sum_{i=0}^{M-1} \mathbf{H}(z^{1/M} W^i), \quad \text{and} \quad (\mathbf{H}(z)) \uparrow_M = \mathbf{H}(z^M) \tag{1.2}$$

Given filters $H_0(z), H_1(z), \ldots, H_{N-1}(z)$, their M-th order analysis polyphase matrix $\mathbf{E}(z)$ is the $N \times M$ matrix defined by

$$\mathbf{h}(z) \triangleq (H_0(z), H_1(z), \ldots, H_{N-1}(z))^T = \mathbf{E}(z^M)\mathbf{d}(z),$$

where $\mathbf{d}(z) = (1, z^{-1}, \ldots, z^{-(M-1)})^T$ is the length M delay vector. Thus, $\mathbf{E}(z)$ has i-th column $\left(z^i \mathbf{h}(z)\right) \downarrow_M$. Similarly, the M-th order synthesis polyphase matrix of the filters $F_0(z), F_1(z), \ldots, F_{N-1}(z)$ is the $M \times N$ matrix $\mathbf{R}(z)$ obeying

$$\mathbf{f}(z) \triangleq (F_0(z), F_1(z), \ldots, F_{N-1}(z)) = \tilde{\mathbf{d}}(z)\mathbf{R}(z^M).$$

Thus the i-th row of $\mathbf{R}(z)$ is $\left(z^{-i}\mathbf{f}(z)\right) \downarrow_M$. If the $H_i(z), F_i(z)$ are respectively the analysis and synthesis filters of a FB, then $\mathbf{E}(z), \mathbf{R}(z)$ are respectively said to be the M-th order analysis and synthesis polyphase matrices of the FB. An easily proved result that we often use is the following:

Lemma 1: Polyphase lemma. Let $\mathbf{e}(z), \mathbf{r}(z)$ be the M-th order analysis and synthesis polyphase matrices of the filters $H(z)$ and $F(z)$ respectively. Thus $\mathbf{e}(z)$ is a row vector and $\mathbf{r}(z)$ is a column vector. Then,

$$\mathbf{e}(z)\mathbf{r}(z) = (H(z)F(z)) \downarrow_M \tag{1.3}$$

Maximal decimation: **All FBs** studied in the paper are **maximally decimated** with **integer decimation** rates, even if this is not explicitly stated. Similarly, references to a 'set of decimators' (or 'decimator–set') always implicitly mean a set of positive integers (not necessarily distinct) obeying (1.1).

10.2 BACKGROUND: EQUIVALENT UNIFORM FBS; PR EQUATIONS

The main focus of the paper is to find conditions on the decimators that permit existence of various types of nonuniform perfect reconstruction (PR) FBs with those decimators. To do this, we must first know what conditions on the filters of the FB guarantee the PR property. This section begins by reviewing the PR conditions for uniform FBs. We then review the transformation of a nonuniform FB with decimators n_i to an equivalent uniform FB with a decimation rate L that is a multiple of all the n_i. This yields the general PR conditions for nonuniform FBs, that will be used in all the later sections. In spite of the possible transformation to uniform FBs, the nonuniform PRFB design problem by no means reduces to the uniform PR design. However, such a reduction *does* occur if the nonuniform FB is allowed to have filters that are LPTV(L) (linear periodically time varying with period L) instead of LTI. With LTI filters, achieving PR is tougher, and is the subject of the later sections.

10.2.1 PR for uniform FBs, and the nonuniform to uniform transform

For the uniform FB of Fig. 2, the problem of achieving PR is very well understood. The following are three equivalent necessary and sufficient conditions on the filters for PR in this case [11]:

1 *Biorthogonality condition.* $(S_i(z)Q_j(z))\downarrow_M = \delta(i-j)$.
2 *AC matrix formulation.* Let $W = e^{-j2\pi/M}$. Then,

$$\begin{bmatrix} A_0(z) \\ A_1(z) \\ \vdots \\ A_{M-1}(z) \end{bmatrix} \triangleq \underbrace{\begin{bmatrix} S_0(z) & \dots & S_{M-1}(z) \\ S_0(zW) & \dots & S_{M-1}(zW) \\ \vdots & \ddots & \vdots \\ S_0(zW^{M-1}) & \dots & S_{M-1}(zW^{M-1}) \end{bmatrix}}_{\text{alias cancellation (AC) matrix } \mathbf{S}(z)} \begin{bmatrix} Q_0(z) \\ Q_1(z) \\ \vdots \\ Q_{M-1}(z) \end{bmatrix} = \begin{bmatrix} M \\ 0 \\ \vdots \\ 0 \end{bmatrix} \tag{2.1}$$

For any uniform FB (PR or otherwise), the $A_i(z)$ defined above are called the 'aliasing gains'. The PR condition (2.1) thus specifies all aliasing gains. It arises from the frequency domain relation between the output $\widehat{X}(z)$ and input $X(z)$ of *any* uniform FB (PR or otherwise):

$$\widehat{X}(z) = \frac{1}{M}\sum_{i=0}^{M-1} A_i(z)X(zW^i) \tag{2.2}$$

3 *Polyphase formulation.* If $\mathbf{E}(z), \mathbf{R}(z)$ are respectively the M-th order analysis and synthesis polyphase matrices of the FB (as in Fig. 2b), then $\mathbf{R}(z) = \mathbf{E}^{-1}(z)$. That this is equivalent to the biorthogonality condition stated earlier follows from the polyphase lemma (Section 1.3), which shows that the ij-th entry of $\mathbf{E}(z)\mathbf{R}(z)$ is precisely the quantity $(S_i(z)Q_j(z))\downarrow_M$ occurring in the biorthogonality condition.

Now any nonuniform FB (as in Fig. 1) is transformable into a uniform FB, which we will call its equivalent uniform FB [1], [4], [5], [6]. This transform is described by Fig. 4, which shows how a single channel with decimator n_k is replaceable by p_k channels with decimators $L = n_k p_k$. Repeating this process on all channels of the nonuniform FB, with L as any common multiple of all its decimators n_i (usually $L = \mathrm{lcm}\{n_i\}$), yields a *uniform* L-channel FB. The nonuniform FB has PR if and only if the equivalent uniform FB has PR. The filters in the uniform FB are various delayed versions of those in the nonuniform one. Inserting these relations between the filters into the PR conditions for uniform FBs gives the PR conditions for nonuniform FBs. These conditions, described next, generalize the uniform FB PR conditions, and are heavily used later.

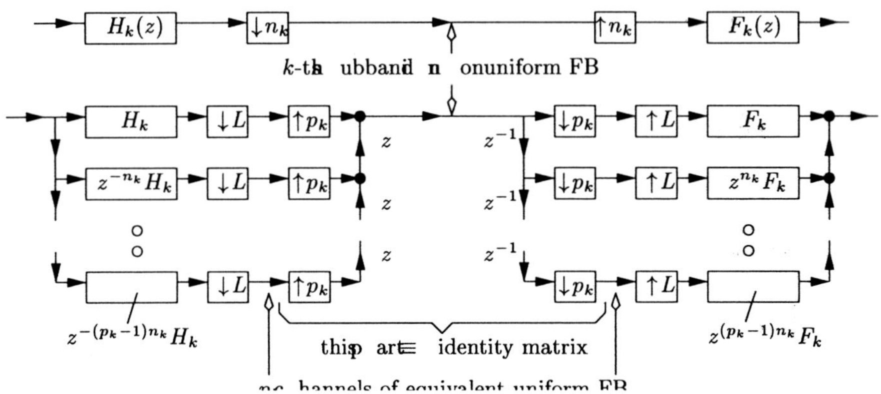

Figure 10.4. Transforming a nonuniform FB to an equivalent uniform FB

10.2.2 The general PR conditions for nonuniform FBs

Biorthogonality condition. The uniform FB biorthogonality condition, when applied to the uniform FB derived from the nonuniform one of Fig. 1, is equivalent to

$$(H_i(z)F_j(z)) \downarrow_{\gcd(n_i,n_j)} = \delta(i-j) \qquad \text{(biorthogonality condition)} \tag{2.3}$$

This has been observed earlier [4], [10]. Appendix A contains a proof for easy reference. The condition gets its name from its time-domain equivalent. To describe this, let $h_i(n), f_i(n)$ respectively be the impulse responses of $H_i(z), F_i(z)$. We define two sets of sequences

$$\{\mu_{ik}(n) = h_i^*(kn_i - n) \;\mid i = 0, 1, \ldots, M-1,\; k = \text{any integer}\} \tag{2.4}$$

$$\{\eta_{jl}(n) = f_j(n - ln_j) \;\mid j = 0, 1, \ldots, M-1,\; l = \text{any integer}\} \tag{2.5}$$

The action of the FB on its input $x(n)$ is now elegantly expressible using these sequences: The j-th subband signal $c_j()$ and the FB output $\widehat{x}()$ are given by

$$c_j(l) = \Big\langle x(n), \mu_{jl}(n) \Big\rangle = \sum_{n=-\infty}^{\infty} x(n) h_j(ln_j - n), \qquad \text{and}$$

$$\widehat{x}(n) = \sum_{j=0}^{M-1} \sum_{l=-\infty}^{\infty} c_j(l)\eta_{jl}(n) = \sum_{j=0}^{M-1} \sum_{l=-\infty}^{\infty} c_j(l) f_j(n - ln_j)$$

Here $\langle a(n), b(n)\rangle = \sum_n a(n)b^*(n)$ is the inner product of the sequences $a(n)$ and $b(n)$ (in the space of all sequences $x(n)$ for which $\sum_n |x(n)|^2$ is finite). Thus, the FB output $\widehat{x}(n)$ is a linear combination of the sequences from (2.5), using weights $c_j(l)$ that are inner products of the input $x(n)$ with the sequences from (2.4). Thus PR (i.e., $\widehat{x}(n) = x(n)$) is achieved if the two sets (2.4),(2.5) form a biorthogonal system, i.e., if

$$\langle \mu_{ik}(n), \eta_{jl}(n)\rangle = \delta(i-j)\delta(k-l)$$

This can indeed be shown to be the 'time domain' equivalent of (2.3).

$$\left[\begin{array}{ccc|ccc|ccc} E_0^k & \cdots & E_{n_k-1}^k & E_{n_k}^k & \cdots & E_{(p_k-1)n_k-1}^k & E_{(p_k-1)n_k}^k & \cdots & E_{L-1}^k \\ z^{-1}E_{(p_k-1)n_k}^k & \cdots & z^{-1}E_{L-1}^k & E_0^k & \cdots & E_{(p_k-2)n_k-1}^k & E_{(p_k-2)n_k}^k & \cdots & E_{(p_k-1)n_k-1}^k \\ z^{-1}E_{(p_k-2)n_k}^k & \cdots & z^{-1}E_{(p_k-1)n_k-1}^k & z^{-1}E_{(p_k-1)n_k}^k & \cdots & E_{(p_k-3)n_k-1}^k & E_{(p_k-3)n_k}^k & \cdots & E_{(p_k-2)n_k-1}^k \\ \vdots & \vdots & \vdots & \vdots & \vdots & \vdots & \vdots & \vdots & \vdots \\ z^{-1}E_{n_k}^k & \cdots & z^{-1}E_{2n_k-1}^k & z^{-1}E_{2n_k}^k & \cdots & z^{-1}E_{L-1}^k & E_0^k & \cdots & E_{n_k-1}^k \end{array}\right]$$

Figure 10.5. Polyphase matrix structure for uniform FBs derived from nonuniform ones

AC matrix formulation [4]. In (2.1), we set $M = L, W = e^{-j2\pi/L}$, and the filters as those of the uniform FB derived as in Fig. 4, from the nonuniform FB of Fig. 1. The i-th row in (2.1) is a sum of filter product terms $S_j(zW^i)Q_j(z)$. We group terms arising from the k-th subband in Fig. 1, i.e., those with $S_j(z) = z^{-ln_k}H_k(z)$ and $Q_j(z) = z^{ln_k}F_k(z)$ for $l = 0, 1, \ldots, p_k - 1$ where $n_k p_k = L$ (see Fig. 4). This yields a sum of form $H_k(zW^i)F_k(z)A_{ik}$, where

$$A_{ik} = \sum_{l=0}^{p_k-1} W^{-iln_k} = \sum_{l=0}^{p_k-1} e^{-j2\pi il/p_k} = \begin{cases} p_k & \text{if } i \text{ is a multiple of } p_k \\ 0 & \text{otherwise} \end{cases}$$

Thus, we can rewrite (2.1) using a new L–row AC matrix $\mathbf{H}(z)$ that has only M columns (one for each analysis filter of the nonuniform FB), as follows:

$$\underbrace{\begin{bmatrix} \mathbf{h}_0(z) \ldots \mathbf{h}_{M-1}(z) \end{bmatrix}}_{\text{AC matrix } \mathbf{H}(z)} \begin{bmatrix} F_0(z) \\ \vdots \\ F_{M-1}(z) \end{bmatrix} = \begin{bmatrix} L \\ 0 \\ \vdots \\ 0 \end{bmatrix}, \quad \text{where} \tag{2.6}$$

$$\mathbf{h}_i(z) = p_i \begin{bmatrix} \mathbf{h}_i'(z) \\ \mathbf{h}_i'(zW^{p_i}) \\ \vdots \\ \mathbf{h}_i'(zW^{(n_i-1)p_i}) \end{bmatrix}, \quad \text{and } \mathbf{h}_i'(z) = \begin{bmatrix} H_i(z) & \underbrace{0 \ldots 0}_{p_i-1 \text{ zeros}} \end{bmatrix}^T \tag{2.7}$$

If $n_i = M$ and $p_i = 1$ for all i (i.e., if the FB is uniform), the form of $\mathbf{H}(z)$ indeed reduces to that of (2.1).

Polyphase formulation. The PR condition is $\mathbf{R}(z) = \mathbf{E}^{-1}(z)$, just as for uniform FBs. However, as the equivalent uniform FB has interdependencies between the filters, its analysis polyphase matrix $\mathbf{E}(z)$ has a special structure [1]: Its rows can be partitioned into groups, where the k-th group corresponds to the k-th subband analysis filter $H_k(z)$ in Fig. 1. This group has $p_k = L/n_k$ rows as shown in Fig. 5. The first row is the L-th order analysis polyphase matrix (vector) of $H_k(z)$. Each subsequent row is formed by shifting length - n_k blocks of the previous row to the right, with the last block multiplied by z^{-1} and circulated back to the left end.[2] These rows are the polyphase vectors of filters $z^{-an_k}H_k(z)$ for $a = 1, 2, \ldots, p_k - 1$. Similarly, the synthesis polyphase matrix $\mathbf{R}(z)$ of the equivalent uniform FB has *columns* arrangeable into groups. The k-th group has a

[2]The submatrix of $\mathbf{E}(z)$ shown in Fig. 5 is *block pseudocirculant* with block size $1 \times n_k$ (generalizing the notion of pseudocirculants [11]).

form like the transpose of that in Fig. 5, with the $E_l^k(z)$ replaced by the entries $R_l^k(z)$ of the L-th order synthesis polyphase vector of the synthesis filter $F_k(z)$, and the z^{-1} factors replaced by z elements.

The paraunitary case. The uniform FB of Fig. 2 is said to be paraunitary (or orthonormal) if $\mathbf{E}^{-1}(z) = \widetilde{\mathbf{E}}(z)$; or in other words, if PR is obtained with $\mathbf{R}(z) = \widetilde{\mathbf{E}}(z)$, or equivalently with $Q_i(z) = \widetilde{S}_i(z)$. By generalization, the nonuniform FB of Fig. 1 is said to be orthonormal if PR is obtained (i.e., (2.3) is obeyed) with $F_i(z) = \widetilde{H}_i(z)$. From the relations between the filters of the nonuniform and the equivalent uniform FB, we see that each of these is paraunitary if and only if the other is. Notice that the two sets of (2.4),(2.5) which form a biorthogonal system in any PRFB, will *coincide*, hence forming an *orthonormal* system, if and only if the FB is paraunitary. This is because $F_i(z) = \widetilde{H}_i(z)$ is equivalent to $\eta_{jl}(n) = \mu_{jl}(n)$ in (2.4),(2.5). A general PRFB that is not necessarily orthonormal is often called a biorthogonal FB, due to the condition (2.3). Two other properties of orthonormal FBs, proved for uniform FBs in [11], are the unit energy and power complementarity properties, stated respectively as

$$\frac{1}{2\pi}\int_0^{2\pi} \left|H_i(e^{j\omega})\right|^2 d\omega = 1, \quad \text{and} \quad \sum_{i=0}^{M-1} \frac{H_i(z)\widetilde{H}_i(z)}{n_i} = 1$$

We can prove these for nonuniform FBs using the result for uniform ones and the transformation of Fig. 4.

10.2.3 Relation between the nonuniform and uniform PR designs

Transforming a nonuniform FB to an equivalent uniform one helps to find the PR conditions on its filters. These two FBs also share several properties (i.e., each has the property iff the other does). Examples are PR and paraunitariness; and rationality, stability, and FIR nature of filters. However, the equivalent uniform FB does not help in designing nonuniform PRFBs. This is due to its special structure: It has groups of filters that are delayed versions of each other. There are no known uniform PRFB design methods that allow imposition of this structure. Notice that the delayed versions of a filter have the *same* magnitude response, while uniform PRFB designs usually approximate ideal *nonoverlapping* analysis filter responses.

Most choices of the analysis filters $H_i(z)$ of Fig. 1 yield an equivalent uniform FB with an *invertible* analysis polyphase matrix $\mathbf{E}(z)$. However, this is not sufficient for existence of LTI synthesis filters ($F_i(z)$ of Fig. 1) resulting in PR: For this we further require that the inverse $\mathbf{R}(z) = \mathbf{E}^{-1}(z)$ have the special structure described in Section 2.2. This added constraint is not always easy to satisfy. If $\mathbf{E}(z)$ is paraunitary, then $\mathbf{R}(z)$, being equal to $\widetilde{\mathbf{E}}(z)$, automatically has the desired structure, and a nonuniform (paraunitary) PRFB is possible. However, again none of the many known parameterizations of uniform paraunitary FBs [11] allow imposition of the special structure of Section 2.2 that $\mathbf{E}(z)$ must have in order to represent a nonuniform FB.

The structural constraints on $\mathbf{E}(z)$ and $\mathbf{R}(z)$ can however be completely given up if the filters in the nonuniform FB are allowed to be LPTV(L) instead of LTI [1]. This is shown by Fig. 6, in which $p_k = L/n_k$ channels of a uniform L-channel (maximally decimated) FB are converted into a single channel with decimator n_k. The analysis and synthesis filters in this channel are LPTV(L). The procedure is repeated for each k using disjoint subsets of channels of the uniform FB. Clearly the nonuniform FB has the PR property if and only if the uniform one does. In the rest of the paper, we assume all

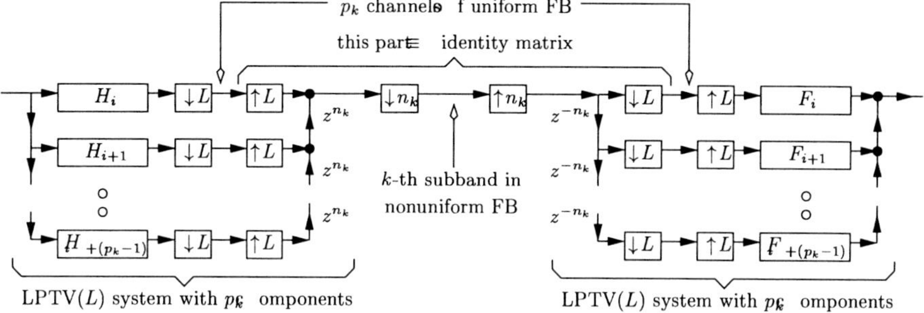

Figure 10.6. Equivalence between uniform FBs and nonuniform FBs with LPTV filters

analysis and synthesis filters of all FBs to be LTI. The nonuniform PR design is then significantly harder.

10.3 PROBLEM STATEMENT, AND UNCONSTRAINED FBS

10.3.1 Problem statement

The nonuniform perfect reconstruction (PR) FB design problem in its full generality can be stated as follows:

1 **Conditions on decimators for PR.** Given a set of positive integers n_i satisfying the maximal decimation condition (1.1), find necessary and sufficient conditions on the n_i for existence of a PRFB in some specified class C of FBs, having the n_i as decimators.

2 **Parameterization of the PRFBs.** When the n_i satisfy such a condition, find *all* possible PRFBs in C having n_i as decimators.

The FB class C here is defined by some constraint on the filters of its constituent FBs. Important examples that we will consider are delay-chains (FBs in which all filters are delays), rational FBs and FIR FBs. Other constraints that the class C can impose are realness of filter coefficients, stability of filters, and paraunitariness (or orthonormality). Note that in general the class definition does not directly by itself impose any constraint on either the number of channels or the nature of the decimators in the FB. However, the requirement that a FB in the class be maximally decimated and have PR could impose various constraints on these parameters. The statement of the problem is to characterize (a) the nature of these constraints, and (b) all PRFBs in C having a general decimator–set that obeys these constraints.

The solution to the problem of course depends on the FB class C. It is completely known for delay-chains, but unknown for rational FBs. Notice that the parameterization problem depends on first finding conditions on the decimators for PR, which can be quite tough in itself. So we will mainly focus on finding conditions for PR. Our aim will be to weaken the sufficient conditions and strengthen necessary ones until we obtain a set of necessary *and* sufficient conditions (the final goal, which we do not always achieve). We will also derive some results on the parameterization problem, especially in connection with tree structures.

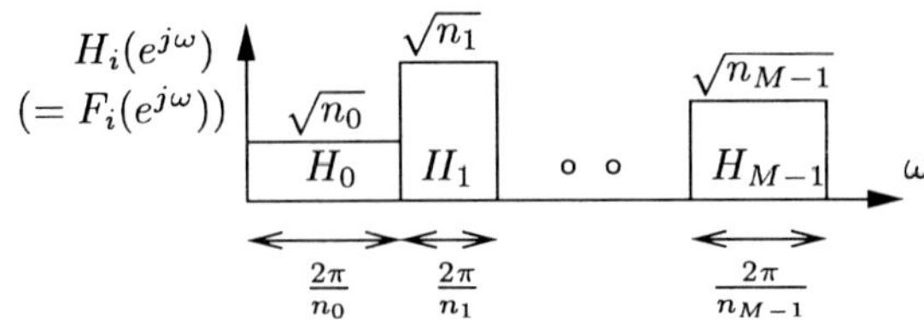

Figure 10.7. Ideal contiguous–stacked complex coefficient brickwall FB

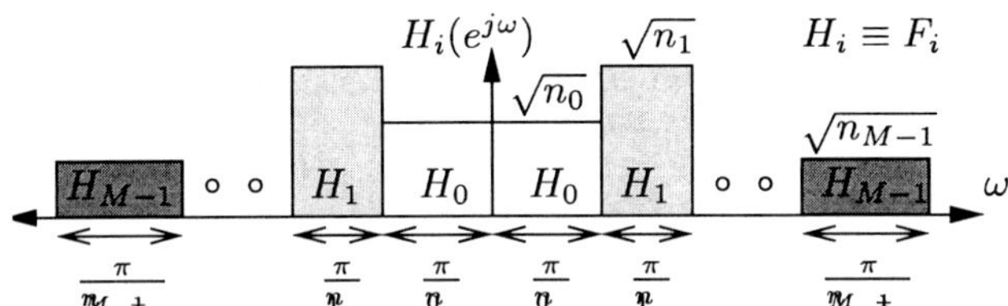

Figure 10.8. Ideal contiguous–stacked real coefficient brickwall FB

10.3.2 FBs with unconstrained complex and real coefficient filters

Let the class C in the above formulation be simply the class of all FBs, with no filter constraints (i.e., allowing ideal brickwall filters etc.). Then a PRFB in C always exists, no matter what the decimators n_i are (of course, provided they obey (1.1)). This is because the FB in Fig. 7, with ideal contiguous–stacked brickwall filters, always has PR. In fact it is a paraunitary FB. We will hence exclude this class C from all further discussion.

Note that the filters of Fig. 7 always have complex coefficients. Now let C be the class of all *real coefficient* FBs (i.e., FBs in which all filters have real coefficients). No other constraint is imposed, so the filters could still be ideal. However, it is now more difficult to find conditions on the decimators for existence of PRFBs in C. Taking a cue from Fig. 7, we can examine *brickwall* FBs, i.e., FBs as in Fig. 1 where the filters $H_i(e^{j\omega})$ have nonoverlapping supports, are constant on their supports and $H_i(z) = F_i(z)$. Since the H_i partition the input spectrum, PR is possible if and only if for each i, the i-th channel perfectly reconstructs all inputs that are bandlimited to the passband of $H_i(e^{j\omega})$. (In fact we then get a paraunitary PRFB, by suitable scaling of the filters.) This equivalently means that $H_i(e^{j\omega})$ has an aliasfree(n_i) support. For the (real coefficient) FB of Fig. 8, the bandpass sampling theorem states that this happens iff the band edges of H_i are at integer multiples of π/n_i [6]. Thus, the FB of Fig. 8 has PR if and only if

$$\sum_{i=0}^{k} \frac{1}{n_i} \text{ is an integer multiple of } \frac{1}{n_{k+1}} \text{ for all } k = 0, 1, \ldots, M-2. \tag{3.1}$$

Thus, a given set of decimators n_i can be used to build a real coefficient PRFB of the form of Fig. 8 if and only if (3.1) holds for some ordering of the n_i. For example, the set $\{2, 3, 6\}$ obeys this condition (with ordering (2,6,3) or (3,6,2)). The set $\{2, 3, 7, 42\}$ violates the condition (it is the only such set with ≤ 4 decimators). However, this does not preclude existence of PRFBs with more complicated stackings of nonoverlapping real coefficient brickwall filters, e.g., as in Fig. 9. Given a set S of decimators, does such a PRFB using

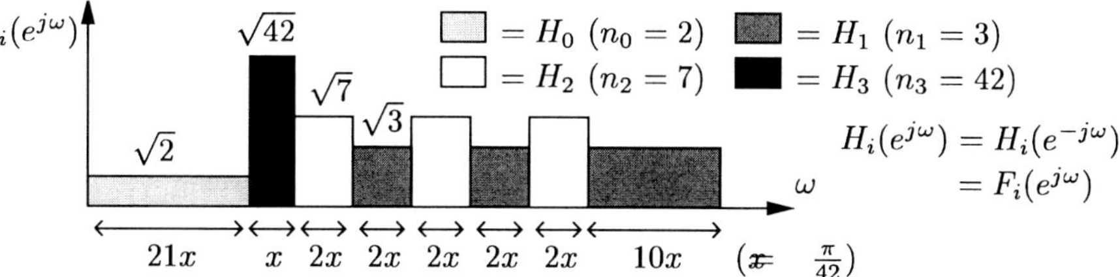

Figure 10.9. Non–contiguous stacked ideal real coefficient brickwall FB

the set S *always* exist? Does its nonexistence imply that there is no PRFB using S with real coefficient filters (ideal or otherwise) at all? The answers are not currently known to the authors.

An important class of FBs studied in the later sections is that of all *rational* FBs, i.e., those in which all filters have rational transfer functions. As Section 6.1 will show, neither of the above decimator–sets $\{2, 3, 6\}$ and $\{2, 3, 7, 42\}$ permit existence of a rational PRFB (since they have pairs of coprime decimators). Thus it is tempting to conclude that the decimators of rational PRFBs are more restricted than those of real coefficient PRFBs. Indeed, intuition suggests that for any decimator–set S, existence of rational PRFBs using S implies that of real coefficient rational PRFBs using S. This is in fact true for all sets S for which rational PRFBs are currently known to exist. However, as we will see later, there are many sets for which it is not known whether either rational PRFBs or PRFBs with real coefficient filters (rational or otherwise) exist. Thus, in general we do not know whether existence of the former implies that of the latter. The constraint of realness of filter coefficients will not be applied or studied further in the rest of the paper.

10.4 TREE STRUCTURES

Cascading uniform PRFBs in a tree structure is the most common method of designing nonuniform PRFBs. As pointed out in Section 1, this method, though useful, is far from providing a complete PR theory of nonuniform FBs, i.e., a full solution to either of the two basic problems posed in Section 3.1. However, tree structures do provide very useful tools in the study of these problems. This section aims at analyzing their role in this study. Section 4.1 defines some basic terminology we will often use later in describing and studying tree structures. Section 4.2 analyzes the method of cascading uniform PRFBs in tree structures, and shows with examples how it falls short of a full PR theory of nonuniform FBs. Section 4.3 presents general methods that use trees to improve upon known PR conditions on the decimators of nonuniform FBs belonging to various FB classes. By 'improving a PR condition' we mean strengthening a necessary condition, or weakening a sufficient one. These methods will be applied to specific conditions later on.

10.4.1 Basics and terminology

A **tree structured FB** is one of the form shown in Fig. 10, built by repeated insertion of FBs into the subbands of other FBs. These constituent FBs of the tree structure will be called its **units**. They could be either uniform or nonuniform FBs, and may themselves be tree structured FBs. The terms **parent, child, root and leaf units** will often be used to describe the relative positions of the units in the tree; their meanings are presumed

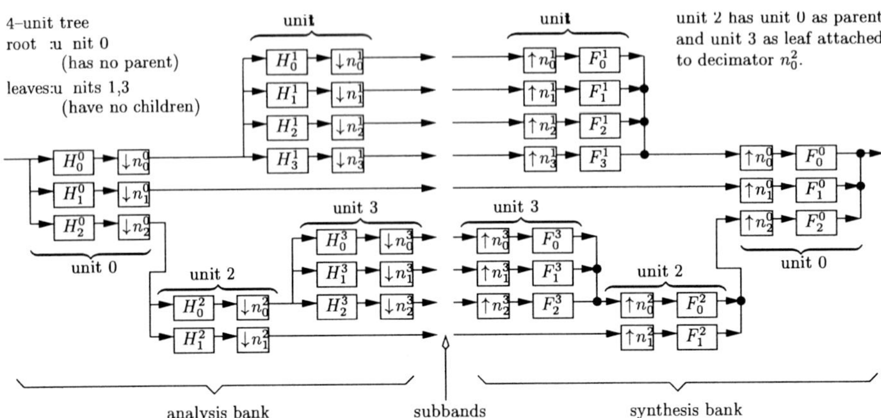

Figure 10.10. Tree structure of filter banks

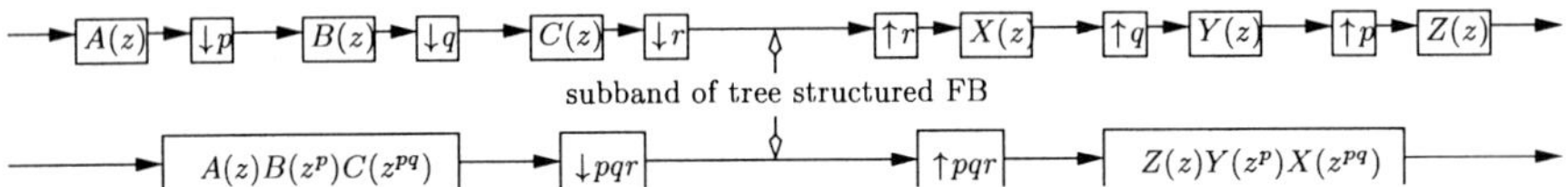

Figure 10.11. Structure of a channel of a tree structured FB

to be self–evident or clear from the examples shown in Fig. 10. We also use the term **descendant**, an obvious extension of 'child'.

Figure 11 shows how the decimators and filters of a tree structured FB are related to those of its units. The same FB may be derivable from many trees, differing in the choice of filters in the units (e.g., in Fig. 11, replace filter $A(z)$ by $A(z)/T(z^p)$ and $B(z)$ by $B(z)T(z)$ for arbitrary $T(z)$) or even in the sets of decimators in the units and the number of units (e.g., combine units 2,3 of Fig. 10 into a single FB). Every FB is derivable from a **trivial tree**, which by definition is one with only one unit, i.e., one whose root is also a leaf. We will also use the notion of a **tree structured set of decimators**. Shown in Fig. 12, this is defined exactly as a tree structured FB, except that the units of the tree are now just sets of decimators rather than FBs. The distinction is made because while a decimator–set S may sometimes be derivable from many tree structures, a FB using S need not always be derivable from all of these. In fact derivability from *all* these trees usually occurs only in very special cases (e.g., when the set S is *dyadic*, Section 7). Often we have the other extreme where the trivial tree is the only one that the FB is derivable from. Two other useful notions are as follows:

Uniform–trees. A uniform–tree structure of FBs or decimator–sets is a tree structure in which all units are uniform. (A uniform decimator–set, like a uniform FB, is one in which all decimators are equal.) Its importance, elaborated in Section 4.2, stems from the fact that design of uniform FBs is so well understood.

Properties preserved by trees. It is fairly clear that a tree structure whose units are PRFBs generates a (tree structured) PRFB. Similarly a tree of rational FBs generates a rational FB. In general we say that a property of FBs is **preserved by tree structures** if it is true that whenever all units of a tree of FBs obey the property, so does the resulting

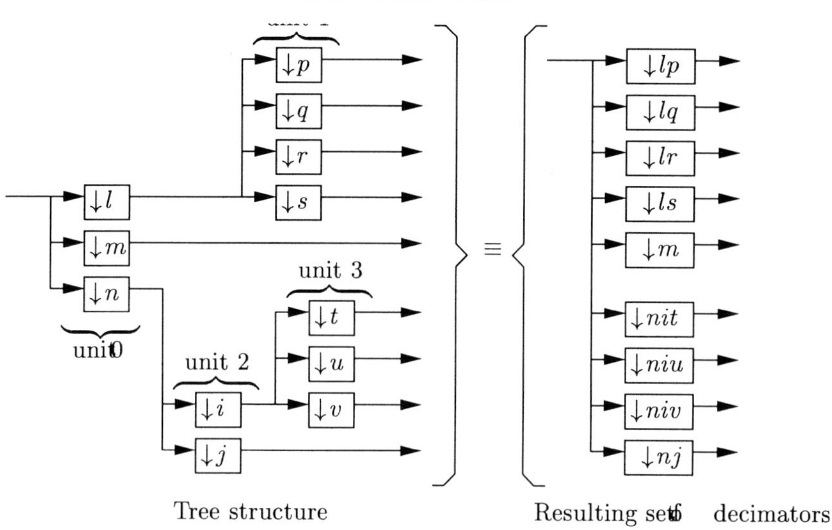

Figure 10.12. Tree structure of sets of decimators

tree structured FB. A similar statement holds for properties of decimator–sets. Two obvious but important properties of FBs preserved by trees are PR and maximal decimation. From Fig. 11, we can infer that the property of having filters that are rational, stable, real coefficient, FIR, or delays, and also the paraunitariness property, are all preserved by tree structures. The property of being a uniform FB is clearly *not* preserved by trees. Other useful nontrivial examples will be presented later (Section 6.3). As Section 4.3 will show, the ability of trees to strengthen known PR conditions on the decimator–sets depends crucially on whether or not certain properties are preserved by trees.

10.4.2 Uniform–trees: An incomplete PR theory for nonuniform FBs

A uniform–tree of FBs or decimator–sets is one in which each unit is uniform (i.e., all its decimators are equal). Its role in the central problem of Section 3.1 can be summarized as follows:

Role of uniform–trees. Derivability of a decimator–set S from a uniform–tree is a *sufficient* condition on S for existence of PRFBs using S and belonging to the specified FB class C, for all C of interest in this work.

This statement follows from the simple fact that a uniform–tree whose units are rational PRFBs generates a rational PRFB, and so on. More generally, the statement holds for every FB class C having two features, namely (a) C contains uniform PRFBs with all decimation rates, and (b) the property of being in C is preserved by tree structures. All C of interest here, e.g., the rational and FIR FB classes, have these features. Thus it is important to have an algorithm to test whether or not a given decimator–set S is derivable from a uniform–tree. Such derivability is assured, for instance, if S has no more than two *distinct* decimators, or if each decimator divides every decimator greater than itself (e.g., when they are all powers of the same integer). Appendix B proves this, and gives complete algorithms to test for derivability from uniform–trees.

Due to the common use of uniform–trees to design nonuniform PRFBs, the term 'tree structure' in the literature sometimes implicitly refers only to uniform–trees. In this work however, trees are always more general, i.e., unless explicitly referred to as uniform–trees, they could have nonuniform units too. This is necessary, for as we now show, uniform–trees do not provide a complete PR theory for nonuniform FBs.

Deficiencies of uniform–trees.

1 *Uniform-tree condition is not necessary for PR:* There are decimator–sets S that are not derivable from uniform–trees, but can be used to build PR *delay-chain* FBs, i.e., FBs in which all filters have the form z^{-k} for integer k. An example [4] is the set $S = \{6, 10, 15, 30, \ldots, 30\}$ (30 occurring 20 times), discussed in detail and generalized in Section 5.3. A delay-chain belongs to every FB class C of interest here (e.g., the FIR class). Thus, the uniform–tree condition is not necessary for any of these classes.

2 *Uniform-tree FBs are not a full parameterization:* Even if a decimator–set S is derivable from a uniform–tree, there may be PRFBs using S which are not derivable from any uniform–tree of FBs. We will now illustrate two examples of such a situation.

Example 1: Based on modifying filters of tree structured FBs. We take the analysis bank of a tree structured PRFB, find all subbands with a fixed decimation rate N, and transform them using an invertible square matrix $\mathbf{E}(z)$. If $\mathbf{h}(z)$ is the vector of analysis filters in the channels being transformed, the transform is equivalent to replacing $\mathbf{h}(z)$ by $\mathbf{h}'(z) = \mathbf{E}(z^N)\mathbf{h}(z)$. We preserve PR by effecting a corresponding change of synthesis filters using the inverse transform $\mathbf{E}^{-1}(z)$. Now if all the subbands being transformed come from the same unit FB in the tree, the transform can be effected by modifying only the filters of this unit, and the tree structure is preserved. More generally, if $\mathbf{E}(z)$ is block–diagonal with each block acting on subbands coming from the same unit of the tree, then again the tree structure can be preserved. However, this is no longer possible in general once we choose $\mathbf{E}(z)$ to avoid this degeneracy. In fact it is then fairly easy to ensure that the special structure of $\mathbf{h}(z)$ imposed by the tree is absent in the new filter vector $\mathbf{h}'(z)$. Thus, the new FB has the same decimator–set S but cannot be derived from the same tree. In particular if we choose a set S with a unique uniform–tree representation which is chosen as the starting tree in the above construction, the new FB is not derivable from a uniform–tree though its decimators are. An example of this kind is shown in [10], using FIR orthonormal FBs having the decimator–set $\{6, 6, 6, 6, 9, 9, 9\}$.

Example 2: Based on PR delay-chains. Consider the set with decimators 6,10,15,30 occurring 2,4,1,6 times respectively. From Section 5.2 we can show that this set can be used to build a PR delay-chain FB, which is clearly not derivable from a uniform–tree as the gcd of its decimators is unity. Now we build a tree in which the root is a uniform 2-channel PR delay-chain and the leaves are two such (identical) FBs, both children of the root. This yields a new (tree structured) PR delay-chain FB in which the gcd of the decimators is 2. Thus if this FB is to be derived from a *uniform–tree*, the root of the tree must be uniform with decimator 2. From its construction, this implies that in fact the new FB is *not* derivable from a uniform–tree. However, its set of decimators *is* derivable from a uniform–tree (in fact, in multiple ways). Note that this example cannot be produced starting from a uniform–tree of FBs in the manner used to create Example 1 above. Thus it shows a *deeper reason* for the incompleteness of FB parameterizations using uniform–trees.

10.4.3 Using trees to improve PR conditions on the decimators

Weakening sufficient conditions

Let S be a general decimator–set (obeying (1.1)). We seek conditions on S that permit existence of a PRFB that uses S as its decimator–set and belongs to some specified FB class C. For all C of interest here, the most elementary but very strong *sufficient* condition on S for this purpose is that S be uniform (i.e., all its decimators be equal), as uniform FBs can always be built. However, using the fact that the FB class definition (i.e., property of being in the class C) is preserved by tree structures, Section 4.2 has obtained a much weaker (and hence improved) sufficient condition, namely that S be derivable from a *uniform–tree*.

The process just described can be easily generalized to improve (i.e., weaken) *any* sufficient condition P on the decimator–set S (rather than merely the condition that S be uniform). The only requirement for this process to work is that the FB class definition be preserved by trees (which holds for all classes of interest here). The improved sufficient condition, denoted by P', states that S be derivable from a tree in which all units obey the original sufficient condition P. Because S is always derivable from the trivial tree, the new condition is indeed weaker, i.e., P implies P'. It is also easy to test for P' once we have a test for the original condition P: We simply list all possible tree representations of S and run the test for P on all units of each of them. (Of course the specific nature of P could make even faster tests possible.)

It may happen that $P' \equiv P$, i.e., the 'weaker' condition is *not strictly weaker*. For example, suppose P itself is preserved by trees. Then if S is derivable from a tree in which each unit obeys P (i.e., P' holds), it implies that S itself obeys P. Thus $P' \equiv P$ here. In fact a little further thought shows that $P' \equiv P$ if and only if P is preserved by trees. Note that by its definition, P' itself is preserved by trees. Thus repeated application of the above method cannot weaken the sufficient condition P any more than the first one does.

The only currently known instance where the above method strictly weakens a sufficient PR condition on decimator–sets is the one mentioned at the beginning of this section, leading to the uniform–tree sufficient condition. (Of course we can create other artificial instances, which lead to sufficient conditions that are stronger, and hence not as useful.) In the next section we will show a method to improve *necessary* conditions, and see that there are more nontrivial examples where this method causes a strict improvement.

Strengthening necessary conditions

We begin by illustrating the general method using a specific necessary condition that follows from Theorem 4 of Section 6.2. The condition states that the decimator–set of a rational PRFB cannot have a subset of $g+1$ decimators within which the gcd of any pair is g. The set $S = \{2, 4, 8, 12, 24\}$ can be seen to obey this condition. Suppose there is a rational PRFB using decimator–set S. We can create tree structures whose units are this and other rational PRFBs. The resulting FB also is a (tree structured) rational PRFB. Hence its decimator–set must obey the above necessary condition too. Thus we can obtain a new and stronger necessary condition on S by applying the original one to all the tree structured decimator–sets created from S as just described. In the present case, this new condition is *strictly stronger:* Using a two–unit tree in which the leaf is uniform with decimator 2 and is attached to the decimator 2 in the root S, we obtain the decimator–set $\{4, 4, 4, 8, 12, 24\}$ which violates the original condition. (Its subset $\{4, 4, 4, 8, 12\}$ has

5 decimators within which the gcd of any pair is 4.)

The new necessary condition P'' created as above from the original condition P will be referred to as the **tree version** of the necessary condition P. It is stronger, i.e., P'' implies P, because in particular the tree chosen in the above construction can be taken as the trivial one with S as its only unit. Generalizing the above example, we summarize the method of strengthening necessary conditions as follows:

Theorem 1: Tree versions of necessary conditions. Let C be a FB class such that the property of being in C is preserved by trees. Let P be a *necessary* condition on a general decimator–set S for existence of a PRFB in C with S as its set of decimators. Consider any decimator–set S'' derivable from a tree structure in which each unit is either identical to S or allows building of PRFBs in C (i.e., obeys some sufficient condition). Let P'' be the condition that all such sets S'' satisfy P. Then P'' is also a necessary condition on S, called the *tree version* of the necessary condition P.

Remarks:

1 We have just defined tree versions of necessary conditions, which are stronger necessary conditions. Earlier we had defined tree versions of sufficient conditions, which are weaker sufficient conditions. Some basic differences exist between these two methods of using trees to improve known conditions. For example, the above definition of the necessary condition P'' involves a known sufficient condition. The weaker this sufficient condition, the stronger P'' becomes. This is notably different from the earlier situation for tree versions of sufficient conditions.

2 *Algorithm to test P''.* The condition P'' on S demands that P hold for several sets S'' derived from S (including S itself) as described above. As there are infinitely many of the S'', we cannot state a general algorithm that tests for P''. One needs specific tests designed using the features of P and the sufficient condition used to define P''. This is again unlike the situation for tree versions of sufficient conditions.

3 *When are tree versions not* strictly *stronger?* Suppose P is preserved by tree structures. Then if S obeys P, all units in the tree generating S'' obey P, and hence so does S''. Thus, S obeys P'' too, i.e., $P'' \equiv P$. Here too, as with tree versions of sufficient conditions, P'' is preserved by trees, and is hence unchanged by forming its tree version. The only difference is that now we cannot in general claim that $P'' \equiv P$ implies that P is preserved by tree structures.

Tree versions of necessary conditions have not been observed earlier. A possible reason for this is that the simplest known necessary conditions for the rational FB class (compatibility and pairwise noncoprimeness, Section 6.1) are preserved by trees, and are hence identical to their tree versions. Section 6.3 shows another necessary condition that is made strictly stronger by forming its tree version.

10.5 DELAY-CHAINS

A delay-chain FB is one in which all filters are delays, i.e., of form z^{-k} for integer k. (We call z^{-k} a delay even though it is actually an 'advance' for $k < 0$.) Such a FB, while quite useless from a practical standpoint, is a useful tool in solving the problems of Section 3.1. This section presents a complete solution to these problems when the class C of FBs under study is that of delay-chain FBs. Because delays trivially obey various filter properties, delay-chain PRFBs belong to every class C studied in this paper: They are FBs with FIR, rational, stable, real coefficient filters, and we will see that they are also paraunitary. Thus, solving the problem of Section 3.1 for the class of delay-chains yields a *sufficient* condition on the decimators for existence of PRFBs in any of these classes. We will see that this condition is strictly weaker than the other sufficient condition studied earlier in Section 4.2, namely derivability from uniform–trees. In fact existence of a delay-chain is the weakest known sufficient condition for all of the earlier mentioned rational FB classes.

10.5.1 PR condition on the set of decimators

In Fig. 1, if $H_k(z) = z^{-l_k}$ for all k, the k-th subband signal is $c_k(n) = x(n_k n - l_k)$, i.e., it contains a certain subset of the input samples $x(n)$. Let $L = \text{lcm}\{n_i\}$, and consider any L consecutive samples of $x(n)$. We see that the k-th subband contains exactly L/n_k of these samples. Due to maximal decimation, we have

$$\sum_k (L/n_k) = L \tag{5.1}$$

Thus, if any of the L chosen input samples occurs in more than one subband, there must be a sample that does not occur in any subband. In this case, PR is clearly impossible no matter what the choice of synthesis filters. On the other hand, if none of the input samples occurs in more than one subband, then (5.1) implies that each of them occurs in exactly one subband. We can then achieve PR by appropriately interleaving the subband samples, which is done by the choice of synthesis filters as $F_k(z) = z^{l_k}$. Thus, PR is possible iff no input sample occurs in more than one subband. This condition means that if $i \neq j$, then $n_i n - l_i \neq n_j m - l_j$, i.e., $l_i - l_j \neq n_i n - n_j m$, for any integers n, m. As n, m range over all integers, the right side here ranges over all multiples of $\gcd(n_i, n_j)$. Thus the PR condition may be summarized as follows:

Theorem 2: Delay-chain PRFBs. In Fig. 1, if $H_k(z) = z^{-l_k}$ for integers l_k, PR is possible iff no input sample occurs in more than one subband. Under this condition, PR is obtained with the unique choice $F_k(z) = \widetilde{H}_k(z) = z^{l_k}$, yielding a PR delay-chain FB, which is thus always paraunitary. The necessary and sufficient condition on the decimators n_i for existence of such a FB is that there exist integers l_i satisfying

$$(l_i - l_j) \not\equiv 0 \pmod{\gcd(n_i, n_j)} \quad \text{if } i \neq j \tag{5.2}$$

10.5.2 Testing the PR condition

Given the decimators n_k, it is required to test for existence of integers $l_0, \ldots, l_{M-1}$ obeying (5.2). Now if (5.2) holds for some integers l_k, then it also holds if each l_k is replaced by $l_k + m_k n_k + C$ for any integers m_k and any fixed integer C. Hence, without loss of generality we can assume that $0 \leq l_k < n_k$ and $l_0 = 0$. This makes the number of possible sets of l_k *finite*, so clearly there is an algorithm for our purpose.

For example, we can try to assign the l_k sequentially, as follows: Suppose we have $l_0, l_1, \ldots, l_{N-1}$ obeying (5.2) for some $N < M$. We assign to l_N all possible values obeying $0 \leq l_N < n_N$ and $(l_N - l_j) \not\equiv 0 \pmod{\gcd(n_N, n_j)}$ for $j = 0, 1, \ldots, N-1$. Each value yields a larger set $l_0, l_1, \ldots, l_N$ obeying (5.2), and we can now repeat the process on this set. If there is no valid choice for l_N, we must restart with another valid set of choices for $l_0, l_1, \ldots, l_{N-1}$. Initializing this recursive scheme using $l_0 = 0$, we can thus list *all* sets $\{l_0, l_1, \ldots, l_{M-1}\}$ obeying (5.2). In particular this finds whether or not there exist such sets. This solves both problems of Section 3.1 for the class C of delay-chain FBs.

To determine only the existence of a PR delay-chain, the above algorithm can often be accelerated using the following result:

Fact 1. Let integers $l_0, \ldots, l_{N-1}$ obey (5.2) for some $N < M$, and let n_N be a common multiple of $n_0, n_1, \ldots, n_{N-1}$. Then there is an integer l_N such that $0 \leq l_N < n_N$ and $l_0, \ldots, l_{N-1}, l_N$ obey (5.2) too.

Proof: From the premise, $l_0, \ldots, l_{N-1}, l_N$ will satisfy (5.2) if and only if

$$(l_N - l_j) \not\equiv 0 \pmod{\gcd(n_N, n_j)} \quad \text{for } j = 0, 1, \ldots, N-1. \tag{5.3}$$

Also $\gcd(n_N, n_j) = n_j$. Thus (5.3) is equivalent to $l_N \neq l_j + nn_j$ for all integers n, for $j = 0, 1, \ldots, N-1$. For each j there are n_N/n_j integers of this form $l_j + nn_j$ in the range $[0, n_N)$. Thus, of the n_N integers in $[0, n_N)$, at most $B = \sum_{k=0}^{N-1} n_N/n_k$ of them are excluded as possible choices of l_N (in fact we can even show that exactly B choices are excluded). As $N < M$, maximal decimation (1.1) means that $B < n_N$, so there are still valid choices of l_N in the interval $[0, n_N)$. □

Thus, suppose there is a decimator n_N such that each $n_j \geq n_N$ is a multiple of all $n_i < n_j$. It then suffices to verify existence of valid delays l_k obeying (5.2) for all $n_k < n_N$. As an extreme case, if every n_j is a multiple of all $n_i < n_j$ (i.e., every n_j divides all $n_i > n_j$), then a delay-chain PRFB always exists. In fact the decimator–set is then derivable from a uniform–tree (Appendix B). Fact 1 is also useful in proving Theorem 3 which follows soon.

Nonuniqueness of delay-chains: When a decimator–set allows building of a PR delay-chain FB, in general this delay-chain is not unique. The non-uniqueness can be much deeper than that caused simply by adding a constant delay to all the filters. For example, when several delay-chains are possible, it could happen that some of them are also derivable from uniform–trees, while some others are not, as seen in Section 4.2.

10.5.3 Delay-chains vs. uniform–trees

Our study of tree structures showed that (a) known PR conditions on decimators can sometimes be strengthened using trees (Section 4.3), and (b) derivability of the decimators from a uniform–tree is a *sufficient* PR condition for all FB classes that we study (Section 4.2). Does this teach us more about delay-chains? Firstly, the condition (5.2) is both necessary and sufficient for existence of PR delay-chains. Hence it remains unaltered by the procedures of Section 4.3. Next, the uniform–tree condition is *not* necessary, as we now show:

Theorem 3: PR delay-chains without uniform–trees. There are infinitely many PR delay-chain FBs that cannot be derived from uniform–trees. Such FBs can be built using every set of decimators of the form

$$S = \{n_0, n_1, n_2, L, L, \ldots, L\}, \quad \text{where} \quad L = \operatorname{lcm}(n_0, n_1, n_2),$$

and $n_0 = m_1 m_2$, $n_1 = m_2 m_0$, $n_2 = m_0 m_1$ where the m_i are pairwise coprime integers greater than unity. (Here L occurs $L(1 - \sum_{i=0}^{2}(1/n_i))$ times in S.)

Proof: By Fact 1, decimators of S allow building of a PR delay-chain FB iff there are integers l_0, l_1, l_2 obeying (5.2) for $i, j \in \{0, 1, 2\}$. This condition is easily ensured, in fact it holds iff $\gcd(n_i, n_j) \geq 2$ for $i, j \in \{0, 1, 2\}$ with strict inequality for at least one $i \neq j$. (We can then make a valid choice of the l_i from the numbers 0,1,2.) Further if

$\gcd(n_0, n_1, n_2) = 1$, the set cannot be derived from a uniform–tree (Appendix B). Both these requirements are satisfied by the choice of n_i stated by the theorem. □

An example of a delay-chain PRFB not derivable from a uniform–tree was first shown in [4]. Its set of decimators $\{6, 10, 15, 30, 30, \ldots, 30\}$ (30 occurring 20 times) is a special case of the construction of Theorem 3 with $(m_0, m_1, m_2) = (5, 3, 2)$. This is not the only way to produce such examples: Delay-chain PRFBs can also be built with the decimator values 6, 10, 15, 30 when the number of their respective occurrences are 2, 4, 1, 6 or 2, 3, 2, 7. The former set of decimators is the *smallest* such example.[3] It can be used as the root of a tree to derive the example of [4], but not the latter example. In all these cases, the decimators have no common factor, ensuring that they are not derivable from uniform–trees. In fact if the decimators of a delay-chain PRFB do have a common factor, the FB can be built from smaller PR delay-chains as follows:

Fact 2. Let all decimators in a PR delay-chain FB have common factor $K > 1$. Then the FB can be derived from a tree structure in which each unit is a PR delay-chain FB and the root is uniform with decimator K.

Proof: Let $x(n)$ be the FB input. For $0 \leq k < K$, let $f_k(n) = x(Kn - k)$, which is the k-th subband signal in a uniform K channel delay-chain PRFB. Now consider the i-th channel of the given PR delay-chain, with decimator n_i, analysis filter z^{-l_i}, and hence, subband signal $x(n_i n - l_i)$. Since n_i is a multiple of K, either *all* its samples lie in the sequence $f_k(n)$, or none of them do (depending on whether or not $l_i \equiv k \pmod{K}$). We now collect all subbands whose samples do lie (entirely) in $f_k(n)$. Due to the PR condition for delay-chains (Theorem 2), these subbands collectively contain all samples of $f_k(n)$ (as none of the other subbands have any of them), and each of these samples occurs in exactly one of these subbands. Further the delays in all these subbands are equal (to k) modulo K. Thus these subbands can be generated by inserting a suitable delay-chain PRFB as a child (in a tree) in the k-th subband signal $f_k(n)$ of a uniform K channel delay-chain PRFB. Repeating this process for $k = 0, 1, \ldots, K - 1$ yields the desired tree structure. □

Remarks:

1 The above result does not generalize easily to other classes of FBs (besides delay-chains). For example, consider the decimators $\{4, 4, 4, 4\}$, having common factor $K = 2$. These decimators can be used to build *rational* and *FIR* PRFBs that are not derivable from *any* tree structure (besides the trivial one).

2 A common factor $K > 1$ among all decimators does not by itself ensure their derivability from a tree whose root is uniform with K as decimator.[4] However, if the decimators also allow building of a delay-chain PRFB, then by Fact 2, there is at least one such tree, as the FB itself is derivable from such a tree.

3 All decimators of a PR delay-chain FB need not have a common factor $K > 1$ (see the example in Theorem 3). However, further conditions on the decimators can force such a common factor to exist, thus making Fact 2 apply. For example, suppose the PR delay-chain has a decimator of value m occurring $m - 1$ times (m is thus the smallest decimator). Then all decimators must have m as a factor. This is provable by a slight extension of the proof of Fact 2. In fact it even generalizes

[3]This is true when size is measured by either the number of decimators, or their lcm, or the largest one. In fact there is no other example with 13 or fewer decimators. This is verifiable by exhaustive search aided by a computer and Fact 2.

[4]The set of decimators $\{4, 6, 6, 10, 10, 10, 10, 60\}$ shows this for $K = 2$. The choice of root prevents the leaves from obeying (1.1).

to rational FBs in place of delay-chains (Theorem 5, Section 7), although this is harder to prove.

10.6 THE CLASS OF RATIONAL FBS

In this section and most of Section 7, the FB class C of interest is that of rational FBs, i.e., FBs in which all filters are rational. We seek necessary and sufficient conditions on a decimator–set S for existence of a rational PRFB using S. The *weakest known sufficient condition* is that of existence of a PR *delay-chain* (Section 5). This is clearly sufficient since delay-chains are rational FBs, but *is it also necessary*? Or is there a decimator–set which does not permit existence of PR delay-chains, but allows building of rational PRFBs (whose filters are not all delays)? This is a major open question in the PR theory of nonuniform FBs.

A possible approach to answer the above question is to try to build a rational PRFB with decimators that do not allow building of PR delay-chains. However, starting with an arbitrary decimator–set such as $S = \{2, 3, 6\}$ does not help, as S violates a known *necessary* condition (called 'compatibility', Section 6.1) on the decimators of a rational PRFB. Such sets must be excluded, and to this end it helps to derive more necessary conditions. This is our main contribution in this section. The previously known necessary conditions for PR are described in Section 6.1. Each subsequent subsection develops a new necessary condition that is strictly stronger than a previously known one. Table 1 (Section 8) presents a comprehensive summary of all known conditions, many of which are new results of the present work. The table studies the interrelationship between the conditions, and lists example decimator–sets illustrating their use.

All the new necessary conditions we develop still collectively remain insufficient for existence of *delay-chain* PRFBs, and thus it is *still not known* whether they are sufficient for existence of rational PRFBs. Our work reduces the 'gap' between the necessary conditions and the sufficient one. Proving that the sufficient condition is in fact necessary would in some sense render obsolete most of the present section. However this appears tough to do, in fact the statement may not even be true. Our work is a step towards the truth.

10.6.1 Previously known necessary conditions on decimators

1 *Pairwise noncoprimeness.* No two decimators of a rational PRFB can be coprime [4]. If $\gcd(n_i, n_j) = 1$ for two decimators n_i, n_j in Fig. 1, the biorthogonality condition (2.3) for PR implies $H_iF_j = 0$ and $(H_iF_i)\downarrow_{n_i} = (H_jF_j)\downarrow_{n_j} = 1$. This is impossible for a rational FB, as $H_iF_j = 0$ forces $H_i \equiv 0$ or $F_j \equiv 0$.

2 *Compatibility.* Every decimator occurring only once must divide some other decimator [1, 5, 4]. In particular, the largest decimator must occur at least twice. As Section 6.4 will show, without this condition the rational FB cannot even be a nonzero LTI system, let alone have PR.

3 *Strong compatibility.* This condition, developed in [4], places a lower bound $b_j \geq 1$ on the number of occurrences N_j of each decimator n_j. The condition is stated as follows:

$$N_j \geq b_j \triangleq \frac{1}{p_j}\left(\min_{p_i \neq p_j} \operatorname{lcm}(p_i, p_j)\right), \quad \text{where } p_i = \frac{L}{n_i}, \tag{6.1}$$

where $L = K(\operatorname{lcm}\{n_i\})$ for any integer $K > 0$. This will be shown in Section 6.4, which in fact proves a new condition strictly stronger than the above.

Note that the bound b_j of (6.1) is independent of the integer K. Also, it only needs verification for *distinct* decimator values, because if $n_i = n_j$ then $N_i = N_j$, $b_i = b_j$. For a *uniform* set of decimators, $p_i = p_j$ for all i, j, so we define $b_j = 1$ here (so that the bound holds). Excluding this case, $b_j = 1$ iff p_j is a multiple of some $p_i \neq p_j$, i.e., iff n_j divides some distinct decimator n_i. So the bound need not be checked for such decimators. Also, strong compatibility implies compatibility, because it demands that any n_j occurring only once (i.e., with $N_j = 1$) must have $b_j = 1$, i.e., must divide some other decimator. In fact strong compatibility is a *strictly* stronger necessary condition than compatibility, as shown by the set of decimators $\{2, 4, 6, 24, 24\}$. However it does not imply pairwise noncoprimeness [1] (shown by $\{2, 5, 10, 10, 10\}$). Likewise, a set could satisfy pairwise noncoprimeness but violate compatibility (and hence strong compatibility), e.g., $\{2, 4, 6, 12\}$.

10.6.2 The pairwise gcd test

Theorem 4: Pairwise gcd test. Among the decimators of a rational PRFB, there cannot be a subset of $g + 1$ decimators such that the gcd of any two elements from the subset is a factor of g. In particular (for $g = 1$), this implies the pairwise noncoprimeness condition (Section 6.1).

Proof: As with pairwise noncoprimeness, the proof uses the biorthogonality condition (2.3) for PR. Let $g + 1$ decimators $n_0, n_1, \ldots, n_g$ be such that the gcd of any pair divides g. From (2.3), $(H_i(z)F_j(z)) \downarrow_{\gcd(n_i,n_j)} = 0$ if $i, j \in \{0, 1, \ldots, g\}$, $i \neq j$. In this case $g/\gcd(n_i, n_j)$ is given to be an integer, so decimating both sides by it,

$$(H_i(z)F_j(z)) \downarrow_g = 0, \quad \text{for } i, j \in \{0, 1, \ldots, g\},\ i \neq j \tag{6.2}$$

Form the g-th order analysis polyphase matrix $\mathbf{E}(z)$ (of size $(g+1) \times g$) of the filters $H_i(z)$, and the g-th order synthesis polyphase matrix $\mathbf{R}(z)$ (of size $g \times (g + 1)$) of the F_i. Thus, from the polyphase lemma (Section 1.3), for $i, j \in \{0, 1, \ldots, g\}$, $(H_i(z)F_j(z)) \downarrow_g$ is the ij-th entry in the product $\mathbf{P}(z) = \mathbf{E}(z)\mathbf{R}(z)$. Hence by (6.2), $\mathbf{P}(z)$ is a $((g + 1) \times (g + 1))$ *diagonal* matrix. Its i-th diagonal entry is the filter $(H_i(z)F_i(z)) \downarrow_g$, with impulse response $c_i(gn)$, where $c_i(n)$ is the impulse response of H_iF_i. From (2.3), $(H_i(z)F_i(z)) \downarrow_{n_i} = 1$, so $c_i(n_in) = \delta(n)$ i.e., $c_i(0) = 1$. Hence $c_i(gn) \not\equiv 0$, i.e., no diagonal element of $\mathbf{P}(z)$ is identically zero. Thus, as these elements are rational filters, there is a z such that $\mathbf{P}(z)$ has full rank $g + 1$. However, this is impossible from the sizes of $\mathbf{E}(z), \mathbf{R}(z)$. □

10.6.3 Tree version of strong compatibility

In Section 4.3, we saw how given a necessary condition P on the decimators for PR, we could form its 'tree version' P'', which is a stronger (though not necessarily *strictly* stronger) necessary condition. We can apply this process to the conditions of Section 6.1. Some thought shows that both the pairwise noncoprimeness and the compatibility conditions are preserved by tree structures, and are hence identical to their tree versions (as seen in Section 4.3). However, the same is not true with strong compatibility: Its tree version is strictly stronger than itself. This is shown by the two–unit tree in Fig. 13. Both units R and S are strong compatible, and S allows building of rational PRFBs (as it is uniform). However the resulting set of decimators is not strong compatible. Hence, though R obeys strong compatibility, it violates its tree version.

A complete algorithm to test this new necessary condition is described in Appendix E. Its derivation involves characterizing trees similar to that in Fig. 13. This is done by the following results:

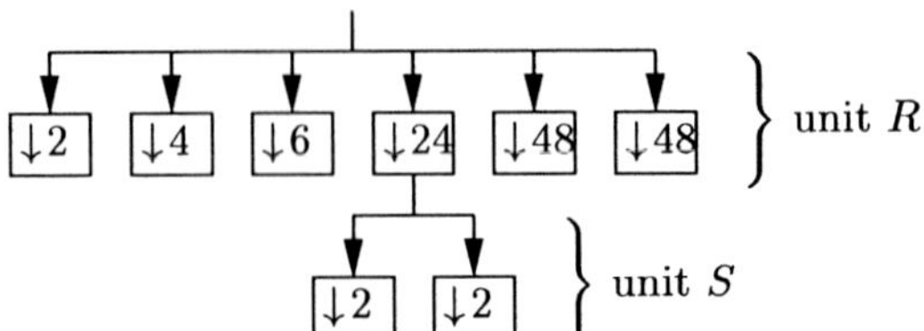

Figure 10.13. Showing that strong compatibility is not preserved by trees

Fact 3. Consider a set T of decimators derived from a 2-unit tree structure having root R and leaf S attached to decimator m_0 of R. Suppose R, S are strong compatible but T is not. Then S is a uniform unit, i.e., all its decimators have equal value K. The decimator m_0 of R does not occur in T, i.e., it occurs only once in R. The decimator m_0K of T obtained at the leaf S also occurs in R. Decimators of this value m_0K are the only ones violating the strong compatibility lower bound on the number of their occurrences in T.

Fact 4. Let a set D of decimators satisfy strong compatibility but violate its tree version. Then there is a tree $\mathcal{T}$ generating a set T of decimators, such that $\mathcal{T}$ and T have the following properties:

1 The tree $\mathcal{T}$ has root D. All leaves of the tree are uniform and are children of its root. All decimators obtained at the leaves have equal value d.
2 If d_i are the decimators of D to which leaves are attached in $\mathcal{T}$, then no decimator in T has value d_i.
3 If $d \notin D$, then $d = \text{lcm}\{d_i\}$. Hence, if $d \notin D$, the d_i are not all equal (for otherwise, $d = d_i \in D$).
4 Decimator $d \in T$ violates the strong compatibility lower bound on the number of its occurrences in T.

Fact 3 is proved in Appendix C and used to prove Fact 4 in Appendix D. Fact 4 gives an algorithm to test whether the set D obeys the tree version of the strong compatibility condition: We find all trees with root D and properties 1,2 and 3 listed in Fact 4. It can be seen that there are only finitely many such trees, and from Fact 4, D violates the condition if and only if one of these trees also obeys property 4. This idea is the basis of the detailed algorithm of Appendix E.

10.6.4 The AC-matrix test

The necessary condition derived here relies heavily on the AC matrix formulation (2.6),(2.7) of the PR condition on the filters of the FB. The algorithm to test the condition is described in Appendix F, and may be taken as the statement of the condition (i.e., this condition, unlike the earlier ones, does not have a short / simple statement). Like the test of Section 6.3, this test also strictly strengthens strong compatibility, but in an independent direction. In this section we derive two lemmas that explain the operation of the test, illustrate the test with examples, and thus justify the algorithm of Appendix F. Deriving the new test also *proves* the necessity of strong compatibility for PR; a result *assumed* in deriving the test of Section 6.3. We further show that (simple) compatibility is necessary even if we allow the rational FB to violate PR but merely insist that it be an *LTI* system (i.e., an *aliasfree* FB) that is not identically zero.

Two key results used by the test

Lemma 2. In Fig. 1, if $\sum_k H_{i_k}(z)F_{i_k}(z) = 0$ for any set of i_k, $0 \le i_k < M$, then the FB cannot have PR.

Proof: If the FB of Fig. 1 has PR, $(H_{i_k}(z)F_{i_k}(z)) \downarrow_{n_{i_k}} = 1$ by biorthogonality (2.3). Let $L = \text{lcm}\{n_i\}$, thus $(H_{i_k}(z)F_{i_k}(z)) \downarrow_L = (1) \downarrow_{(L/n_{i_k})} = 1$. So $\left(\sum_k H_{i_k}(z)F_{i_k}(z)\right) \downarrow_L \ne 0$, violating $\sum_k H_{i_k}(z)F_{i_k}(z) = 0$. □

Lemma 3. Given *rational* filters $B_i(z), C_i(z)$, $0 \le i < N$, let $W = e^{-j2\pi/M}$ and $G_l(z) = \sum_{i=0}^{N-1} B_i(zW^l)C_i(z)$. If $G_l(z) = 0$ for N values of l occurring consecutively in an arithmetic progression, then $G_l(z) = 0$ *for all* values of l in this progression. (The lemma in fact holds for any nonzero complex W.)

Proof: If $N = 1$, the lemma must be taken to mean that if $B_0(zW^l)C_0(z) \equiv 0$ for some l, then it holds for all l. This is clearly true: Rational filters $B_0(z), C_0(z)$ obey $B_0(zW^l)C_0(z) \equiv 0$ iff $B_0 \equiv 0$ or $C_0 \equiv 0$ or both. (Note however that this is in general false if we remove the rationality constraint.) Hence, let $N > 1$. Let the N given consecutive values of l in arithmetic progression be $s, s+d, s+2d, \ldots, s+(N-1)d$. The lemma can then be restated by defining

$$\mathbf{b}(z) = [B_0(zW^s), \ldots, B_{N-1}(zW^s)], \quad \mathbf{c}(z) = [C_0(z), \ldots, C_{N-1}(z)]^T,$$

as follows: If $\mathbf{b}(zW^{nd})\mathbf{c}(z) = 0$ for $n = 0, 1, \ldots, N-1$ then it is true for all integers n. To show this, form the square matrix $\mathbf{B}(z)$ with rows $\mathbf{b}(zW^{nd})$, $0 \le n < N$. By the premise of the lemma, $\mathbf{B}(z)\mathbf{c}(z)$ is the zero vector. This implies linear dependence of the columns of $\mathbf{B}(z)$, and hence of its rows, as it is square. So $\sum_{n=0}^{N-1} \alpha_n(z)\mathbf{b}(zW^{nd}) = 0$ for some rational filters $\alpha_n(z)$ not all identically zero. Let r be the maximum n for which $\alpha_n(z) \not\equiv 0$. Divide the above relation by $\alpha_r(z)$. (This is allowed solely due to the rationality assumption: Otherwise $\alpha_r(e^{j\omega})$ could for instance be zero in an interval.) This yields

$$\mathbf{b}(zW^{rd}) = \sum_{n=0}^{r-1} \beta_n(z)\mathbf{b}(zW^{nd}) \qquad \text{for some rational filters } \beta_n(z).$$

Replacing z by zW^d and postmultiplying by $\mathbf{c}(z)$ shows $\mathbf{b}(zW^{nd})\mathbf{c}(z) = 0$ for $n = r+1$. Using this and repeating the process shows the same for $n = r+2$. Carrying on this way, the result is shown for all $n \ge 0$. For $n < 0$ we use a similar process, now taking r as the *minimum* n obeying $\alpha_n(z) \not\equiv 0$. □

Deriving the test from the AC matrix and Lemmas 2, 3

Examine closely Equations (2.6),(2.7) which are equivalent to PR. Number the rows of the AC matrix from 0 to $L-1$, and the columns from 0 to $M-1$, and let $W = e^{-j2\pi/L}$. The l-th row equation in (2.6) has form

$$\sum_k p_{i_k} H_{i_k}(zW^l)F_{i_k}(z) = 0 \tag{6.3}$$

The summation here ranges over all indices i_k for which the i_k-th column in the AC matrix has a nonzero entry in the l-th row. This happens if and only if l is a multiple of $p_{i_k} = L/n_{i_k}$. Thus, suppose that for all integers m in some set S, the number $l = mp_0$ is not a multiple of any $p_i \ne p_0$. For all these $l = mp_0$, (6.3) holds with the summation being over the same set of filters, i.e., those corresponding to decimator value n_0. Thus, if N_0 is the number of occurrences of decimator n_0, (6.3) takes the form

$$\sum_{i=0}^{N_0-1} p_0\, H_i(zW^l)F_i(z) = 0, \quad \text{for } l = mp_0, \quad \text{for } m \in S \tag{6.4}$$

This is very similar to the system $\sum_{i=0}^{N-1} B_i(zW^l)C_i(z) = 0$ of Lemma 3, with $N = N_0$. The only difference is that here the premise of the lemma may or may not hold, i.e., (6.4) may or may not hold for N_0 values of l occurring consecutively in an arithmetic progression.

The main idea of the test we are developing is to find all progressions for which the premise of Lemma 3 actually holds, and then use the lemma. This may sometimes allow us to deduce that (6.4) actually holds for other values of l too, besides those stated in (6.4) itself. If $l = 0$ turns out to be one of these, then by Lemma 2 we can conclude that PR is impossible, i.e., the given set of decimators fails the test. To perform such a test, we must use the known values of l from (6.4) to find progressions obeying the premise of Lemma 3. From (6.4) it clearly suffices to examine progressions of integers whose common difference d is an integer multiple of p_0. There are infinitely many such progressions, each an infinite sequence. However, since $W^L = 1$, it suffices to consider the progressions modulo L, and to restrict their common difference d as $d < L$. In fact even $d \leq \lfloor L/2 \rfloor$ suffices, as any progression with common difference $L - d$ can be generated in reverse order by one with difference d. We will now show examples of the working of the above test.

Example 1: Compatibility is a special case, and is necessary even for aliasfree FBs. Suppose p_0 occurs only once and is not a multiple of any $p_i \neq p_0$. In other words, n_0 occurs only once and does not divide any other n_i, i.e., compatibility (Section 6.1) is violated. Then (6.4) holds with $N_0 = m = 1$, and a trivial use of Lemma 3 shows that indeed (6.4) also holds for $l = 0$ (i.e., $H_0F_0 \equiv 0$). Thus by Lemma 2, PR is impossible. Hence, passing our new test implies compatibility of the decimators. In fact, even if the rational FB does not have PR but is *LTI* (i.e., aliasfree) with transfer function $T(z)$, (2.6),(2.7) still hold with L replaced by $LT(z)$ in (2.6). Hence, if n_0 violates compatibility as described above, the conclusion $H_0F_0 \equiv 0$ still holds. Thus the FB input–output relation is preserved even if we drop the 0-th channel, making the FB *overdecimated.* As the input of such a FB cannot be recovered from its output using any LTI system, we must have $T(z) \equiv 0$ (else we could use the LTI inverse $1/T(z)$). Thus, compatibility is necessary even if all we demand of the rational FB is that it be aliasfree and not identically zero $T(z) \not\equiv 0$ (as opposed to having PR).

Example 2: A specific set of decimators. Consider the set $\{4, 6, 6, 6, 10, 20, 20, 20\}$. This has 8 decimators with lcm $L = 60$, and $p_5 = p_6 = p_7 = 60/20 = 3$. For $m = 7, 9, 11$, the numbers $l = 3m$ are not multiples of any $p_i \neq 3$. Thus for these l, the l-th row equation in (2.6) reads as

$$\sum_{i=5}^{7} 3H_i(zW^l)F_i(z) = 0 \quad \text{for } l = 3m,\ m = 7, 9, 11 \tag{6.5}$$

The sum has three terms, and the three values of l occur consecutively in an arithmetic progression. Thus by Lemma 3, (6.5) holds for all l in this progression, specifically for $l = 3 \times 5 = 15$, which is not a multiple of any p_i besides $p_5 = 3$ and $p_0 = 15$. The 15-th row equation in (2.6) initially reads as $15H_0(zW^l)F_0(z) + 3\sum_{i=5}^{7} H_i(zW^l)F_i(z) = 0$ (for $l = 15$), but in the light of the above conclusion it now further says that $H_0(zW^l)F_0(z) = 0$. Now another application of Lemma 3 (for the trivial case of $N = 1$) shows that $H_0(z)F_0(z) = 0$ and thus PR is impossible by Lemma 2.

Based on the above discussion and examples, Appendix F shows a complete general algorithm to test the necessary condition derived above using Lemmas 2 and 3. This test, called the AC matrix test, implies not merely compatibility (Example 1) but strong compatibility too (Appendix F), and is in fact strictly stronger than strong compatibility: The decimators of Example 2 are strong compatible and yet fail the test.

10.7 CONDITIONS BASED ON REDUCTIONS TO TREE STRUCTURES

As seen in Section 4.2, a decimator–set may be derivable from trees in many ways, but a FB using the decimator–set may not be derivable from all of these trees. However, all PRFBs using decimators obeying certain conditions must be derivable from certain non-trivial trees associated with the conditions. Fact 2 is a result of this type for delay-chains. Another example is as follows: If the decimator–set comes from a dyadic or 'wavelet' tree (i.e., has form $\{2, 2^2, \ldots, 2^{r-1}, 2^r, 2^r\}$ for some integer $r > 0$), then *all* PRFBs using those decimators are derivable from this tree. This was proved in [10], [3] for rational orthonormal and biorthogonal FBs respectively. It parameterizes all FBs with dyadic decimator–sets, i.e., solves problem 2 of Section 3.1 for such sets. However, it does not reveal any new conditions on decimators for existence of rational PRFBs (problem 1 of Section 3.1). This is because it concerns only dyadic decimator–sets, which, being derivable from uniform–trees, are already known to allow building of PRFBs in every FB class of interest here.

Suppose on the other hand that we have a condition on a more general decimator–set S that allows us to conclude that every rational PRFB using S is derivable from some (nontrivial) tree. Such a condition provides a parameterization result for FBs using such decimator–sets S. Further, it reduces the problem of existence of rational PRFBs using S to that of existence of rational PRFBs using the smaller decimator–sets in the units of the tree. We can obtain a new necessary condition on S for existence of such a FB, by applying all the known conditions on these smaller sets. In this section, we derive three such conditions (Theorems 5,6,7) all of which yield as a special case, the result on dyadic FBs mentioned earlier. We refer the reader to Table 1 (Section 8) for example decimator–sets showing the use of the new necessary conditions generated by these results. Finally, we present two other results (Theorems 8,9) that also pertain to other filter constraints besides rationality such as orthonormality, stability and the FIR property.

Theorem 5: Root extraction test. Let a set of decimators $n_0, n_1, \ldots, n_{M-1}$ obey

$$\sum_{i=k}^{M-1} \frac{1}{n_i} = \frac{1}{N}, \quad \text{where } N = \text{integer multiple of } n_0, n_1, \ldots, n_{k-1}. \tag{7.1}$$

Let there be a rational PRFB using these decimators. Then all n_i for $i \geq k$ are multiples of N, and the FB is always derivable from a two unit tree structure of rational PRFBs in which the root has decimators $n_0, n_1, \ldots, n_{k-1}, N$, as shown in Fig. 14.

This result is a special case of Theorem 6 (which is proved in Appendix G). A corollary obtained with $N = 2$ is that any rational PRFB having a decimator of value 2 must be derivable from a two unit tree of rational PRFBs in which the root is a uniform two band FB. Repeated use of this corollary shows the result of [3] on derivability of rational biorthogonal FBs with dyadic decimators from trees. The corresponding result of [10] for orthonormal FBs does not directly follow: Theorems 5,6 do not themselves show how to ensure orthonormality of all the units of the tree, given that of the overall FB. This can

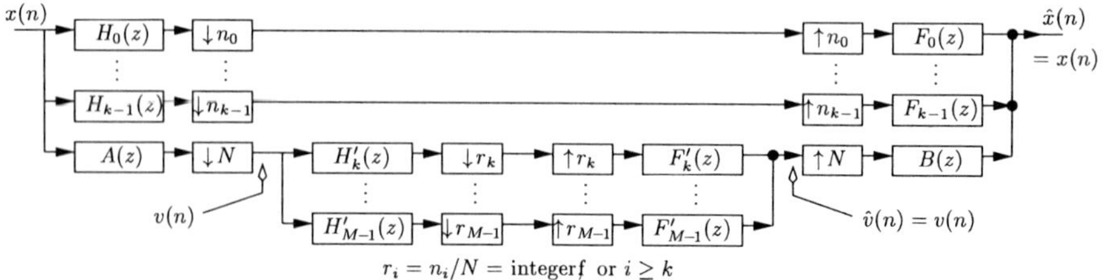

Figure 10.14. Root extraction test (Theorem 5): Showing equivalent tree structure for any rational PRFB with decimators obeying the premise (7.1) of the test

be done using Theorem 8 which follows later. Finally, note that even nonrational PRFBs with decimators n_i obeying (7.1) must be derivable from trees as in Fig. 14, *provided* the n_i have the property that all n_i for $i \geq k$ are multiples of N. In other words, this property no longer follows from (7.1) (as is clear from Section 3.2), but the derivability from a tree follows if the property is made an additional premise (Appendix G).

Theorem 6: Generalized root extraction test. Let the given decimator–set $D = \{n_0, \ldots, n_{M-1}\}$ have disjoint[5] subsets S, T_1, T_2 such that S is nonempty and

$$\sum_{n_i \in S} \frac{1}{n_i} = \frac{1}{N} \text{ for some integer } N, \tag{7.2}$$

$$T_1 = \{n_i : n_i \in D, n_i = \text{factor of } N\}, \tag{7.3}$$

$$\gcd(n_i, n_j) = \text{factor of } N \quad \text{whenever } n_i \in S \cup T_2, n_j \in T_2, i \neq j, \tag{7.4}$$

$$\sum_{n_i \in T_1} \frac{N}{n_i} + |T_2| = N - 1 \quad (\text{where } |T_2| = \text{number of elements in } T_2). \tag{7.5}$$

Then, if a rational PRFB exists with these decimators, all $n_i \in S$ are multiples of N, and the FB is derivable from a 2-unit tree of rational PRFBs. This tree has root decimator–set obtained from D by replacing $S \subset D$ by a single decimator of value N. The leaf–decimator set is derived from S by dividing all its elements by N.

Comments on Theorem 6. This result, proved in Appendix G, is more complicated to state but also more general than Theorem 5. Theorem 5 represents the special case when T_2 is empty and $S \cup T_1 = D$ (in which case (7.2) and (7.3) imply (7.5), due to (1.1)). Note that while one of the sets T_1, T_2 can be empty, (7.5) shows that they cannot *both* be empty except in the trivial case where $N = 1$ and $S = D$. With S, T_1 defined as in (7.2),(7.3), their disjointedness is equivalent to $|S| > 1$, which ensures that each entry of T_1 is less than all entries of S. Disjointedness of T_2 from S, T_1 is a separate requirement that does not follow from (7.2)–(7.5). Both S and T_1 can have multiple occurrences of a given decimator value; in fact from (7.3), every $n_i \in T_1$ occurs as many times in T_1 as it does in D. However, entries of T_2 are all distinct from each other, for else by (7.4), T_2 would have some elements that are factors of N and are hence in T_1 too, violating their disjointedness. Unlike Theorem 5, Theorem 6 is not obeyed by nonrational FBs

[5] Most 'sets' in our work, including D, S, T_1 here, are really 'multisets', i.e., can contain multiple occurrences of the same decimator value. However, *disjointedness* here has its usual set–theoretic meaning. Thus, here if one of the sets S, T_1, T_2 has a decimator of value m, the other two sets cannot have any decimator of value m even if D has several such decimators.

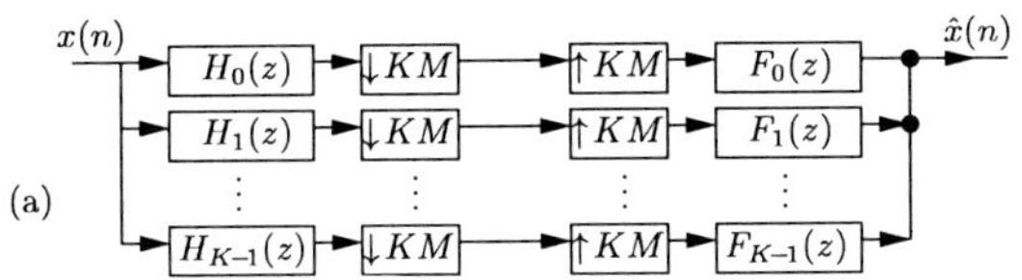

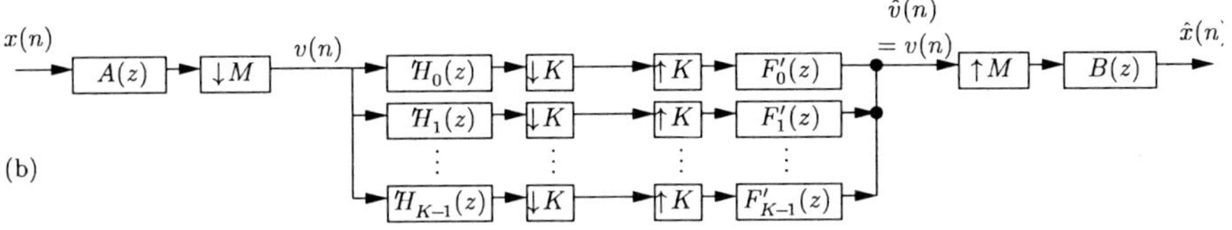

Figure 10.15. Leaf extraction test (Theorem 7). (a) K channels with decimator KM. (b) Equivalent structure under the premise of the test

even if the condition that all $n_i \in S$ be multiples of N is made an additional premise (a counterexample can be created with brickwall FBs).

Theorem 7: AC matrix based leaf extraction test. Consider Fig. 15a, showing a subset of the channels of some maximally decimated FB. Suppose the system in Fig. 15a is not identically zero, and all its filters are rational. Then the following statements are equivalent:

(a) Let $W = \exp(\frac{-j2\pi}{KM})$ and $G_l(z) \triangleq \sum_{i=0}^{K-1} H_i(zW^l)F_i(z)$. Then, $G_l(z) = 0$ for all $l \in \{0, 1, \ldots, KM-1\}$ that are not integer multiples of K (or equivalently, as $W^{KM} = 1$, for all such integers l).

(b) There are rational filters A, B, H_i', F_i' such that the systems of Figs. 15a, 15b are equivalent (i.e., for $i = 0, 1, \ldots, K-1$, $H_i(z) = A(z)H_i'(z^M)$, $F_i(z) = B(z)F_i'(z^M)$) and the H_i', F_i' form a K band PRFB.

Application of Theorem 7. The result is proved in Appendix G, we mainly use the fact that (a) implies (b) in its statement. Suppose precisely K decimators of a rational PRFB have value KM. Examine the k-th row on the left side of the L–row AC matrix equation (2.6) of the FB (where L is a multiple of all decimators in the FB). For $k = l(L/(KM))$ (l a positive integer), this evaluates to the sum of $G_l(z)$ (defined in Theorem 7) and other terms coming from channels whose decimators n_i are such that k is a multiple of L/n_i. If there are no such terms, $G_l(z) = 0$. Even if there are such terms, we have seen in deriving the AC matrix test (Section 6.4) how one can sometimes deduce that they sum to zero (and hence that $G_l(z) = 0$) using filter rationality and the other rows in (2.6). Suppose the decimators are such that such a deduction of $G_l(z) = 0$ is possible for all l that are not integer multiples of K. The condition (a) of Theorem 7 is then obeyed by the K channels with decimator KM for all rational PRFBs with this set of decimators. Thus, Theorem 7 implies that all these PRFBs are derivable from a two unit tree of rational PRFBs, in which the leaf is uniform with decimator K and generates the K channels with decimator KM. Thus for such a decimator–set, existence of rational PRFBs is equivalent to existence of rational PRFBs using the smaller decimator–set in the root of the above–mentioned tree. This technique can be applied to dyadic decimator sets to deduce the result of [3]. However this result follows more easily from Theorem 5. Finally, note that even if we remove all rationality restrictions in the statement of Theorem 7, (b) still implies (a) (Appendix G). The converse (which is more useful) is however no longer true (a counter example can be created using brickwall FBs).

Theorems 5—7 involve a decimator–set D having a subset S whose entries have reciprocals summing to $1/N$ for some integer N. Given a rational PRFB using D, the goal is to derive the subset of its channels corresponding to S from a single channel with decimator N, by attaching a leaf FB using the decimator–set S/N. (S/N is obtained from S by dividing each of its entries by N.) The theorems give various conditions on D under which this can be done for all rational PRFBs using decimator–set D. That S/N is a set of integers is either an assumption or a conclusion. Note that certain *rational* PRFBs are shown to be derivable from trees whose units are all rational PRFBs. If the original PRFB obeys a constraint other than (or besides) rationality, such as orthonormality or the FIR constraint, then can all units of the tree also be chosen to obey this constraint? A partial answer (for certain constraints) lies in the following result:

Theorem 8. Consider any of the following properties of FBs: (a) PR, (b) orthonormality, (c) stable filters, and (d) FIR filters. Suppose a tree structure of rational FBs yield a (necessarily rational) FB which obeys one particular property from this list. Then the filters in each unit of the tree can be modified without changing the overall FB, in such a way that each unit also becomes a FB that satisfies that property.

This result is proved in Appendix H. There is an important point to note about the list of properties in its statement. One could consider adding to the list, *combinations* of the listed properties, i.e., (e) PR and stable filters, (f) orthonormality and stable filters, (g) PR and FIR filters, (h) orthonormality and FIR filters. However, these *have not been listed.* Thus, for instance if the overall FB has PR with stable filters, all Theorem 8 assures is that the individual FBs can be altered to have either PR or stable filters — whether they can have *both* is left undecided. Indeed, it is an *open problem* as to whether or not Theorem 8 holds with any of the properties (e)–(h) added to its list of properties (though we believe that it probably does hold even in this case). That it holds for property (h) has been proved *for dyadic trees* in [10, Th. 2]. This proof can be extended to cover both properties (f) and (h) for all uniform–trees in which no unit has more than one child (dyadic trees being a special case). Further extensions (to arbitrary trees) are unknown.[6]

Our last result is one that, given a PRFB with decimator–set S, deduces existence of another PRFB, which has a possibly different decimator–set S_1 and preserves certain properties of the original FB such as rationality and orthonormality. Thus, given a necessary condition P on S for existence of a PRFB using S with such a property, we can get a stronger necessary condition P_1 by applying P not merely to S but also to S_1. It may turn out that $S_1 \equiv S$, or that S_1 is derivable from a tree using S as root, in which case P_1 is automatically tested once we test for the tree version P'' of P (i.e., P'' is even stronger than P_1). However, this does not always happen, i.e., sometimes we indeed get a new condition. The result is as follows:

Theorem 9: Subset extension test. Consider any subset of channels of any PRFB. Let the decimators in this subset have lcm L and reciprocals that sum to p/L. Then there exist $L-p$ channels with decimation L which when augmented to the chosen subset, extend it into a PRFB. If the original PRFB has any *one* of the following properties: (a) rational filters, (b) orthonormality, (c) orthonormality and rational filters, (d) orthonormality and FIR filters, then the new 'extended' FB can also be chosen to have that property.

---

[6] [10, Th. 4] appears to show Theorem 8 with property (h) for all uniform–trees, but in fact it does not: In its proof, $T_i'(z) = T_i(z)/B(z^b)$ has not been shown to be FIR. Similarly, [3, Sec. 6] seems to account for property (e), but in fact it only covers stability (property (c)).

Table 10.1. Necessary Conditions on Decimators of Rational PRFBs.

Test type	Test # (i)	Test	Section #	Tests implied	Example decimator–set D_i violating test $\#i$
Pairwise gcd	1	Pairwise noncoprimeness	6.1	—	3,4,12,12,12,12,12
based	2	Pairwise gcd test	6.2 (Th. 4)	1	6,10,14,210,210,...,210
	3	Compatibility	6.1	—	6,6,6,9,9,9,9,18
Generalizing	4	Strong compatibility	6.1	3	6,6,6,6,9,9,27,27,45,270,...,270
compatibility	5	Tree version of #4	6.3, App. E	3,4	4,6,6,12,16,16,16,18,18,72,144,144,144
	6	AC matrix test	6.4, App. F	3,4	6,6,6,6,9,9,24,72,72,72,72,72
	7	Root extraction (RE) test	7 (Th. 5)	—	4,4,4,10,20,20,20
Tree reduction	8	Generalized RE test	7 (Th. 6)	7	4,4,6,8,24,24,24,24,24
based	9	AC–based leaf extraction	7 (Th. 7)	—	4,6,6,12,12,12,20,20,30,60,60
	10	Subset extension test	7 (Th. 9)	—	4,6,6,8,12,12,36,48,48,72,72,72,72
	11	Tree version of #2	?	1,2	3,6,6,12,18,30,180,...,180
Tree versions	12	Tree version of #6	?	3,4,6	??
(Section 4.3)	13	Tree version of #7–10	?	7–10	??

Proof: We use the equivalence between the biorthogonality condition (2.3) and the polyphase formulation of the PR condition on the filters. From Section 2.2, Fig. 4 and Appendix A, we can see that for a specific i, j, (2.3) is equivalent to $p_i \times p_j$ equations of the form $(S_i(z)Q_j(z)) \downarrow_L = \delta(i-j)$. Here $L = n_i p_i = n_j p_j$, and these equations come from choosing S_i, Q_j respectively as the delayed versions of H_i, F_j in Fig. 4. The left side of each such equation can be written as an 'inner product' of length L vectors using the polyphase lemma (Section 1.3). In order to arrange all these equations into a single polyphase matrix equation, L was chosen in Section 2.2 as a multiple of *all* decimators n_i. However, if we restrict attention to a subset of channels of the nonuniform FB (as in Theorem 9), it suffices to let L be a multiple of the decimators in this subset. Thus, the subset chosen in the theorem statement corresponds to a matrix equation $\mathbf{E}(z)\mathbf{R}(z) = \mathbf{I}$ where $\mathbf{E}(z), \mathbf{R}(z)$ are of sizes $p \times L$ and $L \times p$ respectively. The theorem then follows by augmenting these matrices into $L \times L$ ones whose product is still the identity. The augmented matrices are the polyphase matrices of the new FB, and the added rows and columns are the L-th order polyphase vectors of the filters in the added channels. Clearly if $\mathbf{E}(z), \mathbf{R}(z)$ are rational, these vectors can also be chosen to be rational. If the original FB is orthonormal, $\mathbf{E}(e^{j\omega})\mathbf{E}^{*T}(e^{j\omega}) = \mathbf{I}$, so we can extend $\mathbf{E}(e^{j\omega})$ into a unitary matrix (for each ω). Further if the original FB is rational or FIR, the extension can be forced to preserve these properties by using the unitary statespace realizations [11, Chap. 14] of $L \times p$ paraunitary systems.[7] $\square$

10.8 SUMMARY AND COMPARISON OF NECESSARY CONDITIONS

Table 1 lists all currently known necessary conditions on the decimators of rational PRFBs, many of which have been developed in this paper. The following remarks are in order:

1 For each of the tests numbered $\#i = 1, 2, \ldots, 11$, we have an example decimator–set D_i violating the test. We have chosen D_i so that the only other listed tests it fails are (a) any tests that imply test $\#i$, as shown in the second-last column of the table, and (b) *possibly* one or more of the tests #11–13, which we have not designed

[7] The extension is made by extending their 'rectangular unitary' *realization matrices* into (square) unitary ones.

algorithms to perform, and hence, cannot currently decide whether or not they are violated. This shows that except for these last three tests, the interdependencies between the tests are exactly as described in the table (in its second-last column). For example, passing the AC matrix test implies nothing about passing the tree version of the strong compatibility test, and vice versa.[8]

2 The above remark applies in particular to the example set D_{11}, which passes all tests #1–10. It fails test #11 because attaching a uniform leaf with decimator 2 to its decimator of value 3 yields a set with a subset of 7 decimators within which the gcd of any pair is 6. Currently we do not have such examples for tests #12,13. Though we have not devised an algorithm to test for the tree version of the pairwise gcd test, the set D_{11} shows that the tree version is strictly stronger than the original test.

3 The AC matrix test (#6) is also strictly strengthened by forming its tree version (#12), as shown by the set $\{3, 4, 8, 12, 12, 24, 24, 24\}$. This set passes test #6, but fails its tree version because attaching uniform leaves with decimator 2 to its decimators of value 12 yields a set that fails test #6. However, this example also fails another test from Table 1, i.e., pairwise noncoprimeness.

4 Each test P of rows #11–13 is the tree version of some test P_1. As seen in Section 4.3, the test P is well defined, but involves applying P_1 to an infinite family of decimator–sets. Devising a finite algorithm for this process can take ingenuity or hard work, as seen in Section 6.3 (and Appendix E) for the tree version of strong compatibility. This is especially true given the complex nature of some of the P_1 (e.g., the AC matrix test). Design of algorithms for these tests is left for future work.

5 There are decimator–sets that obey all the known necessary conditions #1–10 which we have algorithms to verify, and yet do not allow building of delay-chain PRFBs. Examples are the sets $\{6, 6, 6, 6, 9, 12, 36, 36, 36, 36, 36\}$ and $\{4, 6, 6, 12, 12, 12, 16, 24, 48, 48, 48\}$ (all examples have ≥ 11 decimators). Thus, these necessary conditions, taken together, are still not equivalent to the most general known sufficient condition for existence of rational PRFBs, namely, existence of PR delay-chains. We currently do not know whether or not sets of the kind listed above allow building of rational PRFBs. Thus, the main problem of this paper (Section 3.1) remains unsolved for the rational FB class.

10.9 CONCLUDING REMARKS

We have presented several new conditions on the decimators of rational PRFBs, considerably generalizing many earlier known ones. Our work still leaves necessary and sufficient conditions unknown. The weakest known sufficient condition, when obeyed, allows the decimators to be used to build PRFBs as specialized as delay-chains. Thus, if we impose various less restrictive conditions on the filters of a rational FB (e.g., FIR filters, linear phase filters, orthonormality, etc.), we get many more FB classes for which we do not know the necessary and sufficient conditions on the decimators of PRFBs in the class. It has been shown [4] that existence of rational PRFBs implies that of rational *orthonormal* FBs with stable filters (i.e., all analysis filters have all poles inside the unit circle). However, whether this implies existence of *FIR orthonormal* FBs is not known. Even when a decimator–set is known to allow building of rational PRFBs, complete parameterizations

[8]Example D_5 actually also fails test #10, but in a manner that makes the use of test #10 equivalent to using the tree version of another test (see discussion on Theorem 9).

of the possible PRFBs are not known, except in the restricted cases of uniform and dyadic decimator–sets. Partial parameterizations using trees have been presented in Section 7. Other specific open problems encountered in our study are listed below:

1 *Non–tree–structured PRFBs with tree structured decimator–sets:* Section 4.2 has shown two different constructions leading to such FBs. Are there any more?
2 *Forcing properties of a tree structured FB on all the tree units:* Theorem 8 (Section 7) shows that this is possible for rational FBs with certain properties (e.g., PR, FIR filters), but it is not known whether it is possible for certain others (e.g., PR *and* FIR filters).
3 *Real coefficient FBs (Section 3.2):* Do they always exist? Does existence of rational PRFBs with decimator–set S imply that of real coefficient PRFBs (rational or otherwise) using S?
4 *Algorithms for tree versions of necessary conditions:* These have not been designed for certain necessary conditions due to their complicated nature (e.g., AC matrix test), see Table 1 (Section 8).

10.10 APPENDICES

10.10.1 Appendix A: Proof of Nonuniform Biorthogonality Condition (2.3)

The (uniform) biorthogonality condition on the uniform FB derived from a nonuniform one is equivalent to

$$\left(z^{-n_i c} H_i(z) z^{n_i d} F_i(z)\right) \downarrow_L = \delta(c-d) \tag{10.1}$$

$$\left(z^{-n_i a} H_i(z) z^{n_j b} F_j(z)\right) \downarrow_L = 0 \quad \text{if} \quad i \neq j \tag{10.2}$$

Here $c, d, a \in \{0, 1, \ldots, p_i - 1\}$ and $b \in \{0, 1, \ldots, p_j - 1\}$, where $L = n_i p_i = n_j p_j$. We now use the noble identity $\left(X(z^M)Y(z)\right) \downarrow_M = X(z)(Y(z)) \downarrow_M$. This shows that (10.1) is equivalent to $\left((H_i(z)F_i(z)) \downarrow_{n_i} z^{d-c}\right) \downarrow_{p_i} = \delta(c-d)$. If $c = 0$, the left side here for $d = 0, 1, \ldots, p_i - 1$ is the d-th entry in the p_i-th order analysis polyphase vector of $(H_i(z)F_i(z)) \downarrow_{n_i}$. So the equation is clearly equivalent to

$$(H_i(z)F_i(z)) \downarrow_{n_i} = 1. \tag{10.3}$$

Next, having (10.2) hold for the said values of a, b is equivalent to having it hold for all integers a, b. This is because $L = n_i p_i = n_j p_j$, and $(A(z)) \downarrow_L = 0$ is equivalent to $\left(z^{qL} A(z)\right) \downarrow_L = 0$ for any integer q and transfer function $A(z)$ (by noble identity). As a, b take all integer values, $-n_i a + n_j b$ takes values $k \gcd(n_i, n_j)$ for all integers k. Thus, using the noble identity, with $A_{ij}(z) \stackrel{\triangle}{=} (H_i(z)F_j(z)) \downarrow_{\gcd(n_i,n_j)}$, (10.2) is equivalent to

$$\left(A_{ij}(z)\ z^k\right) \downarrow_{L/\gcd(n_i,n_j)} = 0 \quad \text{for all integers } k \quad \text{if} \quad i \neq j \tag{10.4}$$

The left side here includes all entries of the order $L/\gcd(n_i, n_j)$ polyphase vector of $A_{ij}(z)$, so (10.4) is equivalent to

$$A_{ij}(z) = (H_i(z)F_j(z)) \downarrow_{\gcd(n_i,n_j)} = 0 \quad \text{if} \quad i \neq j. \tag{10.5}$$

Thus (10.1),(10.2) are equivalent to (10.3),(10.5) respectively; proving the nonuniform biorthogonality equation (2.3).

10.10.2 Appendix B: Derivability of Decimator–sets from a Uniform-tree

Claim: If a set S of decimators satisfying (1.1) is derivable from a uniform–tree, then it is derivable from a uniform–tree in which the root has decimator g, where g is the gcd of all elements of S.

Proof: (Can be skipped without losing continuity.) Use induction on the number N of units in the tree. Let r be the root decimator in the given uniform–tree. Clearly r divides all decimators in S, so r divides g. Now the root can have at most r children. If it has less than r children, then r is a decimator in S. Hence $g <= r$, implying $g = r$, i.e., the root already has decimator g. This proves the claim for $N = 2$, as the root of a 2-unit tree has 1 child and $1 < r$. If the root has all r children, consider any child along with all its descendants. These units form another uniform–tree. All decimators generated by this new tree are multiples of g/r. Let g' be their gcd, thus $g' = k(g/r)$ for some integer k. The new tree has $\leq N - 1$ units. So by the induction hypothesis, the decimators it generates can be rederived from a uniform–tree with root decimator g'. Since $g' = k(g/r)$, this root unit can then be rederived from a uniform–tree having decimator g/r for root and k for all the g/r leaves, each of which is a child of the root. After making all these replacements on the starting tree, all children of its root now have decimator g/r, hence the root and its children can be replaced by a single uniform unit with decimator g. This proves the claim. □

The above result suggests an algorithm [6] that tries to build a uniform–tree starting from its root:

Root–to–leaves Algorithm. (Tests derivability of a given set S of decimators from a uniform–tree)

1 Find gcd g of all elements of S. If $g = 1$, then S is not derivable from a uniform–tree.
2 Divide all entries of S by g (represents choosing root decimator g). Find *all* possible partitions of the resulting set into g groups each of which is a valid set of decimators obeying maximal decimation (1.1).
3 We can derive S from a uniform–tree if and only if among these partitions, there is at least one in which each group is derivable from a uniform–tree.

The algorithm is recursive. At Step 2, dividing the entries of S by g yields a set S' lower–bounded by unity. In the ensuing partition of S', any unity element in S' is all by itself a valid group viewed as derivable from a uniform–tree for purposes of Step 3. Such a group denotes absence of a child of the root, just as groups with more than one element represent children of the root. There may possibly be *no* valid partition at Step 2, e.g., when $S = \{4, 6, 6, 10, 10, 10, 10, 120, 120\}$. This of course means that there is no uniform–tree.

Note that though Step 2 can always be implemented in principle, doing it with a simple and efficient algorithm can be tricky. An alternative method builds the tree starting from a leaf and avoids this problem. Its basic idea is in identifying a leaf: Given an arbitrary decimator d in a set S derivable from a uniform–tree, it is not clear whether d is obtained at a leaf unit of the tree. However, this *must* be the case if d is the *maximum* element in S, and further the leaf decimator must of course then divide d. Based on this, we have:

Leaf–to–root Algorithm. (Tests derivability of a given set S of decimators from a uniform–tree)

1 If S has no more than two *distinct* decimators, it is derivable from a tree.

2 Find the maximum entry m in S, and the number of times N that it occurs. For each factor k of m such that $1 < k \leq N$, form a smaller set S_k by setting $S_k = S$ and then replacing k of the elements of value m in S_k by one element of value m/k (i.e., form a leaf unit that is uniform with decimator k).
3 The set S is derivable from a uniform–tree if and only if at least one of the S_k above is.

This is another recursive algorithm, more elegant and simpler to implement, though it may be somewhat unclear whether or not it is faster. Its only step still requiring justification is Step 1. This is easily done: Suppose $S = \{m_0, \ldots, m_0, m_1, \ldots, m_1\}$ with m_i occurring N_i times $(i = 0, 1)$. Let $m_i = gd_i$ where $g = \gcd(m_0, m_1)$. Maximal decimation implies that $N_0 d_1 + N_1 d_0 = g d_0 d_1$. As d_0, d_1 are coprime, this means that $N_i = D_i d_i$ for integers D_i, $i = 0, 1$, where $D_0 + D_1 = g$. Thus S is derivable from a uniform–tree in which the root has decimator g, all its children are leaves, and D_i leaves have decimator d_i $(i = 0, 1)$.

Only necessary and only sufficient conditions. Presence of no more than two distinct decimators, as shown above, is an example of a *sufficient* condition for derivability from a uniform–tree. It is by no means necessary. Another such example is the condition that each decimator divides every decimator larger than itself (a special case is when all of them are powers of the same number). This condition neither implies nor is implied by the earlier one, and neither condition is necessary, as exemplified by the set $\{4, 4, 6, 6, 12, 12\}$. Sufficiency of the new condition is proved using the root–to–leaves algorithm: Clearly $g > 1$ at Step 1, as g is the smallest decimator. At Step 2 in formation of the partition, if we sequentially select elements from the smallest upwards, the condition ensures that at some stage the reciprocals of the selected elements will sum to unity. Repeating this process results in a valid partition, and further each of its groups also satisfies the condition. Thus the proof is completed by induction on the number of decimators.

Derivability of a set of decimators from uniform–trees implies existence of various types of PRFBs (including PR delay-chains) using those decimators. Thus, any conditions necessary for such existence are also necessary for derivability from uniform–trees. Their necessity is often provable directly from the above algorithms. For example, without pairwise noncoprimeness (Section 6.1), $g = 1$ at Step 1 of the root–to–leaves algorithm. If compatibility (Section 6.1) is violated, i.e., if a decimator d does not divide any other decimator, then eventually $m = d$ and $N = 1$ at Step 2 of the leaf–to–root algorithm, i.e., there are no sets S_k. As tests for such necessary conditions are inconclusive whenever they are satisfied, they cannot replace the earlier complete algorithms, though they can potentially increase their efficiency.

10.10.3 Appendix C: Proof of Fact 3

Let $R = \{m_0, \ldots, m_{M-1}\}$, $S = \{k_0, \ldots, k_{K-1}\}$. So $T = \{n_0, \ldots, n_{K+M-2}\}$ with $n_i = m_0 k_i$ for $i = 0, 1, \ldots, K-1$ and $n_{K-1+i} = m_i$ for $i = 1, 2, \ldots, M-1$. Let $L = \text{lcm}\{n_i\}$, $p_i = L/n_i$. Let n_i occur N_i times in T. Let b_i be the strong compatibility lower bound on N_i. The proof is in two parts:

Part 1: Uniformity of S. Suppose S is not a uniform unit, we will then show that $b_j \leq N_j$ for all j, i.e., T is strong compatible. Indeed for $j = 0, 1, \ldots, K-1$ we have from (6.1),

$$p_j b_j = \min_{p_i \neq p_j} \mathrm{lcm}(p_i, p_j) \leq \min_{p_i \neq p_j,\, 0 \leq i < K} \mathrm{lcm}(p_i, p_j) \tag{10.6}$$

$$\leq p_j N_j \qquad \text{(as } S \text{ is strong compatible)} \tag{10.7}$$

The minimization on the right side of (10.6) is not over an empty set because S is nonuniform, i.e., $p_i \neq p_j$ for at least one i such that $0 \leq i < K$. The right side of (10.6) thus equals $p_j b_j^S$ where b_j^S is the strong compatibility (lower) bound on the number N_j^S of occurrences of k_j in S. This bound holds by strong compatibility of S, and $N_j^S \leq N_j$. This justifies (10.7), and thus $b_j \leq N_j$ for $0 \leq j < K$. For $j \geq K$, if $n_j = m_0$ then $b_j = 1 \leq N_j$, because n_j divides a distinct decimator $n_0 = n_j k_0$. If $n_j \neq m_0$, then $N_j \geq N_j^R$, the number of occurrences of $n_j = m_{j-(K-1)}$ in R. Let b_j^R be the strong compatibility lower bound on N_j^R. Thus $b_j^R \leq N_j^R \leq N_j$, and with

$$A = \min_{p_{K-1+i} \neq p_j,\, i \geq 1} \mathrm{lcm}(p_{K-1+i}, p_j) \quad \text{and} \quad B = \mathrm{lcm}(L/m_0, p_j), \quad \text{we have} \tag{10.8}$$

$$p_j b_j^R = \min(A, B), \quad \text{while} \quad p_j b_j = \min_{p_i \neq p_j} \mathrm{lcm}(p_i, p_j) \tag{10.9}$$

Thus if $A \leq B$ in (10.8) (e.g., this holds if $m_i = m_0$ for some $i > 0$), then clearly $b_j \leq b_j^R \leq N_j$. Even if $A > B$,

$$p_j b_j^R = \mathrm{lcm}(L/m_0, p_j) \geq \min_{p_i \neq p_j,\, 0 \leq i < K} \mathrm{lcm}(p_i, p_j) \geq p_j b_j,$$

as $p_i = L/(m_0 k_i)$ for $i < K$, and nonuniformity of S again ensures that $\mathrm{lcm}(p_i, p_j)$ is not being minimized over empty set. (Nonuniformity of S is not needed here if $L/(m_0 K) \neq p_j$.) So again $b_j \leq b_j^R \leq N_j$. Thus, $b_j \leq N_j$ for all j, i.e., T is strong compatible, contradicting the premise of Fact 3. Hence S must be a uniform unit, i.e., $k_0 = k_1 = \ldots = k_{K-1} = K$. □

Part 2: Necessary conditions for $b_j > N_j$. We have already shown in Part 1 that if $j \geq K$, then $b_j > N_j$ is possible only if m_0 occurs only once in R and $m_0 K = n_j$. The proof of Fact 3 will be completed if we show a similar statement for $j < K$, i.e., that $b_j > N_j$ is possible only if m_0 occurs only once in R and $m_0 K = n_i$ for some $i \geq K$. To show this, note that for all $j < K$, all the n_j are identical (shown by Part 1), and hence the same holds for the N_j and the b_j. Also $N_j \geq K$. Thus it suffices to show that $b_0 \leq K$ if either $m_0 = m_l = n_{K-1+l}$ for some $l > 0$, or $m_0 K \neq m_i$ for all $i > 0$. If $m_0 = m_l = n_{K-1+l}$ for some $l > 0$, then

$$p_0 b_0 = \min_{p_i \neq p_0} \mathrm{lcm}(p_i, p_0) \leq \mathrm{lcm}(p_{K-1+l}, p_0) = \mathrm{lcm}\left(\frac{L}{m_0}, \frac{L}{m_0 K}\right) = \frac{L}{m_0} = p_0 K,$$

hence $b_0 \leq K$. If on the other hand m_0 occurs exactly once in R, then $m_0 F = m_l = n_{K-1+l}$ for some $F > 1$, $l > 0$ since R is compatible. Thus if $m_0 K \neq m_i$ for all $i > 0$, then

$$p_0 b_0 = \min_{p_i \neq p_0} \mathrm{lcm}(p_i, p_0) \leq \mathrm{lcm}(p_{K-1+l}, p_0) = \mathrm{lcm}\left(\frac{L}{m_0 F}, \frac{L}{m_0 K}\right) \leq \frac{L}{m_0} = p_0 K,$$

hence $b_0 \leq K$ again. This establishes the claim, hence proving Fact 3. □

10.10.4 Appendix D: Proof of Fact 4

From the premise of Fact 4, there is a tree T' in which each unit is either D or allows building of rational FBs (e.g., uniform units), such that T' generates a set of decimators that is not strong compatible. Note that every unit in T' is strong compatible. We now perform a series of operations on T', each yielding a new tree with all the properties of T', until finally we get the tree T with the desired properties as in Fact 4.

If the root of T' has a child that is not a leaf, then this child, along with all its descendants, forms a tree with fewer units than T'. We can assume that this tree generates a strong compatible decimator–set (else we can replace T' by this tree and repeat the process). We then view this tree as a *single* unit. This makes every child of the root of T' a strong compatible leaf. Next, we delete any leaf such that the residual tree generates a decimator–set that is not strong compatible. This yields the desired tree T having all properties of T'. We now show that T and the decimator–set $\mathcal{T}$ it generates have all the properties listed in Fact 4.

Properties 1,2,4: For any leaf S of T, we see that T can be redrawn as a 2-unit tree with strong compatible units R and S. However T itself generates the set $\mathcal{T}$ that is not strong compatible. Thus we can use Fact 3 to conclude the following: (a) All leaves of T are uniform. (b) For any decimator value obtained at a leaf of T, decimators of $\mathcal{T}$ with that value are the only ones in $\mathcal{T}$ that violate the strong compatibility lower bound on the number of their occurrences in $\mathcal{T}$. (c) Property 2 of Fact 4 holds. Now (b) implies that all decimators obtained at the leaves have the same value d. Also, (a) implies that T has root D: Otherwise the root allows building of rational PRFBs, and hence, so does T (as all children of its root are uniform leaves); violating the fact that $\mathcal{T}$ is not strong compatible. This completes the proof of property 1. Property 4 follows from this and conclusion (b) listed above. Thus we have shown properties 1,2,4 of Fact 4. □

Property 3: Let k_i be the decimator value of the leaf attached to $d_i \in D$ to form T. As $d_i k_i = d$, we have $d = C\,\mathrm{lcm}\{d_i\}$ where $C = \gcd\{k_i\}$. We must show that if $d \notin D$, then $C = 1$. In fact, this may be *false*. Our approach is to assume that $d \notin D$, and then create a new tree T^* generating a decimator–set $\mathcal{T}^*$ with all the properties of Fact 4. This is done by replacing every leaf decimator k_i with k_i/C. (If $k_i = C$ this means deleting the leaf.) Clearly property 1 of Fact 4 continues to hold, with the decimators obtained at the leaves now having value $d^* = d/C = \mathrm{lcm}\{d_i\}$. To prove property 2, let decimator d_i of D have a leaf attached to it in T^*. Then it also has a leaf (uniform with decimator k_i) attached in T. As $d_i \notin \mathcal{T}$ (by property 2 for T), the only way to have $d_i \in \mathcal{T}^*$ is that d_i be the newly formed decimator d/C. This however means that $k_i = C$ (as $d = d_i k_i$), i.e., the leaf attached to d_i in T has been deleted in T^*, contradicting the assumption on d_i. Thus $d_i \notin \mathcal{T}^*$, i.e., T^* obeys property 2. Next we prove property 3. As already seen, if $k_j = C$ for some j, then $d^* = d/C = d_j \in D$. Thus, if $d^* \notin D$, then $k_j > C$ for all j, i.e., decimators d_i with leaves attached in T are the same as those with leaves attached in T^*. So property 3 holds for T^* from $d^* = d/C = \mathrm{lcm}\{d_i\}$. Lastly, we show property 4, i.e., that d^* violates the strong compatibility lower bound b^* on the number N^* of its occurrences in $\mathcal{T}^*$. Let N be the number of occurrences of d in $\mathcal{T}$, and let b be the strong compatibility lower bound on N. Let L be any common multiple of the decimators of $\mathcal{T}$. We must show that $b^* > N^*$. Since T obeys property 4, we have $b > N$. Also, by construction of T^* and the hypothesis $d \notin D$, we have $N^* \geq N/C$. The inequality is strict only if $d/C \in \mathcal{T}$, but this would imply (by definition (6.1) of b) that $b \leq \left(\frac{d}{L}\right) \mathrm{lcm}(\frac{L}{d}, \frac{LC}{d}) = C$. Since $N \geq k_i \geq C$, we get $b \leq N$, a contradiction. Thus $d/C \notin \mathcal{T}$, and hence $N^* = N/C$. Lastly, $b^* = \left(\frac{d}{LC}\right) \mathrm{lcm}(\frac{LC}{d}, \frac{L}{m})$ for some $m \in \mathcal{T}^*$, $m \neq d/C$. Thus $m \in D$ and $m \in \mathcal{T}$ too, and $m \neq d$ by the hypothesis $d \notin D$. So $b \leq \left(\frac{d}{L}\right) \mathrm{lcm}(\frac{L}{d}, \frac{L}{m}) \leq C \left(\frac{d}{LC}\right) \mathrm{lcm}(\frac{LC}{d}, \frac{L}{m}) = Cb^*$. Hence, $b^* \geq b/C > N/C = N^*$ (using $b > N$). Thus $b^* > N^*$ as required. □

10.10.5 Appendix E: Testing Tree Version of Strong Compatibility

Given a decimator–set D, let $V = \{v_0, v_1, \ldots, v_{K-1}\}$ be the set of *distinct* decimator values in D, with v_i occurring N_i times in D. Let L be any multiple of all the v_i, i.e., of $\text{lcm}\{v_i\}$, and let $p_i = L/v_i$. Then D satisfies the tree version of strong compatibility if and only if Routine 1 below returns the value 'TRUE' for all $v_i \in V$ *and* Routine 2 returns value 'TRUE'.

Routine 1: (To be performed for all $v_i \in V$)

1 Initialization: Set $M = N_i$, $A = V$ and delete v_i from A.
2 If A is empty, return(TRUE). Else, let $j = l$ minimize $\text{lcm}(p_i, p_j)$ over all j such that $v_j \in A$. If $M < \text{lcm}(p_i, p_l)/p_i$, return(FALSE).
3 If v_l does not divide v_i, return(TRUE). Else, add $N_l(v_i/v_l)$ to M and delete v_l from A. This represents attaching to every decimator of value v_l, a leaf that is uniform with decimator v_i/v_l. Then go to Step 2.

Routine 2:

1 Find all subsets S of V having at least two but less than $K-1$ elements, such that the lcm $l(S)$ of all elements of S does not divide any $v_j \in V$.
2 For each S of Step 1, let $\sigma(S)$ be the sum of all the numbers $N_i(l(S)/v_i)$ for all $v_i \in S$. Let $b(S)$ be the minimum of $\left(\frac{l(S)}{L}\right)\text{lcm}(\frac{L}{l(S)}, \frac{L}{v_i})$ over all $v_i \notin S$. This step represents attaching to every decimator whose value v_i lies in S, a leaf unit that is uniform with decimator $l(S)/v_i$, so that all decimators thus obtained at the leaves have value $l(S)$. In the resulting tree structured set of decimators, $\sigma(S)$ is the number of occurrences of decimator $l(S)$ and $b(S)$ is the strong compatibility lower bound on $\sigma(S)$.
3 If $\sigma(S) \geq b(S)$ *for all* S above, return(TRUE). Else return(FALSE).

The action of the routines is independent of which multiple of $\text{lcm}\{v_i\}$ we choose L to be. To explain how the above test works, refer to the statement of Fact 4. Routine 2 lists all trees T obeying properties 1,2,3 of Fact 4 such that $d \notin D$ (see property 3), and returns a 'FALSE' value if any of these obey property 4. The set S of Step 1 represents choice of the d_i of property 2. We demand that S must have at least two elements, and that $l(S) \neq v_j$ for all $v_j \in V$, to ensure that property 3 holds with $d \notin D$. In fact we further demand that $l(S)$ must not divide any $v_j \in V$, for if it does, $b(S) = 1$ at Step 2. We also exclude sets S with $\geq K-1$ elements, for then T generates a set with at most two *distinct* decimators. Such a set, being derivable from a uniform–tree (Appendix B), is always strong compatible, i.e., $\sigma(S) \geq b(S)$ will hold at Step 3.

Routine 1 becomes a test for strong compatibility if we delete Step 3 in it. Hence we can assume strong compatibility of the given set of decimators. Thus the only task remaining is to examine whether there is a tree T obeying all properties of Fact 4 with $d \in D$ in property 3. This is achieved by the addition of Step 3. To see this, let there be such a tree T, with $d = v_i$, producing a set T of decimators. The quantity $b = \text{lcm}(p_i, p_l)/p_i$ of Step 2 is the lower bound on N_i, which holds by assumption of strong compatibility. Now the number N_T of occurrences of v_i in T is at least N_i. Further if $v_l \in T$, then the strong compatibility lower bound on N_T does not exceed b, and hence cannot be violated. Thus $v_l \notin T$, i.e., all decimators of value v_l must have leaves attached to them to convert them into decimators of value v_i. This justifies Step 3.

In the special case when $L \triangleq \text{lcm}\{v_j\} \in V$, Routine 2 can be skipped (it always returns 'TRUE'), and Routine 1 needs execution only for $v_j = L$ (it returns 'TRUE' for all other v_j). This is provable from the fact that for $v_j = L$, $p_j = 1$. In general, Routine 1 appears to be the important part of the test: There are relatively fewer decimator–sets for which violation of the test is detected by Routine 2 but not by Routine 1 (examples of such sets being $\{2,3,24,24,36,36,36\}$ and $\{2,4,6,48,48,72,72,72\}$).

10.10.6 Appendix F: Algorithm for the AC Matrix Test

In the given set of decimators, let $v_0, v_1, \ldots, v_{K-1}$ be the *distinct* decimator values, with v_j occurring N_j times. Let L be any common multiple of the v_j, and let $p_j = L/v_j$. The algorithm is then as follows:

1 *Initialization.* Create a matrix $\mathbf{U}$ with rows numbered 0 to $L-1$ and columns 0 to $K-1$, where the lj-th entry u_{lj} is 1 if l is a multiple of p_j, and zero otherwise. Thus $\mathbf{U}$ is initialized to describe the positions of the zero and nonzero entries in the AC matrix (2.6),(2.7). In particular, $u_{0j} = 1$ for all j.

2 Set $\mathbf{U}' = \mathbf{U}$ (saving the current value of $\mathbf{U}$ in $\mathbf{U}'$). For all l, j such that u_{lj} is the only entry in the l-th row having value unity, set $u_{lj} = 2$. This identifies sets of filters having the same decimator value v_j, and satisfying an equation of the form $\sum_i B_i(zW^l)C_i(z) = 0$.

3 For each $d = kp_j$ for integer k obeying $1 \leq kp_j \leq \lfloor L/2 \rfloor$, let $c_s^d(n) = s + nd$ for $s = 0, p_j, 2p_j, \ldots, d - p_j$. If $u_{lj} = 2$ for $l \equiv c_s^d(n) \pmod L$ for N_j consecutive integers n, set $u_{lj} = 2$ for $l \equiv c_s^d(n) \pmod L$ for *all* integers n. Do this for each $j = 0, 1, \ldots, K-1$. (This represents use of Lemma 3.)

4 If $u_{0j} = 2$ for any j, the given set of decimators fails the AC matrix test. (This is where we apply Lemma 2.) If $\mathbf{U}' = \mathbf{U}$, the set passes the test. If neither of these happens, go to Step 2.

Passing the above test is a necessary condition on the decimators of any rational PRFB, as the discussion of Section 6.4 proves. The test outcome is independent of which common multiple of the v_j we choose L to be. The above algorithm may be made more efficient in many ways (e.g., we can declare the test as passed if $\mathbf{U}' = \mathbf{U}$ after Step 2); our main purpose here is to state a correct (rather than highly efficient) algorithm.

Lastly, we prove that the above test implies strong compatibility. Consider any fixed $j \in \{0, 1, \ldots, K-1\}$, and find the smallest $l > 0$ such that u_{lj} is not set to value 2 at Step 2. This is the smallest nonzero multiple of p_j that is also a multiple of some $p_i \neq p_j$, i.e., it is $\min_{p_i \neq p_j} \operatorname{lcm}(p_i, p_j) = p_j b_j$ where b_j is as in (6.1). Thus, after Step 2, $u_{lj} = 2$ for $l = kp_j$ for $k = 1, 2, \ldots, b_j - 1$. So if $N_j < b_j$, Step 3 will use the sequence $c_0^{p_j}(n)$ to set $u_{lj} = 2$ for all $l = np_j$. In particular it sets $u_{0j} = 2$, which means that the test is failed (see Step 4). Hence if the test is passed, we have $N_j \geq b_j$ for all j, which is the strong compatibility condition (6.1).

10.10.7 Appendix G: Proofs of Theorems 6,7

Proof of Theorem 6: We will prove the claim of the theorem after replacing its premises (7.2)–(7.5) about the decimator–set D by the following premise: The set D has two nonempty disjoint subsets S, T such that

$$\sum_{n_i \in S} \frac{1}{n_i} = \frac{1}{N} \text{ for some integer } N, \tag{10.10}$$

$$|T| = N - 1, \quad \text{and} \tag{10.11}$$

$$\gcd(n_i, n_j) = \text{factor of } N \quad \text{whenever} \quad n_i \in S \cup T,\ n_j \in T,\ i \neq j \tag{10.12}$$

This suffices because from a rational PRFB obeying (7.2)–(7.5), we can create one obeying (10.10)–(10.12) by inserting in each of its channels with decimator $n_i \in T_1$, a uniform rational PRFB with decimator N/n_i. This process preserves the channels corresponding to the decimator subset S, and creates $\left(\sum_{n_i \in T_2}(\frac{N}{n_i})\right)$ new decimators each of value N. The set T consists of T_2 and these new decimators; thus (10.11) follows from (7.5), and (10.12) from (7.4) and the fact that the new decimators have value N. Having proved the claim using (10.10)–(10.12), we remove the inserted uniform leaf FBs to prove it under the original premise (7.2)–(7.5).

Part 1: Proof under additional assumption that all $n_i \in S$ are multiples of N. Let us be given a rational PRFB with decimator–set D and filters as in Fig. 1, such that D has disjoint subsets S, T obeying (10.10)–(10.12). Let $\mathbf{E}(z), \mathbf{R}(z)$ respectively be the N-th order analysis and synthesis polyphase matrices of the analysis and synthesis filters corresponding to channels with decimators $n_i \in T$. Let $\mathbf{e}_i(z)$ be the N-th order analysis polyphase vector of $H_i(z)$ where $n_i \in S$. From (10.11), $\mathbf{E}(z), \mathbf{R}(z)$ have sizes $(N-1) \times N$ and $N \times (N-1)$ respectively. We use (10.12) with the PR condition (2.3) and the polyphase lemma, as in Section 6.2. This shows that $\mathbf{e}_i(z)\mathbf{R}(z) = \mathbf{0}$, and that $\mathbf{E}(z)\mathbf{R}(z)$ is a $(N-1) \times (N-1)$ diagonal matrix, none of whose diagonal entries is identically zero. This implies (using rationality of the filters) that $\mathbf{R}(z)$ has $N-1$ linearly independent columns. All the $\mathbf{e}_i(z)$, being 'orthogonal' to all these columns, must be 'proportional', i.e., $\mathbf{e}_i(z) = H_i'(z)\mathbf{a}(z)$ for some rational filters $H_i'(z)$ and vector $\mathbf{a}(z)$. Let $A(z)$ be the filter with $\mathbf{a}(z)$ as its N-th order analysis polyphase vector. Computing $H_i(z)$ from $\mathbf{e}_i(z)$ shows that $H_i(z) = A(z)H_i'(z^N)$. A similar argument shows that for all i such that $n_i \in S$, $F_i(z) = B(z)F_i'(z^N)$ for some rational $B(z), F_i'(z)$. Thus, under the additional assumption that all decimators in S are multiples of N, we see that the given rational PRFB is derivable from a two unit tree of rational FBs. The units of the tree have decimator–sets exactly as desired, and using Theorem 8, their filters can further be modified so that they also have PR. This completes Part 1 of the proof.

Part 2: Extending Part 1 to nonrational FBs in the setting of Theorem 5. When the original premises (7.2)–(7.5) of Theorem 6 are obeyed in the special manner that results in the premise of Theorem 5, the effect on (10.10)–(10.12) is to cause $D = S \cup T$ and $n_j = N$ for all $n_j \in T$. Now in Part 1, the diagonal elements of $\mathbf{E}(z)\mathbf{R}(z)$ are $(H_j(z)F_j(z)) \downarrow_N$ where $n_j \in T$ (by polyphase lemma). Thus, in the above special case, by (2.3), in fact $\mathbf{E}(z)\mathbf{R}(z)$ is the identity. Hence we can choose the $A(z), B(z)$ of Part 1 to have N-th order analysis and synthesis polyphase vectors $\mathbf{a}(z), \mathbf{b}(z)$ respectively, such that the $N \times N$ matrices $\begin{bmatrix} \mathbf{E}(z) \\ \mathbf{a}(z) \end{bmatrix}$ and $\begin{bmatrix} \mathbf{R}(z) & \mathbf{b}(z) \end{bmatrix}$ have product equal to identity. This possible even without any rationality restriction on the filters (of course A, B are then nonrational in general). These matrices now become the polyphase matrices of the root FB. Thus, the root automatically has PR, and hence so does the leaf (since the overall FB has PR), without the need to use Theorem 8 (which requires filter rationality). Thus, for the special case of Theorem 5 (as distinct from the general setting of Theorem 6), we have extended Part 1 to nonrational FBs.

Part 3: Proving the additional premise used in Part 1, using filter rationality. For each i such that $n_i \in S$ we insert a q_i channel uniform rational PRFB within the i-th channel of the given PRFB, where $q_i = \text{lcm}(N, n_i)/n_i$. This forms q_i new decimators of value $n_i q_i$. Let S' be the set of these decimators. Then, the newly formed tree–structured rational PRFB also has a decimator–set satisfying the premises (10.10)–(10.12), with S replaced by S' and T unchanged. Indeed, (10.10),(10.11) obviously hold, while (10.12) follows from the observation that if $\gcd(n_i, n_j)$ is a factor of N and q_i contains precisely the factors of N that are not present in n_i (i.e., $q_i = \text{lcm}(N, n_i)/n_i$) then $\gcd(n_i q_i, n_j)$ is also a factor of N. Further S' also obeys the additional assumption that its elements are multiples of N, by the choice of the q_i. Let $q_i > 1$ and consider two analysis filters $C_i^l(z)$, $l = 0, 1$ of the q_i band leaf FB inserted in the channel with decimator $n_i \in S$. The corresponding analysis filters of the new tree–structured FB are $H_i(z)C_i^l(z^{n_i})$. However, using Theorem 6 (which Part 1 has proved for the new FB), these filters have the form $A(z)D_i^l(z^N)$ for some rational $D_i^l(z), A(z)$ where $A(z)$ is independent of l, i. Taking ratios of these filters (a crucial step that requires filter rationality) shows that

$$\frac{C_i^0(z^{n_i})}{C_i^1(z^{n_i})} = \frac{D_i^0(z^N)}{D_i^1(z^N)}, \tag{10.13}$$

which implies that each equals $X_i(z^{\text{lcm}(N,n_i)})$ for some rational $X_i(z)$. Replacing z by z^{1/n_i} and using the definition of q_i, we have $\frac{C_i^0(z)}{C_i^1(z)} = X_i(z^{q_i})$. This means that the q_i-th order analysis polyphase vectors $\mathbf{e}_i^l(z)$ of $C_i^l(z)$, $l = 0, 1$, are linearly dependent, as $\mathbf{e}_i^0(z) = \mathbf{e}_i^1(z)X_i(z)$. Thus, the inserted q_i band uniform leaf FB with the filters $C_i^l(z)$, while assumed to have PR, has an analysis polyphase matrix that is not invertible (since it contains the rows $\mathbf{e}_i^l(z), l = 0, 1$). This contradiction disproves the assumption that $q_i > 1$. Hence $q_i = 1$, or in other words, n_i is a multiple of N. □

Proof of Theorem 7: We first write the input–output relations, analogous to (2.2), of the systems of Fig. 15:

$$\hat{X}(z) = \frac{1}{KM} \sum_{l=0}^{KM-1} X(zW^l)G_l(z) \qquad \text{for Fig. 15a} \tag{10.14}$$

$$\hat{X}(z) = \frac{1}{M} \sum_{l=0}^{M-1} A(zW^{Kl})B(z)X(zW^{Kl}) \quad \text{for Fig. 15b} \tag{10.15}$$

Here G_l are as defined in statement (a) of Theorem 7, and (10.15) uses the PR property of the FB formed by the H_i', F_i'. That (b) implies (a) in Theorem 7 follows directly by comparing (10.14) and (10.15), and holds even without any rationality requirements on the filters. We now prove that (a) implies (b) (for which the rationality is essential). Form the M-th order AC matrix $\mathbf{H}(z)$ (of size $M \times K$) using analysis filters $H_i(z)$, i.e., let the q-th row of $\mathbf{H}(z)$ be $(H_0(zW^{Kq}), H_1(zW^{Kq}), \ldots, H_{K-1}(zW^{Kq}))$ for $q = 0, 1, \ldots, M-1$. Let $\mathbf{f}(z) = (F_0(z), F_1(z), \ldots, F_{K-1}(z))^T$. Thus, the condition (a) is equivalent to $\mathbf{H}(zW^l)\mathbf{f}(z) = \mathbf{0}$ for $l = 1, 2, \ldots, K-1$. Replacing z by zW^{-l}, $\mathbf{H}(z)\mathbf{f}(zW^{-l}) = \mathbf{0}$. Now the $K-1$ columns $\mathbf{f}(zW^{-l})$, $l = 1, 2, \ldots, K-1$ are linearly independent. For otherwise, there are rational filters $\alpha_l(z)$ such that $\sum_{l=j}^{K-1} \alpha_l(z)\mathbf{f}(zW^{-l}) = 0$ for all z, where $1 \le j < K$ and $\alpha_j(z) \not\equiv 0$. Dividing this by $\alpha_j(z)$ and replacing z with zW^j shows that $\mathbf{H}(z)\mathbf{f}(z) = \mathbf{0}$ too. This would mean that $G_l(z) = 0$ *for all* integers l. This shows, by (10.14), that the system of Fig. 15a is identically zero, contradicting the premise of the theorem. Thus, the $K-1$ columns $\mathbf{f}(zW^{-l})$, $l = 1, 2, \ldots, K-1$ are linearly independent, and each row of $\mathbf{H}(z)$ is 'orthogonal' to all these columns (i.e., their product is identically zero). Hence all these rows must be 'proportional' to each other, i.e., $\mathbf{h}_1(z) = C(z)\mathbf{h}_0(z)$ for some scalar filter $C(z)$, where $\mathbf{h}_i(z)$ is the i-th row of $\mathbf{H}(z)$. This means that $H_i(zW^K)/H_0(zW^K) = H_i(z)/H_0(z) \triangleq D_i(z)$, i.e., $D_0(z) = 1$ and for $i = 1, 2, \ldots, K-1$, $D_i(e^{j\omega}) = D_i(e^{j(\omega + \frac{2\pi}{M})})$, i.e., $D_i(e^{j\omega})$ is periodic with period $\frac{2\pi}{M}$. So

$D_i(e^{j\omega}) = P_i(e^{j\omega M})$, i.e., by rationality, $D_i(z) = P_i(z^M)$. Thus, $H_i(z) = A(z)H_i'(z^M)$ where $A(z) = H_0(z)$ and $H_i'(z) = P_i(z)$, showing that the analysis banks of Figs. 15a and 15b can be made equivalent. Next, replacing z with zW^{-l} in condition (a) of the theorem shows that the condition holds even if each H_i is interchanged with F_i. Hence the same process can be repeated for the synthesis banks.

The above process may not ensure PR for the K band FB formed by the H_i', F_i' (which we will refer to as the leaf FB). However, G_l now takes the form $G_l(z) = A(zW^l)B(z)\sum_{i=0}^{K-1} H_i'(z^M W^{Ml})F_i'(z^M) = A(zW^l)B(z)G_l'(z^M)$, where $G_l'(z) = \sum_{i=0}^{K-1} H_i'(zW_K^l)F_i'(z)$ and $W_K = W^M = \exp(\frac{-j2\pi}{K})$. Thus, condition (a) implies that $G_l'(z) = 0$ for $l = 1, 2, \ldots, K-1$. (The alternative $A(zW^l)B(z) = 0$ is infeasible as it makes the systems identically zero.) Now the input–output relation of the leaf FB is $\hat{V}(z) = \frac{1}{K}\sum_{i=0}^{K-1} V(zW_K^l)G_l'(z)$ (analogous to (2.2)). Thus the leaf FB is LTI with (rational) transfer function $U(z) = G_0'(z)/K$. Hence, dividing all the $H_i'(z)$ by $U(z)$ and multiplying $A(z)$ by $U(z^M)$ gives a new system with all the properties desired in condition (b). This proves that (a) implies (b). □

10.10.8 Appendix H: Proof of Theorem 8

It suffices to prove the result for 2–unit trees, as we can continue by induction. A general 2–unit tree is specifiable as follows: The triples of (analysis filter, synthesis filter, decimator) are $(H_i(z), F_i(z), m_i), i = 0, 1, \ldots, M-1$ for the root and $(A_i(z), B_i(z), k_i), i = 0, 1, \ldots, K-1$ for the leaf, which is attached to decimator m_0 of the root. Thus the filters allowing and requiring modification are H_0, F_0 and the leaf filters A_i, B_i. The overall FB is unaffected iff the modifications preserve all the products $H_0(z)A_i(z^{m_0})$ and $F_0(z)B_i(z^{m_0})$.

Realizing stability, FIR filters: Let all the $H_0(z)A_i(z^{m_0})$ be stable. Then for every unstable pole $z = p$ of $A_j(z)$, there are m_0 unstable poles in $A_j(z^{m_0})$, one at each m_0-th root of p. To cancel these, we must have $H_0(z) = H_0'(z)C(z^{m_0})$ where H_0, H_0' have the same set of poles and $C(z) = (1 - z^{-1}p)$, so that $C(z^{m_0})$ is FIR with m_0 zeroes at the right places. Hence, replacing H_0 by H_0' and the A_i by A_iC removes the unstable pole of A_j and preserves the analysis filters of the overall FB. Thus all A_i can be made stable. Similarly if H_0 has an unstable pole p, each $A_i(z^{m_0})$ must have a zero at p, and hence for each i, $A_i(z) = A_i'(z)(1 - p^{m_0}z^{-1})$ where A_i, A_i' have the same set of poles. Thus, replacing A_i by A_i' and $H_0(z)$ by $H_0(z)(1 - p^{m_0}z^{-m_0})$ removes the unstable pole of H_0. Thus all filters can be made stable while preserving the overall FB. Similarly, if all the $H_0(z)A_i(z^{m_0})$ are FIR, the above argument can be repeated for *all* poles (rather than just the unstable ones), and all analysis filters can be made FIR.

Realizing PR, orthonormality: If the overall FB has PR, from (2.3) we get

$$\begin{aligned}(H_0(z)A_i(z^{m_0})F_0(z)B_j(z^{m_0}))\downarrow_{\gcd(m_0k_i, m_0k_j)} &= \\ ((H_0(z)F_0(z))\downarrow_{m_0} A_i(z)B_j(z))\downarrow_{\gcd(k_i,k_j)} &= \delta(i-j) \qquad (10.16)\end{aligned}$$

With rational filters $X(z), Y(z)$ defined such that $XY = (H_0F_0)\downarrow_{m_0}$, let $A_i' = A_iX$, $B_i' = B_iY$ for all i. Thus from (10.16), $\left(A_i'(z)B_j'(z)\right)\downarrow_{\gcd(k_i,k_j)} = \delta(i-j)$, i.e., replacing each A_i by A_i' and B_i by B_i' causes the leaf FB to obey (2.3) and hence to have PR. The overall FB is preserved on replacing $H_0(z)$ by $H_0'(z) = H_0(z)/X(z^{m_0})$ and $F_0(z)$ by $F_0'(z) = F_0(z)/Y(z^{m_0})$. Since now both the leaf and the overall FB have PR, the root must have PR too. Thus the root and leaf have been modified as desired. Further if the overall FB is orthonormal, then it has PR with $F_0(z)B_i(z^{m_0}) = \widetilde{T}_i(z)$ where

$T_i(z) = H_0(z)A_i(z^{m_0})$ (and of course, $F_i = \widetilde{H_i}$ for $i > 0$). Using $\widetilde{PQ} = \widetilde{P}\widetilde{Q}$, this means that (10.16) holds with F_0, B_i replaced by $\widetilde{H_0}, \widetilde{A_i}$ respectively. So we repeat with these substitutions, the earlier arguments used to make the root and leaf PR, and choose X such that $Y = \widetilde{X}$, i.e., such that $X\widetilde{X} = \left(H_0\widetilde{H_0}\right)\downarrow_{m_0} \triangleq W(z)$. (This is possible by spectral factorization, as $W(z)$ is rational and $W(e^{j\omega}) \geq 0$.) This ensures that the root and leaf are modified to be PR with $F_0' = \widetilde{H_0'}$ and $B_i' = \widetilde{A_i'}$. In other words, for all FBs, PR is obeyed and the synthesis filter corresponding to a given analysis filter D is $\widetilde{D}$. Thus both the root and leaf have been modified to be orthonormal rational FBs.

ACKNOWLEDGEMENT

Work supported in parts by the National Science Foundation Grant MIP 0703755 and ONR Grant N00014-99-1-1002.

REFERENCES

[1] S. Akkarakaran and P.P. Vaidyanathan, New results and open problems on nonuniform filter banks, in Proc. IEEE ICASSP, Phoenix, AZ, Mar. 1999.

[2] T. Chen and L. Qiu, General multirate building structures with application to nonuniform filter banks, IEEE Trans. Ckts. Syst.–II, **45** (1998), 948-958.

[3] S. Dasgupta and A. Pandharipande, On biorthogonal nonuniform filter banks, preprint.

[4] I. Djokovic and P.P. Vaidyanathan, Results on biorthogonal filter banks, Appl. Comp. Harmonic Anal., **1** (1994), 329-343.

[5] P.-Q. Hoang and P.P. Vaidyanathan, Non-uniform multirate filter banks: Theory and design, in Proc. IEEE ISCAS, Portland, Oregon, May 1989, pp.371-374.

[6] J. Kovačević and M. Vetterli, Perfect reconstruction filter banks with rational sampling factors, IEEE Trans. Sig. Proc., **41** (1993), 2047-2066.

[7] J. Li, T.Q. Nguyen, and S. Tantaratana, A simple design method for near-perfect-reconstruction nonuniform filter banks, IEEE Trans. Sig. Proc., **45** (1997), 2105-2109.

[8] K. Nayebi, T.P. Barnwell,III, and M. Smith, Nonuniform filter banks: A reconstruction and design theory, IEEE Trans. Sig. Proc., **41** (1993), 1114-1127.

[9] R.G. Shenoy, Multirate specifications via alias–component matrices, IEEE Trans. Ckts. Syst.–II, **45** (1998), 314–320.

[10] A.K. Soman and P.P. Vaidyanathan, On orthonormal wavelets and paraunitary filter banks, IEEE Trans. Sig. Proc., **41** (1993), 1170-1183.

[11] P.P. Vaidyanathan, Multirate Systems and Filter Banks, Englewood Cliffs, NJ: Prentice-Hall, 1993.

[12] M. Vetterli and J. Kovačević, Wavelets and Subband Coding, Englewood Cliffs, NJ: Prentice-Hall, 1995.

INDEX